Oil and Gas Industry: Advanced Technologies

Oil and Gas Industry: Advanced Technologies

Editor: Jane Urry

Callisto Reference
www.callistoreference.com

Callisto Reference,
118-35 Queens Blvd., Suite 400,
Forest Hills, NY 11375, USA

Visit us on the World Wide Web at:
www.callistoreference.com

ISBN: 978-1-63239-863-5 (Hardback)

The publisher's policy is to use permanent paper from mills that operate a sustainable forestry policy. Furthermore, the publisher ensures that the text paper and cover boards used have met acceptable environmental accreditation standards.

Printed in the United States of America.

Cataloging-in-publication Data

Oil and gas industry: advanced technologies / edited by Jane Urry.
p. cm.
Includes bibliographical references and index.
ISBN 978-1-63239-863-5
1. Petroleum industry and trade--Technological innovations. 2. Petroleum industry and trade. 3. Gas engineering.
4. Petroleum engineering. 5. Petroleum--Prospecting. I. Urry, Jane.

TP751 .O35 2017
665.7--dc23

Table of Contents

Table of Contents

Preface

The oil and gas industry is one of the largest and most advanced sectors of engineering. This book on the oil and gas industry discusses topics related to well logging, reservoir engineering and drilling. Recent developments in the field of oil and gas engineering or petroleum engineering are related to technologies concerning oil and gas recovery and extraction as well as improved computational methods for fluid dynamics. This book discusses the fundamentals as well as modern approaches of this field. This book elucidates the concepts and innovative models around prospective developments with respect to the oil and gas sector. This text presents researches and studies performed by experts across the globe. This book, which is informed by the latest techniques that are available in oil and gas exploration and extraction, will be of great help to researchers in the fields of petroleum engineering, fluid dynamics and geology. The extensive content of this book on oil and gas industry provides the readers with a thorough understanding of the subject.

This book was inspired by the evolution of our times; to answer the curiosity of inquisitive minds. Many developments have occurred across the globe in the recent past which has transformed the progress in the field.

This book was developed from a mere concept to drafts to chapters and finally compiled together as a complete text to benefit the readers across all nations. To ensure the quality of the content we instilled two significant steps in our procedure. The first was to appoint an editorial team that would verify the data and statistics provided in the book and also select the most appropriate and valuable contributions from the plentiful contributions we received from authors worldwide. The next step was to appoint an expert of the topic as the Editor-in-Chief, who would head the project and finally make the necessary amendments and modifications to make the text reader-friendly. I was then commissioned to examine all the material to present the topics in the most comprehensible and productive format.

I would like to take this opportunity to thank all the contributing authors who were supportive enough to contribute their time and knowledge to this project.

Editor

Preface

The oil and gas industry is one of the biggest and most [illegible] sectors of engineering. This book on oil and gas covers the essential topics related to well logging, reservoir engineering and drilling. Recent developments [illegible] petroleum engineering are related to [illegible] of enhancing oil and gas recovery [illegible] as well as improved [illegible] methods for fluid dynamics. This book discusses the fundamentals as well as modern approaches of this field. This book elucidates the concepts and innovative models around prospective developments with respect to the oil and gas sector. This book presents researches and studies performed by experts across the globe. This book, which is informed by the latest techniques that are available in oil and gas exploration and extraction, will be of great help to researchers in the fields of petroleum engineering, fluid dynamics and geology. The extensive content of this book on oil and gas industry provides the readers with a thorough understanding of the subject.

This book was inspired by the evolution of our times; to answer the curiosity of inquisitive minds. Many developments have occurred across the globe in the recent past which has transformed the progress in the field.

This book was developed from a mere concept to drafts to chapters and finally compiled together as a complete text to benefit the readers across all nations. To ensure the quality of the content we instilled two significant steps in our procedure. The first was to appoint an editorial team that would verify the data and statistics provided in the book and also select the most appropriate and valuable contributions from the plenty of contributions we received from authors worldwide. The next step was to appoint an expert of the topic as the Editor-in-Chief, who would head the project and finally make the necessary amendments and modifications to make the text reader-friendly. I was then commissioned to examine all the material to present the topics in the most comprehensible and productive format.

I would like to take this opportunity to thank all the contributing authors who were supportive enough to contribute their time and knowledge to this project.

Editor

1

Dynamic Gelation of HPAM/Cr(III) under Shear in an Agitator and Porous Media

Yu Haiyang[1]*, Wang Yefei[1], Zhang Jian[2], Lv Peng[1] and Shi Shenglong[1]

[1] School of Petroleum Engineering, China University of Petroleum (East China), Qingdao 266555 - China
[2] Technology Research Dept., CNOOC Research Center, Beijing 100027 - China
e-mail: yuhaiyang_upc@live.cn - wangyf@upc.edu.cn - zhangjian@cnooc.com.cn - 412106317@qq.com - 297338429@qq.com

* Corresponding author

Résumé — Gélification dynamique de HPAM/Cr(III) sous cisaillement dans un agitateur et en milieu poreux — La fermeture des eaux et le contrôle du profil constituent l'une des technologies les plus importantes pour l'amélioration de la récupération du pétrole. Afin d'assurer le succès de cette technologie, la clé est de déterminer précisément le temps de gélification et la résistance du gel au cours de l'écoulement du gel en milieu poreux. Le système HPAM (*Hydrolyzed PolyAcrylaMide*) et le système redox (bichromate de sodium et sulfite de sodium) sont communément utilisés ; leur temps de gélification statique en ampoules et en milieu poreux ont été déterminés, ainsi que le temps de gélification dynamique dans un agitateur et en milieu poreux. Le taux de cisaillement a été considéré comme l'un des principaux facteurs affectant le temps de gélification. Les résultats ont montré que le temps de gélification statique en milieu poreux était bien plus long que celui en ampoules. Le temps de gélification initial (IGT) en milieu poreux était deux ou trois fois plus élevé que celui en ampoules, alors que le temps de gélification final en milieu poreux était de six fois celui en ampoules. Sous cisaillement dans un agitateur, le procédé de gélification a été divisé en quatre phases : induction, augmentation brusque, stabilité et diminution. Avec l'augmentation du taux de cisaillement, le temps de gélification a été prolongé et la résistance du gel a diminué. Un taux de cisaillement critique de la gélification a été observé, au-dessus duquel aucun gel ne se formait. Le cisaillement n'avait quasiment aucune influence sur la résistance du gel au cours du stade d'induction, mais dans le procédé d'augmentation brusque, le cisaillement pouvait rapidement dégrader la résistance du gel. Le temps de gélification dynamique en milieu poreux était bien plus long que celui de la gélification statique en milieu poreux et en ampoules. Lorsque les concentrations de HPAM et de RS (*Redox System*) étaient augmentées, l'IGT de la gélification dynamique en milieu poreux était réduit.

Abstract — Dynamic Gelation of HPAM/Cr(III) under Shear in an Agitator and Porous Media — *Water shutoff and profile control is one of the most important technologies to enhance oil recovery. To ensure the success of this technology, the key is to accurately determine gelation time and gel strength during gel flow in porous media. The HPAM (Hydrolyzed PolyAcrylaMide) system and redox system (sodium bichromate and sodium sulfite) is widely used, whose static gelation time in ampoule bottles and porous media was determined, as well as the dynamic gelation time in an agitator and porous media. The shear rate was considered one of the major factors affecting gelation time. The results showed that the static gelation time in porous media was much longer than that in ampoule bottles.*

The Initial Gelation Time (IGT) in porous media was two or three times that in ampoule bottles, while the final gelation time in porous media was six times that in ampoule bottles. Under shearing in an agitator, the gelation process was divided into four phases: induction, sudden increase, stability and decrease. With the increase in shear rate, gelation time was prolonged and gel strength decreased. There was a critical gelation shear rate, above which there was no gel formed. Shear had almost no influence on gel strength during the induction stage but in the process of sudden increase, shear could degrade gel strength sharply. The time of dynamic gelation in porous media was much longer than that of static gelation in porous media and ampoule bottles. When HPAM and RS (Redox System) concentrations increased, the IGT of dynamic gelation in porous media was shortened.

INTRODUCTION

With the development of oilfields, the strips of high permeability emerge in the process of water flooding; and it is very important to enhance the volumetric sweep efficiency after water flooding. The most direct and effective way is profile control and water shutoff using crosslinked polymer gel. A gel with a certain gelation time and gel strength can effectively block the strips of high permeability, cause fluid diversion and enlarge sweep efficiency. It is crucial how to determine the gelation time and gel strength during gelant flow.

There are many methods to obtain the static gelation time, such as the code of gel strength [1], the change in viscosity with time [2], the measurement of the storage modulus and loss modulus, application of electrical conductivity and so on. These methods have advantages in themselves. According to Sydansk's gel-strength code, the gel strength is divided into several levels to determine gelation time quickly and effectively. Mokhtari *et al.* [3] divided gelation time into Initial Gelation Time (IGT) and Final Gelation Time (FGT) with the method of measuring apparent viscosity *versus* time, and the history of viscosity change during the process of gelation could be recorded. Gao *et al.* [4] measured the storage modulus and loss modulus and obtained the gelation time by the Winter-Chambon criterion [5]. Basta *et al.* [6] determined gelation time by the relationship between conductivity and time. Boey *et al.* [7] obtained the gel point where the tangent phase was found to be independent of the frequency.

Before it is injected into formation, gelant has a variety of shear rates. It is very meaningful to study the effect of shear on the gelant. There has been some research on the effect of shear on gelation time, application performances and re-formation after shear [8-12]. This was conducted under the condition of static gelation. Chauveteau *et al.* [13] studied the gelation process of polymer and Cr(III) under stable shear, and the results indicated that it was quite different between gelation under shear and static gelation. Bhaskar *et al.* [14] studied the effects of different shear rates on the gelation process of a redox system (sulfourea and sodium sulfite), and with the increase in shear rate, the reaction speed was slowed and the gel strength decreased. The results of Kolnes *et al.* [15] indicated that the viscosity of a gel system could be restored after slowly stable shearing for a long time. The impact of shear on gelation is achieved in at least two ways: one is that shear accelerates the reaction rate by increasing the mobility of polymer and crosslinker molecules. The other is that shear destroys the structure of gel and degrades the strength.

It is very important to study the process of gel flow in porous media. McCool *et al.* [16] studied the gelation of chromium acetate and polymer during flow in porous media by simulating formation fracture with a long conduit. Gel could be formed in porous media and gelation time was obtained by measuring the resistance coefficient. The results of Seright [17] indicated that a resistance coefficient could be generated by gel in the flow process and firstly increased and then decreased and be stable along the flow direction. The research of Dolan *et al.* [18] confirmed that the gelation time in porous media was much longer than that in static conditions.

In this paper, with the HPAM (Hydrolyzed Poly Acrylamide) system and redox system, we discuss the static gelation time in ampoule bottles and porous media, and the dynamic gelation time in an agitator and porous media. Furthermore, the relationship between the static gelation time and dynamic gelation time was established, and the critical gelation shear rate was determined with the agitator.

1 EXPERIMENTAL SECTIONS

1.1 Materials

The polymer employed in this research was classical partially hydrolyzed polyacrylamide (HPAM), whose molecular weight is 1.2×10^7 and the degree of hydrolysis is 22%. The crosslinker was a Redox System (RS), including sodium bichromate and sodium sulfite, and the mass ratio was 1:2. The Synthetic Water (SW) used in this study contained 6 921 ppm Na^+, 412 ppm Ca^{2+}, 148 ppm Mg^{2+} and 11 853 ppm Cl^-.

The experiments of core flooding were operated with sand packs with 2.50 cm Inner Diameter (ID) and 10.00 cm length or 100.00 cm length; the sand packs were filled with clean micro-glass beads of different meshes and saturated by SW. All of the gelation experiments were performed at 75°C.

1.2 Static Gelation in Ampoule Bottles

HPAM solutions of different concentrations were prepared with SW as gelant, and then the crosslinkers (RS) of different concentrations were added. 15 mL gelant was placed into an ampoule bottle made of thermo-resistant glass. The bottle has a screw cap, and it is 10 cm long with an ID of 2 cm. The air in the headspace above the solution in the ampoule was replaced with nitrogen to remove any oxygen that could cause gel degradation. A series of samples of each formula were prepared for the following static gelation experiments.

1.3 Static Gelation in Porous Media

A series of sand packs with 2.50 cm ID and 10.00 cm length were prepared. After being saturated by SW, Pore Volumes (PV) and permeabilities (K) were calculated, then the sand pack was injected into one PV gelants (HPAM + RS) at the injection speed of 1 mL/min. Then, the sand packs were sealed and placed in a 75°C oven. At regular intervals the breakthrough pressure gradient of each sand pack was measured by SW flooding at the speed of 1 mL/min.

1.4 Dynamic Gelation under Shear in an Agitator

70 mL gelant was placed in an agitator and stirred at a certain rotation speed at 75°C. The viscosity of the gelant was measured by a Brookfield DV-II viscometer at 75°C at regular intervals. At the same time, the effect of the shear rate on dynamic gelation was considered by changing the Revolutions Per Minute (RPM). After being stirred for different times, the gelant remained standing in a 75°C oven and the viscosity was measured.

1.5 Dynamic Gelation in Porous Media

Gelant in the normal conditions was continuously injected into the sand pack of 2.50 cm ID and 100.00 cm length, which was saturated by SW and placed in a 75°C oven. During the whole process, the change in injection pressure with time was recorded. At the same time, the influencing factors were considered, such as polymer and crosslinker concentrations.

2 RESULTS AND DISCUSSION

2.1 Static Gelation in Ampoule Bottles

Gelation time is one of the most important parameters. In this paper, the gelation time, which is divided into Initial Gelation Time (IGT) and Final Gelation Time (FGT), was determined by measuring viscosity [3]. IGT is the time of sudden increase in viscosity. FGT is the time when the reaction is nearly finished. The gelation times of different formulas were determined and are shown in Figure 1. The results showed that the process of gelation could be divided into three phases: slow induction, sudden increase and stability. With the increase in concentrations of HPAM and RS, both IGT and FGT decreased. The value of the slope of the sudden increase phase represented the reaction rate. Increase in the concentration of the system could accelerate the reaction rate and shorten IGT, as well as FGT.

2.2 Static Gelation in Porous Media

After gelant was injected into porous media, the increase in the following water flooding pressure mainly resulted from plugging by crosslinked gelant and the pressure increased with the degree of crosslinking. When the pressure was finally constant, the crosslinking reaction was completed. The IGT is the time when the water flooding pressure suddenly increases, and the FGT is the first moment when the pressure remains unchanged. The IGT and FGT of static gelation in porous media were obtained by measuring the change in pressure with time; the results are shown in Figure 2. In order to eliminate the effect of permeability on water flooding pressure, vadose force (F, unit N) was introduced, which is defined as the required force for fluid flowing through the special core. Numerically, F was equal to the product of water flooding pressure and permeability. From Figure 2, the IGT and FGT of static gelation in porous media were obtained. Furthermore, with increasing concentration of the system, the IGT and FGT decreased. The change in polymer concentration mainly influenced the strength of the gel in porous media; meanwhile, the change in crosslinker concentration mainly influenced gelation time in porous media. The explanation is that the crosslinker RS is an inorganic ion, whose main role is crosslinking polymer molecules and shortening gelation time. However, polymer molecules could enhance the dimensional network structure of gel.

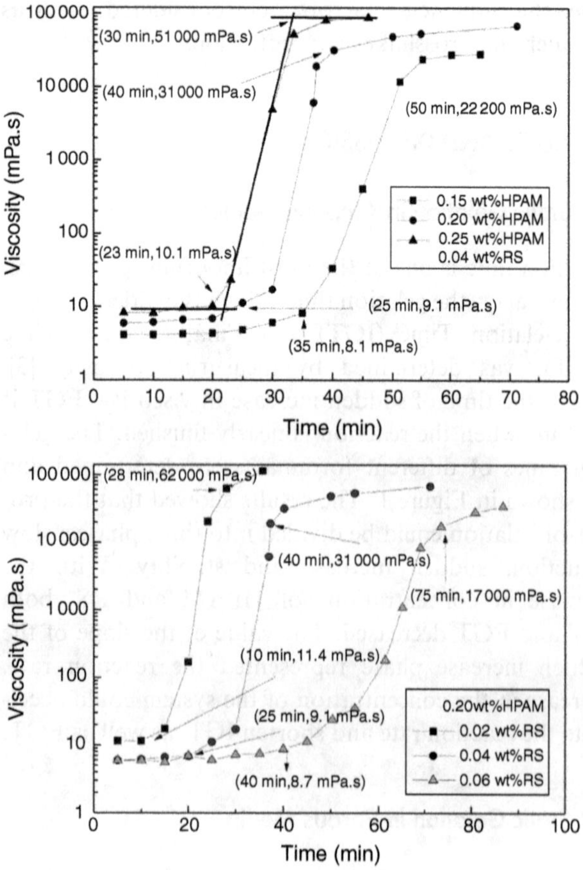

Figure 1

The static gelation in ampoule bottles of system HPAM/Cr(III) at 75°C.

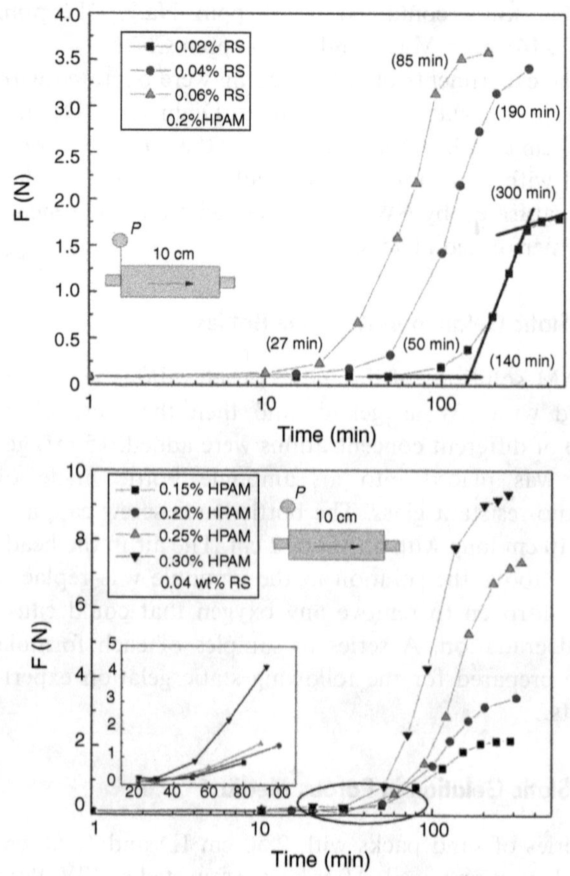

Figure 2

The change of vadose force with time in the process of static gelation o system HPAM/Cr(III) in porous media with 2.50 cm ID and 10.00 cm length at 75°C.

Comparing static gelation in ampoule bottles with static gelation in porous media, it was indicated that both IGT and FGT in porous media were much longer than those in ampoule bottles. Also, IGT in porous media were about three times the IGT in ampoule bottles, while FGT in porous media were about six times the FGT in ampoule bottles. This might result from gelant adsorption and retention in porous media.

2.3 Dynamic Gelation under Shear in an Agitator

In the process of water shutoff and profile control, before it is injected into formation, the gelant has a series of shear rates, which influence the gelation time and gel strength. It is very important to analyze the effect of shear on gelation time and gel strength. The classical shear of the agitator was studied with the HPAM system and RS in the following experiments.

2.3.1 The Effect of the Shear Rate on Gelation Time

The rotation speed of the agitator, with four paddles (2.5 cm length, 0.8 cm width and 0.1 cm thickness; the angle between the paddle and horizontal plane is 45°) was in the range of 45 $\sim$ 600 rpm. The shear rate was calculated with the method of Metzner and Otto [19]. The vessel for gelant was a cylinder with ID 5.86 cm and height 2.6 cm. The volume of gelant was 70 mL in every agitation. The results of the effect of shear on gelation time are displayed in Figure 3. Under different shear rates, the gelation time and gel strength were different. When the shear rate was less than 15.17 s^{-1}, IGT increased with increasing shear rate but was shorter than that in static conditions, and with the increase in shear rate, ultimate gel strength decreased sharply. When the shear rate was up to 18.96 s^{-1}, the system was cross-linked slightly, and the IGT was longer than that in static conditions. Furthermore, FGT in static conditions was faster than that at any shear rate. So, it was considered that when the shear rate was higher than 18.96 s^{-1}, no gel existed. A tiny shear could accelerate the reaction between polymer molecules and the crosslinker and was good for the formation of macromolecular aggregates in the stage of induction [6], while the structure of macromolecular aggregates was destroyed by severe shear and IGT was prolonged. The process of gelation under a shear rate (less than 15.17 s^{-1}) was divided into four phases: induction, sudden increase, stability and decrease [13].

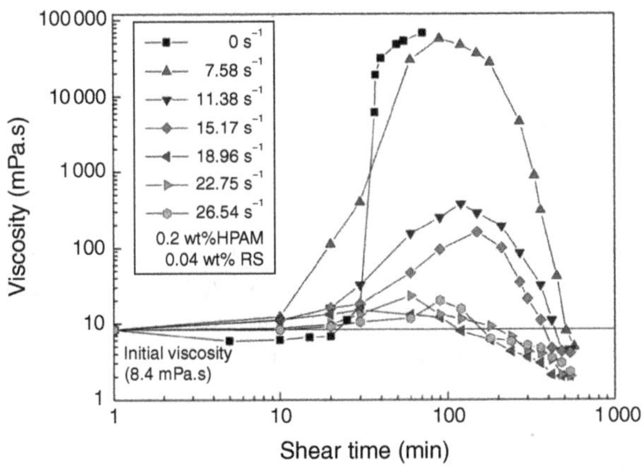

Figure 3

The change of viscosity with shear time under different shear rates with the system of HPAM/Cr(III) at 75°C.

The results of the effect of the concentrations of the polymer and crosslinker on gelation time are shown in Figure 4. When the concentrations of polymer and crosslinker increased, the amount of polymer molecules, crosslinker ions and the probability of a crosslinking reaction increased. This factor resulted in shortened gelation time and increased ultimate gel strength, even under shear conditions.

2.3.2 Static Gelation after Shearing

The dynamic gelation process under shear was divided into induction, sudden increase, stability and decrease stages, shown in Figure 5. Viscosity change in the gelant at different gelation stages after shearing for a certain time was a very important factor to study. For this, the experiment was designed as follows: gelant of 0.2 wt% HPAM and 0.04 wt% RS was sheared at the rate of 11.38 s^{-1} at 75°C and the shear time lasted for 1 min. Then, the shearing was stopped and the gelant was kept gelling under static conditions. The viscosity of static gelation after shearing for 1 min was measured. The shear time was increased and the experiment was repeated. The relationship between the viscosity of static gelation after different shearing times and shear time was established; the data are shown in Figure 5. For the viscosities of dynamic gelation under shear and static gelation after shearing for different times, shear had almost no effect on gel strength at the stage of induction, but it had an enormous influence on gel strength at the stage of sudden increase. At the stage of sudden increase, the viscosity of dynamic gelation under shear increased

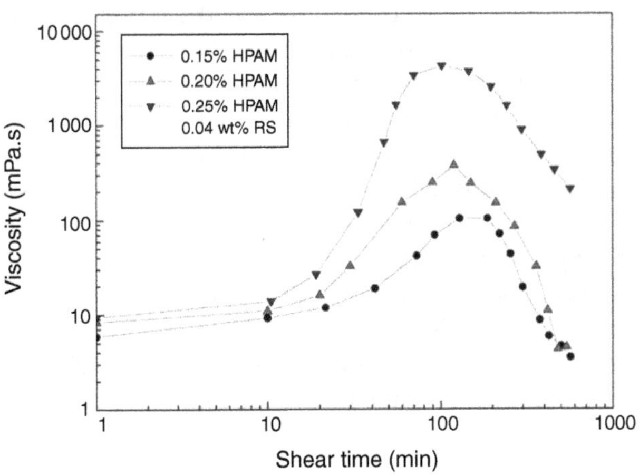

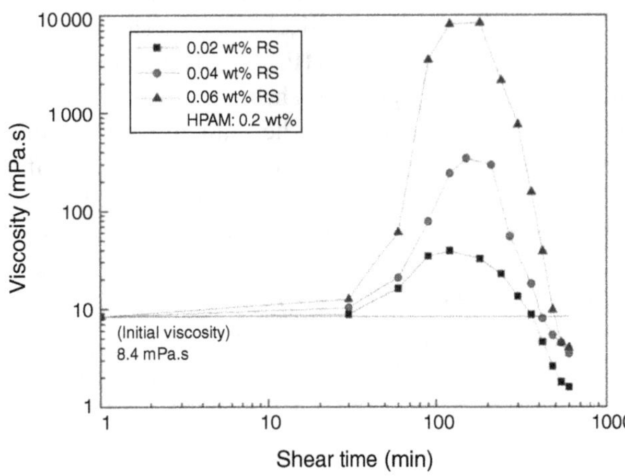

Figure 4

The effect of concentration of polymer and crosslinker on gelation time with the system of HPAM/Cr(III) under shear rate 11.38 s^{-1} at 75°C.

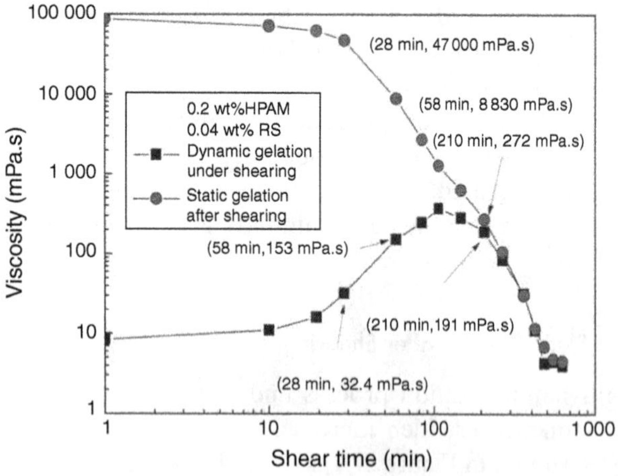

Figure 5

The change of viscosity in the process of static gelation after shearing with the system of HPAM/Cr(III) at 75°C.

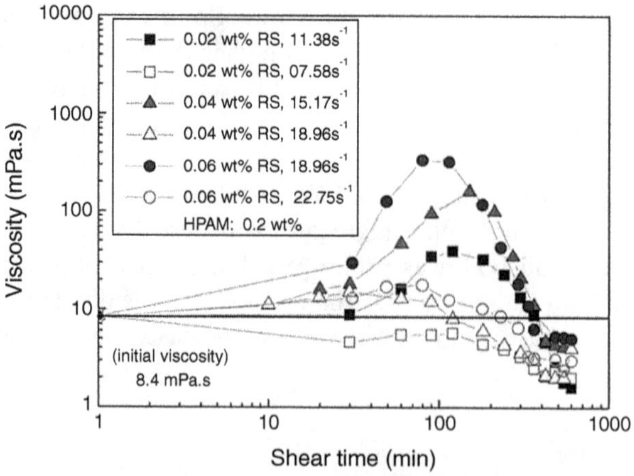

Figure 6

Determination of critical gelation shear rates with different formulas at 75°C.

linearly with shear time and the viscosity of static gelation after shearing for different times decreased linearly with shear time in semilog coordinates. The slopes were 0.022 and −0.024 in semilog coordinates, respectively. At the stage of stability and decrease, the viscosities had almost no difference. Macromolecular aggregates were destroyed by shear at the stage of induction, while the three-dimensional network structure was damaged by shear at the stage of sudden increase.

2.3.3 Critical Gelation Shear Rate

With increasing shear rate, gel strength decreased. When the shear rate reached a certain value, the HPAM system and RS could not form gel. There was a critical value of the shear rate, named the critical gelation shear rate, beyond which the system of HPAM/Cr(III) could not form gel. The critical gelation shear rates of different formulas were measured, and the data are shown in Figure 6. The critical gelation shear rates of different formulas were 11.38 s^{-1} (0.2 wt% HPAM and 0.02 wt% RS), 15.17 s^{-1}(0.2 wt% HPAM and 0.04 wt% RS) and 18.96 s^{-1}(0.2 wt% HPAM and 0.06 wt% RS), respectively. This indicated that the critical gelation shear rate increased with increasing crosslinker concentration. Therefore, the higher the gel strength, the higher the critical gelation shear rate.

2.4 Dynamic Gelation in Porous Media

When the gelant was transported in porous media at 75°C, there were two processes. One was the process of reaction between polymer molecules and crosslinkers, which resulted in the formation of gel and buildup of injection pressure. The other one was breakage of the three-dimensional network structure by the following injected gelant and porous media, which led to a pressure drop. So, the gelation time was longer than that in static porous media. When the gelant began to form gel, the injection pressure rose. The change in pressure with time is shown in Figure 7. With increase in time, the injected volume increased and input pressure rose. The process of input pressure was divided into the induction and sudden increase stages. The IGT could be determined by the input pressure, as shown in Figure 7. However, the FGT was difficult to determine. Under low concentrations of polymer and crosslinker, the pressure P_2 increased by a certain degree, which represented the gelant migration in porous media. In the process of dynamic gelation in porous media, there were mainly two effect factors: the crosslinking reaction, making input pressure rise, and shear degradation, reducing input pressure. When the concentrations of polymer and crosslinker increased, the IGT decrease was shortened, as shown in Figure 8. Compared with static gelation in porous media and in ampoule bottles, the IGT of dynamic gelation in porous media was much longer, which was about four times the IGT of static gelation in porous media, and eight to ten times the IGT of static gelation in ampoule bottles.

CONCLUSIONS

Based on the experimental analysis of the HPAM system and Redox System (RS: sodium bichromate and sodium

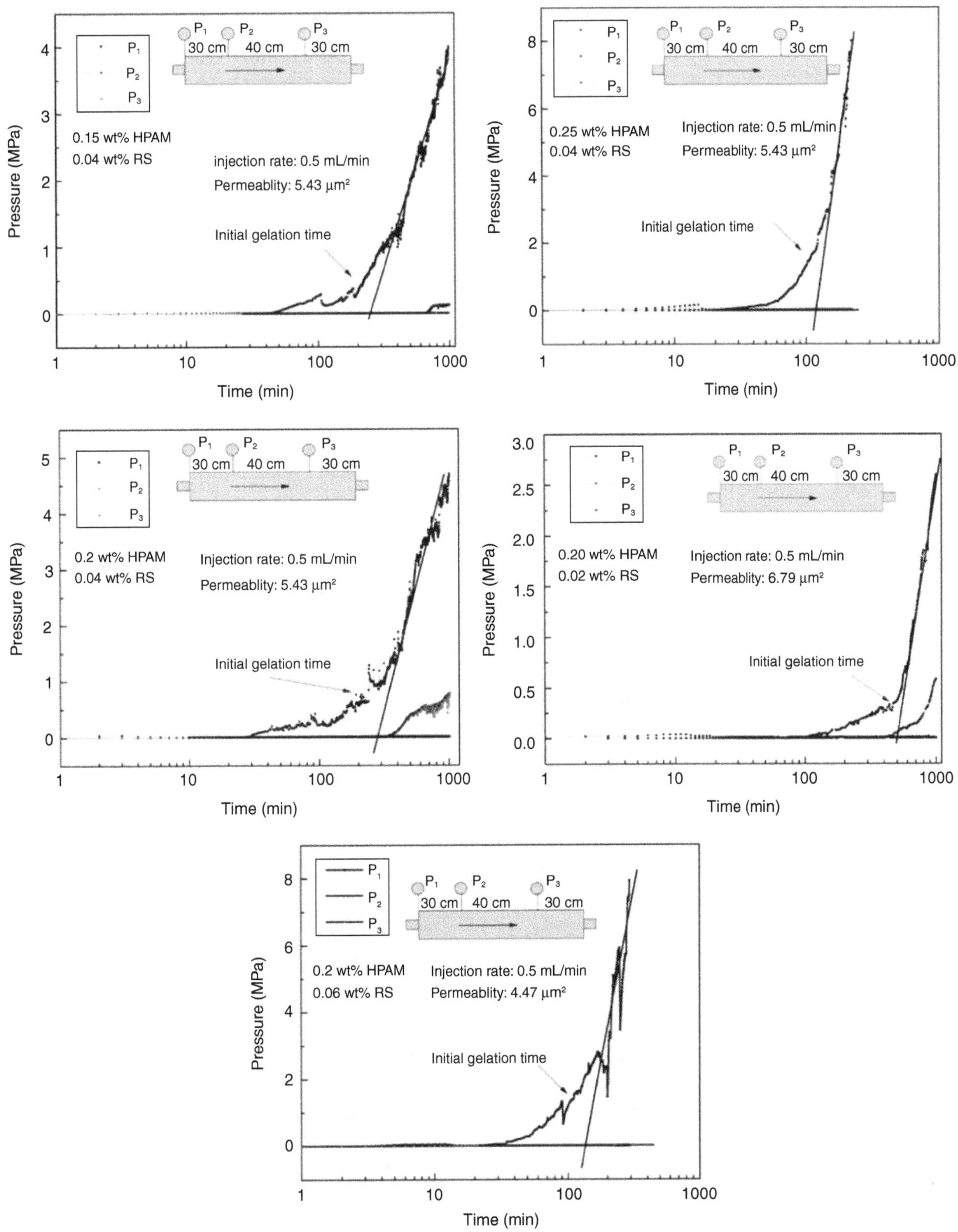

Figure 7

The change of pressure with time in the process of gelant system of HPAM/Cr(III) migration in porous media at 75°C.

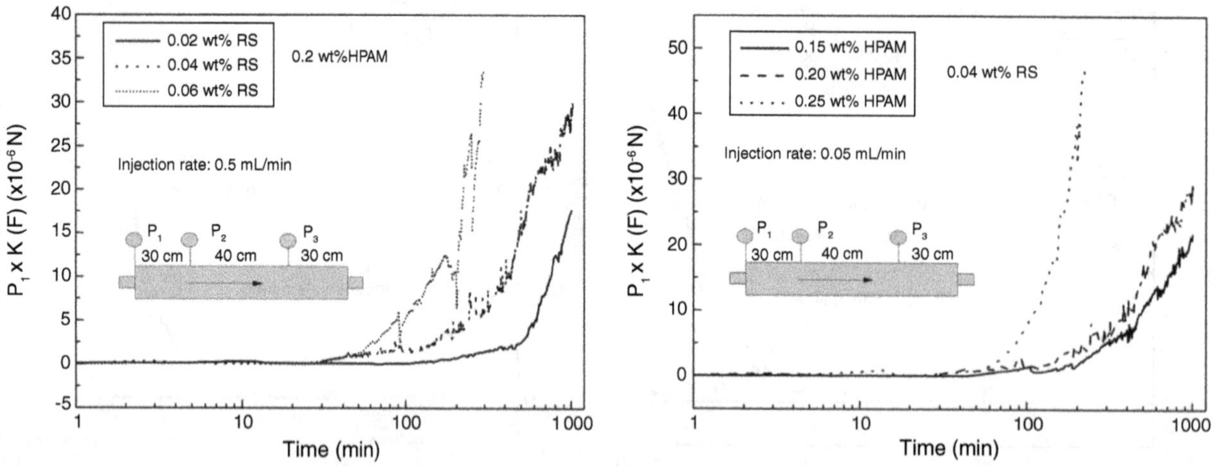

Figure 8

The effect of polymer and crosslinker on dynamic gelation in porous media at 75°C.

sulfite) at 75°C, the most important conclusions are summarized as follows:

- the time of static gelation in porous media was much longer than that in ampoule bottles. IGT in porous media were about three times those in ampoule bottles, while FGT in porous media were six times those in ampoule bottles;
- there was a critical gelation shear rate in the process of gelation under shear, beyond which the system could not form gel, and it was enlarged from 11.38 s^{-1} to 18.96 s^{-1} with increasing RS concentration from 0.02 wt% to 0.06 wt%;
- shear had almost no influence on gel strength during the induction stage, but in the stage of sudden increase, shear could sharply degrade gel strength. The viscosity of a system with 0.2 wt% HPAM and 0.04 wt% RS standing for 40 min after shearing decreased linearly with time in semilog coordinates and the slope was −0.024 in semilog coordinates;
- the gelation time of dynamic gelation in porous media was much longer than that of static gelation in porous media and ampoule bottles. IGT of dynamic gelation in porous media were about four times the IGT of static gelation in porous media and eight to ten times the IGT of static gelation in ampoule bottles.

ACKNOWLEDGMENTS

This work was sponsored by the National S&T Major Project (Grant No.: 2011ZX05024-004) on chemical flooding in Bohai heavy oil reservoir.

REFERENCES

1 Sydansk R. (1988) A New Conformance-Improvement-Treatment Chromium(III) Gel Technology, *SPE/DOE Enhanced Oil Recovery Symposium*, Tulsa, Oklahoma, 17-20 April, *SPE/DOE paper* 17329.

2 Reddy B.R., Eoff L., Dalrymple E.D., Black K., Brown D., Rietjens M. (2003) A Natural Polymer-Based Cross-Linker System for Conformance Gel Systems, *SPE J.* **8**, 2, 99-106.

3 Mokhtari M., Ozbayoglu E.M. (2010) Laboratory Investigation on Gelation Behavior of Xanthan Crosslinked with Borate Intended to Combat Lost Circulation, *SPE Production and Operations Conference and Exhibition*, Tunis, 8-10 June, *SPE paper* 136094-MS.

4 Gao S., Guo J., Wu L., Wang S. (2008) Gelation of konjac glucomannan crosslinked by organic borate, *Carbohydr. Polym.* **73**, 3, 498-505.

5 Chambon F., Winter H. (1985) Stopping of crosslinking reaction in a PDMS polymer at the gel point, *Polym. Bull.* **13**, 6, 499-503.

6 Basta M., Picciarelli V., Stella R. (1991) Electrical conductivity in the kinetic gelation process, *Eur. J. Phys.* **12**, 210-213.

7 Boey F., Qiang W. (2000) Determining the gel point of an epoxy-hexaanhydro-4-methylphthalic anhydride (MHHPA) system, *J. Appl. Polym. Sci.* **76**, 8, 1248-1256.

8 Jousset F., Green D.W., Willhite G.P., McCool C.S. (1990) Effect of High Shear Rate on *In-Situ* Gelation of a Xanthan/Cr(III) System, *SPE/DOE Enhanced Oil Recovery Symposium*, Tulsa, Oklahoma, 22-25 April, *SPE paper* 20213-MS.

9 Tseu J.S., Liang J.T., Hill A.D. (1992) Re-formation of Xanthan/Chromium gels after shear degradation, *SPE Reserv Eng.* **7**, 21-28, *SPE paper* 18506-PA.

10 Bartosek M., Mennella A., Lockhart T.P., Causin E., Rossi E., Passucci C. (1994) Polymer gels for conformance treatments: propagation of Cr (III) crosslinking complexes in porous media, *SPE/DOE Ninth Symposium on improved oil Recovery*, Tulsa, Oklahoma, USA, 17-20 April, *SPE paper* 27828-MS.

11 Broseta D., Marquer O., Zaitoun A., Baylocq P., Fery J.-J. (2000) Shear Effects on Polyacrylamide/Chromium (III) Acetate Gelation, *SPE Reserv. Eval. Eng.* **3**, 3, 204-208.

12 McCool S., Li X., Willhite G.P. (2007) Effect of Shear on Flow Properties During Placement and on Syneresis After Placement of a Polyacrylamide Chromium Acetate Gelant, *SPE International Symposium on Oilfield Chemistry*, Houston, Texas, USA, 28 Feb.-2 March, *SPE paper* 106059-MS.

13 Chauveteau G., Tabary R., Renard M., Omari A. (1999) Controlling *In-Situ* Gelation of Polyacrylamides by Zirconium for Water Shutoff, *SPE International Symposium on Oilfield Chemistry*, Houston, Texas, 16-19 Feb., *SPE paper* 50752-MS.

14 Bhaskar R.K., Stinson J.A., Willhite G.P., Thiele J.L. (1988) The Effects of Shear History on the Gelation of Polyacrylamide/Chromium VI Thiourea Solutions, *SPE Reserv. Eng.* **3**, 4, 1251-1256.

15 Kolnes J., Stavland A., Thorsen S. (1991) The Effect of Temperature on the Gelation Time of Xanthan/Cr (III) Systems, *SPE International Symposium on Oilfield Chemistry*, Anaheim, California, 20-22 Feb., *SPE paper* 21001-MS.

16 McCool C.S., Li X., Willhite G.P. (2009) Flow of a Polyacrylamide/Chromium Acetate System in a Long Conduit, *SPE J.* **14**, 1, 54-66, *SPE*-106059-PA.

17 Seright R.S. (1997) Use of Preformed Gels for Conformance control in Fractured system, *Old Prod. Facil.* **12**, 1, 59-65.

18 Dolan D.M., Thiele J.L., Willhite G.P. (1998) Effects of pH and Shear on the Gelation of a Xanthan-Cr (III) System, *Old Prod. Facil.* **13**, 2, 97-103.

19 Metzner A., Otto R. (1957) Agitation of non-Newtonian fluids, *AIChE J.* **3**, 1, 3-10.

2

Performance Analysis of Fractured Wells with Stimulated Reservoir Volume in Coal Seam Reservoirs

Zhao Yu-long[1]*, Zhang Lie-Hui[1], Feng Guo-Qing[1], Zhang Bo-Ning[1] and Kang Bo[2]

[1] State Key Laboratory of Oil and Gas Reservoir Geology and Exploitation, Southwest Petroleum University, Chengdu, Sichuan 610500, P.R. China
[2] Chengdu North Petroleum Exploration and Development Technology LTD, Chengdu, Sichuan 610500, P.R. China
e-mail: 373104686@qq.com

* Corresponding author

Abstract — *CoalBed Methane (CBM), as one kind of unconventional gas, is an important energy resource, attracting industry interest in research and development. Using the Langmuir adsorption isotherm, Fick's law in the matrix and Darcy flow in cleat fractures, and treating the Stimulated Reservoir Volume (SRV) induced by hydraulic fracturing as a radial composite model, the continuous linear source function with constant production is derived by the methods of the Laplace transform and Duhamel theory. Based on the linear source function, semi-analytical solutions are obtained for a fractured vertical well producing at a constant production rate or constant bottom-hole pressure. With the help of the Stehfest numerical algorithm and computer programing, the well test and rate decline type curves are obtained, and the key flow regimes of fractured CBM wells are: wellbore storage, linear flow in SRV region, diffusion flow and later pseudo-radial flow. Finally, we analyze the effect of various parameters, such as the Langmuir volume, radius and permeability in the SRV region, on the production performance. The research results concluded in this paper have significant importance in terms of the development, well test interpretations and production performance analysis of unconventional gas.*

Résumé — Analyse des performances de puits fracturés avec un volume de réservoir stimulé dans des gisements de charbon — Le gaz de houille (*CoalBed Methane*, CBM), en tant que gaz non conventionnel, est une ressource d'énergie importante, qui attire l'intérêt industriel en matière de recherche et de développement. En utilisant l'isotherme d'adsorption de Langmuir, la loi de Fick dans la matrice et le flux de Darcy dans les fractures clivées, et en traitant le volume de réservoir stimulé (*Stimulated Reservoir Volume*, SRV) induit par la fracture hydraulique comme un modèle composite radial, la fonction de source linéaire continue avec une production constante est obtenue par les méthodes de la transformée de Laplace et de la théorie de Duhamel. Sur la base de la fonction de source linéaire, des solutions semi-analytiques sont obtenues pour un puits vertical fracturé produisant à un taux de production constant ou à une pression de fond constante. En s'aidant de l'algorithme numérique de Stehfest et de la programmation informatique, des courbes de test de puits et de déclin de taux de production sont obtenues, et les régimes d'écoulement clés des puits CBM fracturés sont les suivants : stockage de puits de forage, flux linéaire dans la zone SRV, flux de diffusion et flux pseudo-radial ultérieur. Enfin, nous analysons l'effet de différents paramètres, tels que le volume de Langmuir, le rayon et la perméabilité dans la zone SRV, sur les performances de production. Les résultats mentionnés dans cet article ont une importance significative en termes de développement, d'interprétations de test de puits et d'analyse des performances de production des gaz non conventionnels.

NOMENCLATURE

Field Variables

c_{tf1}	Total compressibility of the SRV region (MPa^{-1})
c_{tf2}	Total compressibility of the outer region (MPa^{-1})
C	Wellbore storage coefficient (m^3/MPa)
D	Diffusion coefficient (m^2/h)
H	Formation thickness (m)
k_{1f}	Permeability of the macro-fracture system in the SRV region (D)
k_{2f}	Permeability of the macro-fracture system in the outer region (D)
L	Reference length, in this paper L can be half fracture length, x_f
M	Molar mass of natural gas (mol/kg)
M_{12}	Mobility ratio between the inner and outer regions (-)
p	Reservoir pressure (MPa)
p_i	Initial reservoir pressure (MPa)
p_L	Langmuir pressure constant (MPa)
p_{1f}	Pressure in the SRV region (MPa)
p_{2f}	Pressure in the outer region (MPa)
p_{sc}	Standard pressure, $p_{sc} = 0.1$ MPa
q^*	Gas diffusion rate from the coal particles to the matrix micro-pore, defined in Equation (3)
q_{sc}	Well production rate in the standard conditions (m^3/d)
$\hat{q}$	Production rate from an instantaneous line source (m^3/d)
q_{scL}	Production rate from a continuous line source (m^3/d)
R	Universal gas constant, $R = 0.008314$ (MPa·m^3)/(K·kmol)
R_m	External radius of the matrix block (m)
r	Radial distance, $r = \sqrt{x^2 + y^2 + z^2}$
r_e	Radial of the outer boundary (m)
r_m	Radial distance of the SRV region (m)
S_{kin}	Skin factor (-)
S_c	Instantaneous point source in real space
t	Production time (h)
t_a	Pseudo-time, defined in Equation (12)
T	Reservoir temperature (K)
T_{sc}	Standard temperature, $T_{sc} = 273$ K
V	Average matrix gas concentration (m^3/m^3)
V_E	Volume of gas adsorbed per unit volume of the coal grain in equilibrium at pressure p (m^3/m^3)
V_i	Volume of gas adsorbed per unit volume of the coal grain in equilibrium at initial pressure p_i (m^3/m^3)
V_L	Langmuir volume (m^3/m^3)
x, y, z	Coordinates (m)
Z_g	Gas deviation factor (-)

Dimensionless variables

r_D	Dimensionless radius, defined in Equation (21)
r_{eD}	Dimensionless outer boundary, $r_{eD} = \frac{r_e}{L}$
r_{Dm}	Dimensionless matrix radius, defined in Equation (20)
t_D	Dimensionless time, defined in Equation (14)
C_D	Dimensionless wellbore storage coefficient, $C_D = \dfrac{C}{2\pi\phi_{1f}c_{tf1}L^2}$
$\bar{\psi}_{SD}$	Pseudo-pressure response of the well with a skin effect in Laplace space
$\Delta\bar{\psi}_{1fD}$	Dimensionless fracture pseudo-pressure of the inner region in Laplace space
$\bar{\psi}_{wD}$	Pseudo-pressure response of a CBM well with storage in Laplace space
ψ'_{wD}	Pseudo-pressure derivative response, $\psi'_{wD} = \dfrac{d\psi_{wD}}{d(t_D/C_D)}$
$\bar{q}_D$	Dimensionless rate in Laplace space
q_D	Dimensionless rate, defined in Equation (34)
t_{Dd}	Dimensionless decline time, defined in Equation (35)
q_{Dd}	Dimensionless decline rate function, defined in Equation (36)
q_{Ddi}	Dimensionless decline rate integral, defined in Equation (37)
q'_{Ddi}	Dimensionless decline rate integral derivative, defined in Equation (38)
V_D	Dimensionless micro-pores volumetric concentration, $V_D = V - V_i$
V_{ED}	Dimensionless equilibrium volumetric concentration, $V_{ED} = V_E - V_i$

Greek Variables Equations

$\Delta\bar{\psi}_{1f}$	Pseudo-pressure of the inner fracture system in Laplace space
μ	Gas viscosity (cp)
μ_{gi}	Gas viscosity, evaluated in the initial pressure conditions (cp)
α	Conversion factor, and $\alpha = 3.684 \times 10^{-3}$ for practical units
$\emptyset_{1f}$	Porosity of the fracture system in the SRV region, fraction
$\emptyset_{2f}$	Porosity of the fracture system in the outer region, fraction
ρ_{sc}	Fluid density under the standard conditions (g/m^3)
$\psi(p)$	Real gas pseudo-pressure (MPa^2/cp)
ψ_{fi}	Pseudo-pressure in initial fracture conditions (MPa^2/cp)
ψ_{wf}	Pseudo-pressure of bottom-hole pressure (MPa^2/cp)
$\psi_L(p_f)$	Langmuir pseudo-pressure constant (MPa^2/cp)
ω_1	Storativity ratio of the inner region (-)
ω_2	Storativity ratio of the outer region (-)
λ	Dimensionless transfer constant, defined in Equation (17)
τ	Sorption time (h)
Λ	Parameters defined in Equation (22)
σ	Ad- and desorption coefficients, defined in Equation (23)
ε	Variable

Subscripts

a	Pseudo
D	Dimensionless variable
f	Fracture system
mD	Matrix dimensionless
wD	Wellbore dimensionless
sc	Standard conditions
1	Inner region
2	Outer region

Superscripts

–	Laplace transform

Operator

$L^{-1}[\,]$	Laplace inverse transform operator

INTRODUCTION

Coal seams, together with tight gas sands, Devonian shales and gas hydrates, are classified as unconventional gas reservoirs. Coal seams, similar to shale reservoirs, are known to be a source rock for natural gas. The gas stored in coal seams is primarily composed of methane, which is usually called CoalBed Methane (CBM). The contribution of CBM to unconventional gas reservoirs has played an important role in gas production all over the world. In the US, for example, the maximum annual production of CBM was 55.67 billion cubic meters in 2008, accounting for 9.75% of its domestic gas production. Although the subsequent annual production from 2009 to 2011 was continuously declining, the production of CBM is still expected to grow dramatically in the future with the economic recovery in the US (Source: US Energy Information Administration, Washington, DC; http://www.eia.gov/dnav/ng/hist/rngr52nus_1a.htm).

Due to the large internal surface area of coal seams, a significant amount of gas is stored as adsorbed gas as well as free gas in the network of cleats, and the characteristics of the coal composition, reservoir temperature and pressure, and the development of cleats vary significantly from one coal bed to another. Therefore, each CBM reservoir has distinct transport behaviors and different producing methods. Well testing is a powerful technique to recognize the *in situ* characterization and evaluate the stimulation effect of oil and gas reservoirs, and has been widely implemented to determine the gas transport and sorption characteristics of coal beds (Bayles and Reznik, 1986). Therefore, a vital task for researchers is to establish various models to evaluate the properties of underground reservoirs.

The physical and mathematical models to describe gas transport in coal reservoirs are well known and have been studied in the past. King *et al.* (1986) described the mathematical and numerical developments for a series of finite difference models that simulate the simultaneous flow of water and gas through dual-porosity coal seams. Bumb and Mckee (1988) derived an approximate analytical solution for single-phase gas flow when gas exists both as free gas and adsorbed on the matrix. Ertekin and Sung (1989) established a single-phase mathematical model in radial-cylindrical coordinates to describe both unsteady-state sorption phenomena in the coal matrix and the coupled laminar gas transport in the fracture network. Anbarci and Ertekin (1990) derived a single-phase flow equation for coalbed methane that includes non-equilibrium sorption phenomena in the coal matrix and laminar gas transport in the cleat system with kinds of outer and inner boundary conditions. Rushing *et al.* (1991) developed a model to analyze the transient pressure response from slug tests in wells completed in multiple, hydraulically fractured coal seams. Engler and Rajtar (1992) presented an analytical solution for horizontal wells producing at constant rate in anisotropic, semi-infinite reservoirs by the Laplace and Fourier integral transform methods. Anbarci and Ertekin (1992) presented a solution package to describe the pressure transient behavior of coal reservoirs in the presence of a hydraulically fractured well. Kamal and Six (1993) presented a new analysis method for pressure transient testing in coal degasification projects. Sarkar and Rajtar (1994) developed a new semi-analytical solution to account for wellbore storage and skin effects on the pressure response of a horizontal well.

Later, Claudia and Joseph (2004) analyzed the production data by analytical solutions, which include analytical solutions for coalbed wells that took into account gas desorption from coal using isotherm data, analytical solutions for conventional gas reservoirs and Arp's decline curve analysis. Meng and Luke (2007) presented a rigorous dual-porosity model for gas reservoir flow incorporating adsorption behavior and obtained an analytical solution for diffusion within a matrix block. Clarkson (2009) and Clarkson *et al.* (2013) reported the production data and pressure transient analysis for coalbed methane wells. Hu *et al.* (2009) developed a new mathematical model with the consideration of Klinkenberg effects on coalbed gas flow properties. Nie *et al.* (2012) established a mathematical model of a horizontal well in a coalbed reservoir with the consideration of pseudo-steady diffusion in the matrix and Darcy flow in the network of cleats to analyze the transient pressure and rate decline behaviors. Based on Fick's law in the matrix and Darcy flow in cleats and hydraulic fractures, Wang *et al.* (2013) presented a new semi-analytical model considering the effects of boundary conditions to investigate the transient pressure behavior for asymmetrically fractured wells in coal reservoirs.

In general, coal seam reservoirs have dual porosity, which is macro-pores in the natural fracture network and those in the coal matrix (Claudia and Joseph, 2004; Meng and Luke, 2007; Clarkson, 2009; Nie *et al.*, 2012; Clarkson *et al.*, 2013; Wang *et al.*, 2013). The production of CBM in coal beds can be cataloged into three stages:

- Darcy flow of CBM through the fracture network to the wellbore;
- desorption of the gas from the matrix-cleat surfaces;
- diffusion of CBM through the network of capillaries and micro-pores within the coal matrix to the matrix-cleat interface (Anbarci and Ertekin, 1990, 1992).

Hydraulic fracturing is an effective well stimulation technique employed for production of oil and gas from damaged wells or from low-permeability reservoirs, and it has become a primary technique for CBM development (Rushing *et al.*, 1991; Anbarci and Ertekin, 1992; Wang *et al.*, 2013). With the development of drilling and fracturing equipment, Massive Hydraulic Fracturing (MHF) has been widely used in CBM wells, which activates and connects existing discrete or micro-seismic natural fractures. Due to the natural fracture network and unique rock mechanics properties of the coal beds, MHF of CBM wells always generates a fracture network around the well, which is usually called Stimulated Reservoir Volume (SRV). The SRV is always defined as the volume of a reservoir containing the fracture network and the main hydraulic fractures with which it is effectively stimulated for improved well performance.

Due to the well testing analysis is an effective method to understand the *in situ* characterization of coal beds and evaluate fracturing effectiveness, and most of the above literature on this subject models the CBM transport characteristics without taking into account the effect of the SRV on the fluid flow performance (Bayles and Reznik, 1986; King *et al.*, 1986; Bumb and Mckee, 1988; Anbarci and Ertekin, 1990, 1992; Sarkar and Rajtar, 1994; Claudia and Joseph, 2004; Meng and Luke, 2007; Clarkson, 2009; Nie *et al.*, 2012; Clarkson *et al.*, 2013; Wang *et al.*, 2013). Therefore, we expect to establish the corresponding mathematical model for the fractured CBM well with the SRV captured. In this paper, we considered the Langmuir adsorption isotherm and pseudo-steady and transient diffusion in the matrix, and Darcy flow both in the outer natural fracture region and inner SRV region.

1 DESCRIPTION OF PHYSICAL AND MATHEMATICAL MODEL

1.1 Characteristics of CBM Reservoirs

The characteristics of CBM reservoirs differ from conventional gas reservoirs in several areas. Unlike conventional gas reservoirs, coal is both the reservoir rock and the source rock. Most of the gas adsorbs on the surface of the coal grains on which desorption will occur during the process of production. The coal seam was treated as a dual porosity system, with the cleats as macro-porosity of the system and the spherical matrix elements representing the micro-pore structure (Claudia and Joseph, 2004; Meng and Luke, 2007; Clarkson, 2009; Cai and Yu, 2011; Nie *et al.*, 2012;

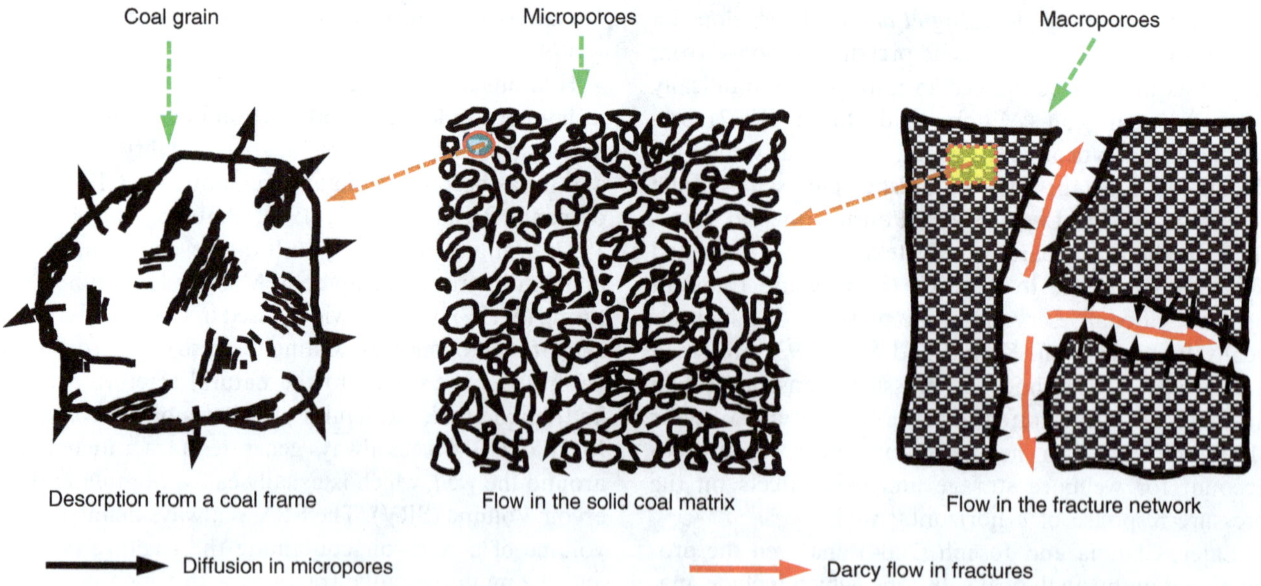

Figure 1

Schematic representation of methane flow through the coal seam.

Clarkson *et al.*, 2013; Wang *et al.*, 2013; Zhao *et al.*, 2012, 2013a, 2014). The seepage of the CBM through the porous media can be described as desorption and seepage flow (the schematic is shown in *Fig. 1*).

The equilibrium volumetric concentration which represents the amount of gas adsorbed in the matrix element is obtained using Langmuir's isotherm (Langmuir, 1916; Zhao *et al.*, 2012, 2013a), which is:

$$V_{\mathrm{E}} = V_{\mathrm{L}} \frac{p}{p_{\mathrm{L}} + p} \tag{1}$$

1.2 Continuous Linear Source Function

Although the natural fractures in a CBM reservoir are developed, the connectivity between the fractures is poor, which causes an ultra-low permeability for this reservoir as the effective permeability is a function of the conductivity, connectivity and density of fractures. Hydraulic fracturing is a common way to develop the low-permeability reservoir by injecting a mixture of liquid, sand and chemical into the formation at high pressure to create a high conductivity path around the well, which sometimes causes many induced small cracks around the main fractures (called induced fractures). As there is the development of natural fractures in a CBM reservoir, so we can use the dual-porosity model to describe it. Some of the natural fractures, including connected and disconnected, may be connected by the induced hydraulic fractures, which causes a fracture network around the well called SRV. The induced fractures can effectively connect the natural fractures around the well, which significantly enhances the permeability of the fracture system. In order to describe a reservoir with developed natural fractures and the SRV around the well simply, we use a composite model and the inner region can be treated as the SRV caused by hydraulic fracturing (*Fig. 2*).

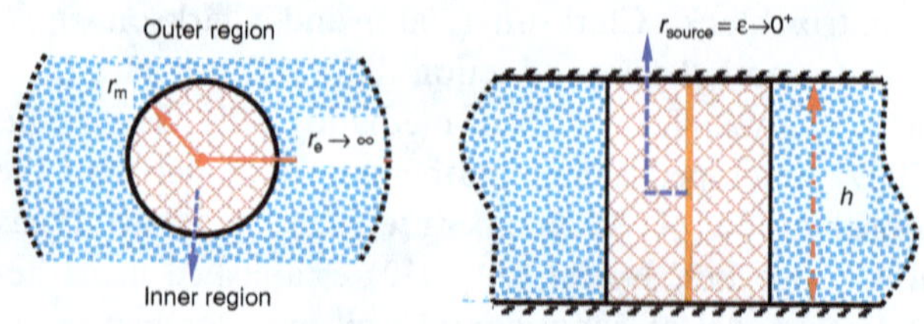

Figure 2

A schematic of a continuous line source in a radial composite reservoir.

In order to make the problem more tractable, we made the following assumptions:

- a uniform thickness of h with a radius of the SRV region of r_{m} and a constant initial pressure p_{i}, and the upper and lower boundaries are sealed;

- the porosity, compressibility and permeability in the inner and outer regions are $\emptyset_{1f}$, c_{f1}, k_{r1}; $\emptyset_{2f}$, c_{f2}, k_{f2}, respectively;
- the diffusion in the micro-pores obeys Fick's law (transient and pseudo-steady diffusion);
- the CBM flow in the macro-structure is only in the radial direction and obeys Darcy's law;
- the adsorbed gas is stored on the cleat surface and obeys the Langmuir isothermal adsorption equation;
- a fully fractured well with symmetrical wings located in the center of the inner region with the half length of x_f;
- the well produces at a constant rate (q_{sc}) or a constant bottom-hole pressure (p_{wf}).

In order to obtain the pressure response expression of a fully penetrated fractured well in a CBM reservoir, we firstly derive the continuous line source function and then integrate the source function along the fractures (Zhao *et al.*, 2012, 2013a,b). The transport equation that describes single-phase, unsteady-state flow in the cleat fracture system can be obtained by performing a mass balance on a radial element volume (Ertekin and Sung, 1989; Anbarci and Ertekin, 1990, 1992; Sarkar and Rajtar, 1994).

For the Inner Region:

$$\frac{1}{r}\frac{\partial}{\partial r}\left(r\frac{k_{1f}}{\mu}\frac{p_{1f}M}{Z_gRT}\frac{\partial p_{1f}}{\partial r}\right)=\frac{\phi_{1f}c_{t1f}}{3.6}\frac{p_{1f}M}{Z_gRT}\frac{\partial p_{1f}}{\partial t}+q_1^* \quad (2)$$

q^* is the gas diffusion rate from the coal particles to the matrix micro-pore. As the matrix porosity is very small and few structural fractures developed, we can ignore the effect of compressibility of the matrix and fracture systems on the diffusion rate. Providing the diffusion follows the first law of diffusion, it yields:

$$q_1^*=\frac{\rho_{sc}}{3.6}\frac{\partial[(1-\phi_{1f})V]}{\partial t}=\frac{Mp_{sc}(1-\phi_{1f})}{3.6RT_{sc}}\frac{\partial V}{\partial t} \quad (3)$$

For the Outer Region:

$$\frac{1}{r}\frac{\partial}{\partial r}\left(r\frac{k_{2f}}{\mu}\frac{p_{2f}M}{Z_gRT}\frac{\partial p_{2f}}{\partial r}\right)=\frac{\phi_{2f}c_{t2f}}{3.6}\frac{p_{2f}M}{Z_gRT}\frac{\partial p_{2f}}{\partial t}+q_2^* \quad (4)$$

$$q_2^*=\frac{\rho_{sc}}{3.6}\frac{\partial[(1-\phi_{2f})V]}{\partial t}=\frac{Mp_{sc}(1-\phi_{2f})}{3.6RT_{sc}}\frac{\partial V}{\partial t} \quad (5)$$

$$\frac{\partial V}{\partial t}=\begin{cases}\left.\dfrac{3D}{R_m}\dfrac{\partial V}{\partial r_m}\right|_{r_m=R_m} & \text{for transient diffusion}\\[2ex] \dfrac{6D\pi^2}{R_m^2}(V_E-V) & \text{for pseudo-steady diffusion}\end{cases} \quad (6)$$

For a well with an infinite or closed outer boundary, we have:

$$p_{2f}(r,t)|_{r\to\infty}=0 \ \text{for infinite} \quad (7)$$

$$\left.\frac{\partial p_{2f}}{\partial r}\right|_{r=r_e}=0 \ \text{for closed} \quad (8)$$

According to characteristics of the Dirac delta function (Zhao *et al.*, 2012), when the line source withdraws the fluid with an instantaneous rate ($\hat{q}$), the inner boundary condition becomes:

$$\lim_{\varepsilon\to0^+}\frac{k_{1f}}{\mu}h\left(r\frac{\partial p_{1f}}{\partial r}\right)_{r=\varepsilon}=-\alpha\hat{q}\frac{Zp_{sc}T}{p_{1f}T_{sc}}\frac{k_{2f}}{\Lambda L^2} \quad (9)$$

The interface conditions defining pressure and flow rate continuity in the reservoir are expressed as follows:

$$p_{1f}(r,t)|_{r=r_m}=p_{2f}(r,t)|_{r=r_m} \quad (10)$$

and

$$p_{1f}\left.\frac{\partial p_{1f}}{\partial r}\right|_{r=r_m}=\frac{p_{2f}}{M_{12}}\left.\frac{\partial p_{2f}}{\partial r}\right|_{r=r_m} \quad (11)$$

Dimensionless Parameters

A pseudo-time function to account for the time dependence of gas viscosity and total system compressibility is defined as follows (Agarwal, 1979; Escobar *et al.*, 2013):

$$t_a=\int_{t_o}^{t}\frac{1}{\mu(t)c_g(t)}dt \quad (12)$$

For compressible fluids, the pseudo-pressure, $\psi(p)$, introduced by Agarwal (1979), is given by:

$$\psi(p)=2\int_0^p\frac{p}{\mu(p)Z(p)}dp \quad (13)$$

Dimensionless time is defined as:

$$t_D=\frac{3.6k_{2f}t_a}{\Lambda L^2}$$

The dimensionless storage of the inner and outer regions are:

$$\omega_1=\frac{(\phi_1\mu_{gi}c_{t1})_f}{\Lambda} \quad (15)$$

$$\omega_2=\frac{(\phi_2\mu_{gi}c_{t2})_f}{\Lambda} \quad (16)$$

The dimensionless transfer constant is defined as:

$$\lambda = \begin{cases} \dfrac{k_{2f}\tau}{\Lambda L^2} & \text{for transient diffusion} \\ \dfrac{k_{2f}\tau}{6\Lambda L^2} & \text{for pseudo-steady diffusion} \end{cases} \quad (17)$$

The desorption time constant is defined as:

$$\tau = \begin{cases} \dfrac{R_m^2}{D} & \text{for transient diffusion} \\ \dfrac{R_m^2}{\pi^2 D} & \text{for pseudo-steady diffusion} \end{cases} \quad (18)$$

And the other dimensionless parameters and groups are defined as follows:

$$M_{12} = \frac{k_{1f}/\mu_{gi}}{k_{2f}/\mu_{gi}} \quad (19)$$

$$r_{Dm} = \frac{r_m}{R_m} \quad (20)$$

$$r_D = \frac{r}{L} = \sqrt{(x_D - x_{wD})^2 + (y_D - y_{wD})^2} \quad (21)$$

$$\Lambda = \begin{cases} (\phi_f \mu_{gi} c_{tf})_{1+2} + \dfrac{6k_{f2}h}{\alpha q_{sc}} & \text{for transient diffusion} \\ (\phi_f \mu_{gi} c_{tf})_{1+2} + \dfrac{2k_{f2}h}{\alpha q_{sc}} & \text{for pseudo-steady diffusion} \end{cases} \quad (22)$$

Define the ad- and desorption coefficients σ as:

$$\sigma = \frac{V_L \psi_L(p_f)}{[\psi_L(p_f) + \psi(p_f)][\psi_L(p_f) + \psi(p_i)]} \frac{\alpha q_{sc} p_{sc} T}{k_{f2} h T_{sc}} \quad (23)$$

According to previous research results (King *et al.*, 1986; Anbarci and Ertekin, 1990, 1992; Sarkar and Rajtar, 1994; Claudia and Joseph, 2004; Clarkson, 2009; Clarkson *et al.*, 2013), the implicit expression of Λ can be treated as a constant at the initial pressure $[\psi(p_f) = \psi(p_i)]$.

Using the above dimensionless groups and combining Equations (2)-(6) with the corresponding boundary conditions when the well produces at a constant rate, detailed derivation of the instantaneous line source solution is shown in Appendix, which is:

$$\Delta\bar{\psi}_{1f} = \frac{p_{sc} T}{T_{sc}} \frac{3.6 k_{2f}}{\Lambda L^2} \frac{\alpha \hat{q}}{k_{1f} h} [K_0(\gamma_1 r_D) + A_C I_0(\gamma_1 r_D)] \quad (24)$$

where:

$$A_C = \frac{M_{12}\gamma_1 K_1(\gamma_1 r_{mD}) K_0(\gamma_2 r_{mD}) - \gamma_2 K_0(\gamma_1 r_{mD}) K_1(\gamma_2 r_{mD})}{M_{12}\gamma_1 I_1(\gamma_1 r_{mD}) K_0(\gamma_2 r_{mD}) + \gamma_2 I_0(\gamma_1 r_{mD}) K_1(\gamma_2 r_{mD})}$$

Using the Duhamel theory (Everdingen and Hurst, 1949), the continuous line source function can be obtained easily by:

$$\Delta\psi_{1f} = \frac{p_{sc} T}{T_{sc}} \frac{3.6 k_{2f}}{\Lambda L^2} \frac{\alpha}{k_{1f} h} \int_0^t q_{scL}(\tau) S_c(t_D - \tau) d\tau \quad (25)$$

where:

$$S_c = L^{-1}[\overline{S_c}]$$

$$\overline{S_c} = [K_0(\gamma_1 r_D) + A_C I_0(\gamma_1 r_D)]$$

Taking the Laplace transform with t_D in Equation (25), the continuous linear source function in composite dual-porosity CBM reservoir is:

$$\Delta\overline{\psi}_{1f} = \alpha \frac{p_{sc} T}{T_{sc}} \frac{\tilde{q}_{scL}}{k_{1f} h} [K_0(\gamma_1 r_D) + A_C I_0(\gamma_1 r_D)] \quad (26)$$

Hereto, the continuous linear source function in a radial composite CBM reservoir is obtained and some missing parameter groups are defined in Appendix.

2 SOLUTIONS OF A FRACTURED WELL IN A CBM RESERVOIR

2.1 Well Producing at a Constant Production Rate

In this section, the pressure transient behavior of a fractured vertical well with the SRV around the well is studied. The dual-porosity property of the coal beds is assumed and the unsteady-state desorption/diffusion formulation is used. The radial composite model consists of inner and zones with different properties. The fracture density of the inner zone is much better than the outer ones, which represents the SRV induced by the hydraulic fracturing. Figure 3 is the schematic of a fully penetrated fractured well in a CBM reservoir with the SRV around the well: to make the problem more tractable, the following assumptions are made:

- the reservoir is horizontal, with a uniform thickness of h, and the upper and lower boundaries are closed;
- a vertical fractured well is located at the center of the reservoir and the half fracture length is x_f;

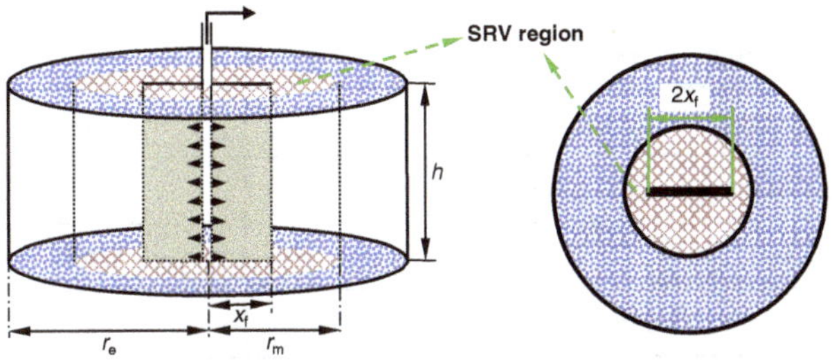

Figure 3

A schematic representation of a composite system with a vertically fractured well.

- the SRV region induced by hydraulic fracturing is a circular area with radius r_m;
- the well produces at a constant rate q_{sc}, and the conductivity of the fracture is infinite.

According to the results of Gringarten *et al.* (1974), the transient pressure response solutions of wells with infinite conductive fractures can be obtained in the uniform flux fracture case by measuring the pressure drop at $x_D = 0.732$ in the fracture. The solution for a vertical uniform flux fractured well can be easily obtained by integrating the continuous linear source function obtained before to x_{wD} over the interval from -1 to 1 (Zeng and Zhao, 2010).

Define the dimensionless pressure as follows:

$$\overline{\psi}_{1fD} = \frac{k_{2f} h T_{sc}}{\alpha q_{sc} p_{sc} T} \Delta \overline{\psi}_{1f} \tag{27}$$

According to the relationship between the linear source strength (q_{scL}) and the well production rate (q_{sc}), which is $q_{sc} = 2\ q_{scL} x_f$. Combine it with Equation (26) and Equation (27), and integrate the x_{wD} from -1 to 1, yielding:

$$\overline{\psi}_{1fD} = \int_{-1}^{1} \frac{1}{M_{12} s} [K_0(\gamma_1 r_D) + A_C I_0(\gamma_1 r_D)] \mathrm{d}x_{wD} \tag{28}$$

The solution of Equation (28) does not consider near-wellbore damage from the hydraulic fracturing fluid or drilling; when considering the effect of the skin factor (S_{kin}), Equation (28) becomes:

$$\overline{\psi}_{SD} = \int_{-1}^{1} \frac{1}{M_{12} s} [K_0(\gamma_1 r_D) + A_C I_0(\gamma_1 r_D)] \mathrm{d}x_{wD} + \frac{S_{kin}}{s} \tag{29}$$

When the well produces at a constant rate, wellbore storage has a significant effect on the bottom-hole pressure. According to the Duhamel principle (Everdingen and Hurst, 1949), the solution in Laplace space for modeling a fractured well in a CBM reservoir with the SRV can be obtained by:

$$\overline{\psi}_{wD} = \frac{\overline{\psi}_{SD}}{1 + C_D s^2 \overline{\psi}_{SD}} \tag{30}$$

2.2 Well Producing at a Constant Bottom-Hole Pressure

The above derivations and solutions are all based on the assumption that the well produces at a constant rate. However, for a CBM well producing at constant bottom hole pressure, the dimensionless production rate can also be derived with the similar method described above. Redefine the following dimensionless groups to replace the corresponding ones used above:

$$\Lambda = \begin{cases} (\phi_f \mu_{gi} c_{tf})_{1+2} + \dfrac{6 p_{sc} T}{T_{sc}(\psi_i - \psi_{wf})} & \text{for transient diffusion} \\ (\phi_f \mu_{gi} c_{tf})_{1+2} + \dfrac{2 p_{sc} T}{T_{sc}(\psi_i - \psi_{wf})} & \text{for pseudo-steady diffusion} \end{cases} \tag{31}$$

$$\sigma = \frac{V_L \psi_L(p_f)}{[\psi_L(p_f) + \psi(p_f)][\psi_L(p_f) + \psi(p_i)]} (\psi_i - \psi_{wf}) \tag{32}$$

According to previous research results (Everdingen and Hurst, 1949; Ozkan and Raghavan, 1991a, b), the solution of a CBM well producing at constant bottom-hole pressure is:

$$\overline{q}_D = \frac{1}{s^2 \overline{\psi}_{wD}} \tag{33}$$

where:

$$q_D = \frac{\alpha q_{sc} p_{sc} T}{k_{2f} h T_{sc}(\psi_{fi} - \psi_{wf})} \tag{34}$$

The standard log-log type curves for rate decline analysis (Blasingame *et al.*, 1991; Blasingame and Lee, 1994) can be obtained by defining the following expressions:

Dimensionless decline time:

$$t_{Dd} = \frac{t_D}{0.5(r_{eD}^2 - 1)[\ln(r_{eD}) - 0.5]} \tag{35}$$

Dimensionless decline rate function:

$$q_{Dd} = q_D[\ln(r_{eD}) - 0.5] \tag{36}$$

TABLE 1
Synthetic data used for the discussion of the results

Initial reservoir pressure, p_i (MPa)	15	Reservoir temperature, T (K)	320
Formation thickness, h (m)	60	Half fracture length, x_f (m)	50
Gas specific gravity, γ_g, fraction	0.65	Well production rate, q_{sc} (m^3/d)	3×10^4
The radius of inner region, r_m	100	Equivalent radial of control area, r_e (m)	2500
Bottomhole pressure, p_{wf} (MPa)	4	Wellbore storage coefficient, C_D	10^{-5}
Skin factor, skin	0.1	The sorption time constant, τ (h)	2×10^5
Langmuir pressure, P_L (MPa)	1.5	Langmuir Volume, V_L (m^3/m^3)	40
Inner region		Outer region	
Permeability, k_{1f} (mD)	0.1	Permeability, k_{2f} (mD)	0.02
Porosity, $\emptyset_{1f}$	0.02	Porosity, $\emptyset_{2f}$	0.01
Total compressibility, C_{tf1} (MPa^{-1})	1.0E-02	Total compressibility, C_{tf2} (MPa^{-1})	1.0E-02

Dimensionless decline rate integral:

$$q_{Ddi} = \frac{\int_0^{t_{Dd}} q_{Dd}(\tau)d\tau}{t_{Dd}} \tag{37}$$

Dimensionless decline rate integral derivative:

$$q'_{Ddi} = -\frac{dq_{Ddi}}{d\ln(t_{Dd})} \tag{38}$$

3 WELL PERFORMANCE ANALYSIS

The dimensionless bottom-hole pseudo-pressure (ψ_{wD}) and the derivative ($d\psi_{wD}/dt_D$) can be obtained using the Stehfest numerical algorithm (Stehfest, 1970) to convert the $\overline{\psi}_{wD}$ back into ψ_{wD}. So, we can obtain the standard log-log type curves of well test analysis of ψ_{wD} and $\psi'_{wD} \times t_D/C_D$ *versus* t_D/C_D.

The standard rate decline curves of q_D *versus* t_D can also be obtained in the same way as the well testing type curves described above, and then by transforming them into real space through Equations (14) and (34) we can obtain the well production performance in real space. By substituting q_D and t_D into Equations (35-38) the standard log-log transient rate decline curves of q_{Ddi} and q'_{Ddi} *versus* t_{Dd} can be obtained (Blasingame *et al.*, 1991; Blasingame and Lee, 1994).

3.1 Type Curves for CBM Fractured Wells

Well testing as well as transient rate decline curves are very interesting to petroleum engineers, as they can reflect the characteristics of the reservoirs and fluid flow behavior, and evaluate the effect of well stimulation. Also, using the type curve matching technology, the engineers can recognize the flow characteristics for a real reservoir and obtain the related property parameters, such as permeability, skin factor, control radial, fracture length, etc.

In this paper, we use the synthetic data in Table 1 to analyze the pressure response and the rate decline of the mathematical model derived in this paper; the results are shown in Figures 4-12.

Figures 4, 7, 9 and 11 show pseudo-pressure and pseudo-pressure derivative type curves of a fractured CBM well with the induced SRV. It can be found that the curves can be classified into the following flow periods.

Stage 1 Early wellbore storage, which is characterized by a unit slope line in both the pseudo-pressure and pseudo-pressure derivative curves (Stage 1 in *Fig. 4*).

Stage 2 A transition flow period between wellbore storage and the linear flow period in the SRV (Stage 2 in *Fig. 4*); the derivative curves in this stage are like a hump.

Stage 3 Linear flow in the SRV region, which is characterized by a half slope in both pseudo-pressure and pseudo-pressure derivative curves on a log-log scale. As the duration of this flow period is short, sometimes it will be masked by the wellbore storage effect of large C_D. Comparing this flow period of pseudo-steady diffusion

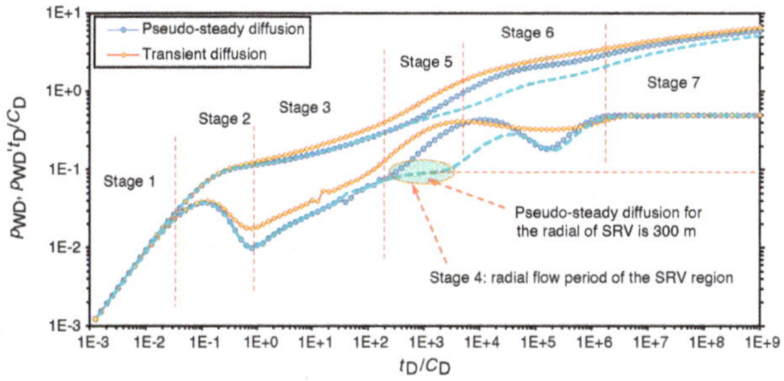

Figure 4

Comparison of the pseudo-steady and transient diffusion models on well test type curves.

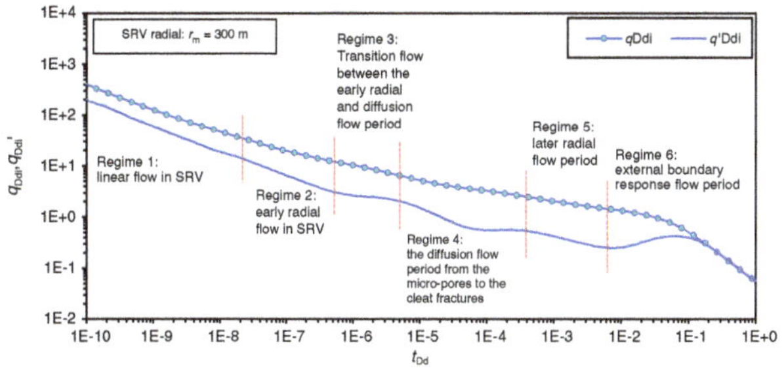

Figure 5

Transient rate decline type curves of a CBM fractured well with the SRV with pseudo-steady diffusion.

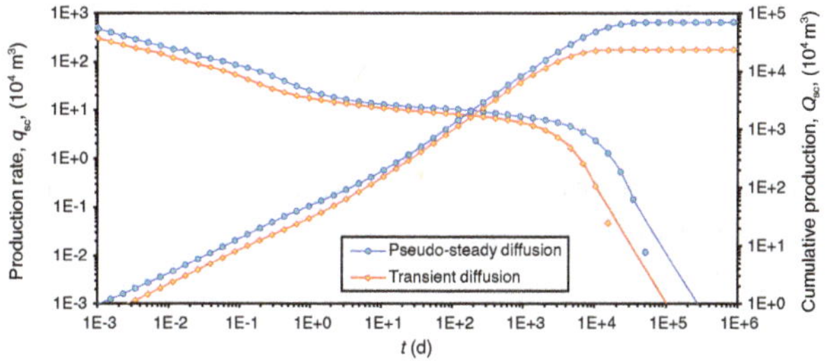

Figure 6

Comparison of the pseudo-steady and transient diffusion models on well production performance.

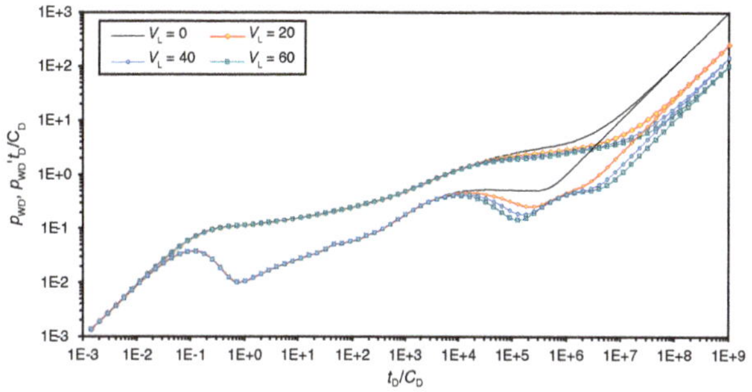

Figure 7

Effect of Langmuir Volume (V_L) on well testing type curves.

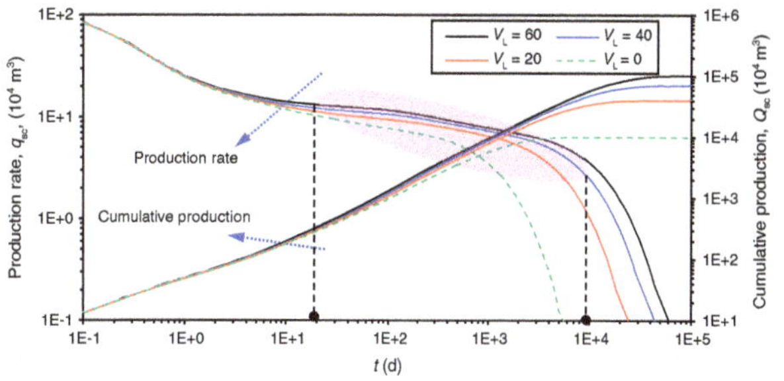

Figure 8

Effect of Langmuir Volume (V_L) on well production performance.

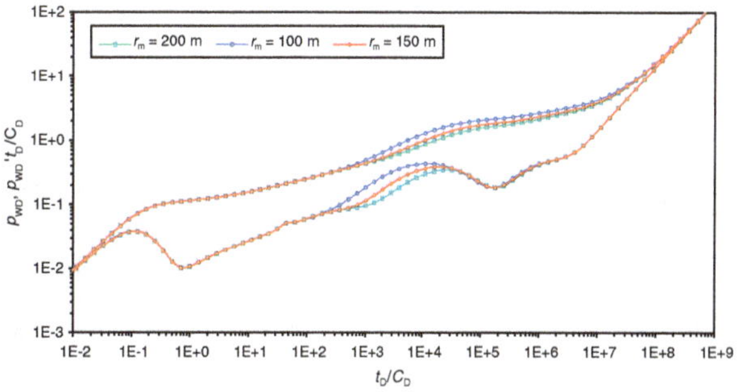

Figure 9

Effect of the SRV radial (r_m) on well testing type curves.

with the transient diffusion (Stage 3 in *Fig. 4*), it can be seen that the transient diffusion needs a higher pressure drop than the pseudo-steady diffusion curves when the well produces at a constant production rate. The gas adsorbed on the matrix particle surface can desorb and diffuse to the cleat fractures instantaneously for the pseudo-steady diffusion model. However, for the transient diffusion model, the gas diffusion needs some time to complete, so the pressure drop is relatively small for steady-state diffusion compared with the transient diffusion model. The above reason can also explain why the production rate and cumulative production for transient diffusion are lower than the pseudo-steady model, as shown in Figure 6.

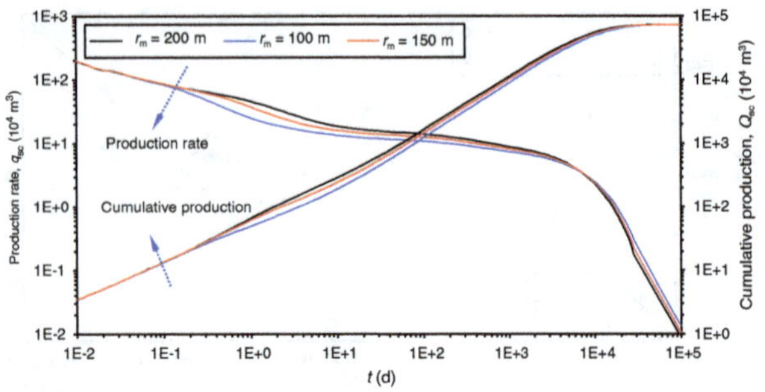

Figure 10

Effect of the SRV radial (r_m) on well production performance.

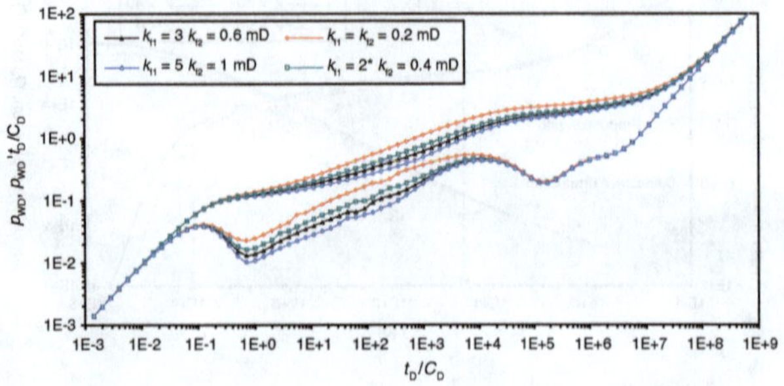

Figure 11

Effect of permeability in the SRV (k_{f1}) on well testing type curves.

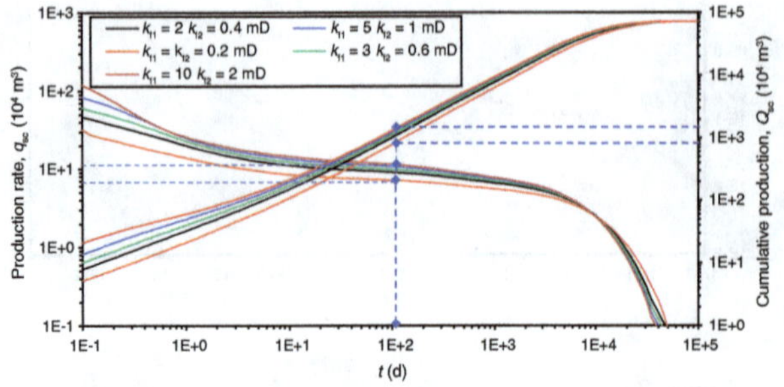

Figure 12

Effect of permeability in the SRV (k_{f1}) on well production performance.

Stage 4 Radial flow period in the SRV region; this flow period theoretically exists when the ratio of the SRV radial to the half fracture length is not big enough. As the half fracture length used in this paper is 50 m and the SRV radial is 100 m, the duration of this flow period is too short to be observed. In order to prove the existence of this flow period and analyze its characteristics, we plot another type curve with the assumption that the SRV radial is 300 m (this situation is shown with the dashed line in *Fig. 4*), which is characterized by a horizontal line with the value of $1/(2\times M_{12})$ (Stage 4 in *Fig. 4*).

Stage 5 The diffusion flow period from the micro-pores to the cleat fractures (Stage 5 in *Fig. 4*). In this flow period, the diffusion/desorption rate becomes more significant compared with the fluid supply in the cleat fractures; the production is mainly dominated by the depletion of micro-pores in the matrix. When the diffusion model is pseudo-steady, the derivative curves have a V-sharp below the straight line. It is the same for the interporosity flow period in natural fractured reservoirs. However for the transient diffusion model, the trough is much gentler.

Stage 6 The later pseudo radial flow period, for which the derivative curves are characterized by a horizontal line (its value is 0.5, Stage 6 in *Fig. 4*). During this flow period, the fluid transfer between micro-pores and cleat fractures has gone up to a state of dynamic balance. If the outer boundary is not infinite but closed, the derivative curves will gradually be upward with the slope of unit (this flow period can be obtained in *Fig. 7, 9 and 11)*.

Figure 5 shows a fractured CBM well with pseudo-steady diffusion producing at a constant bottom-hole pressure in a coalbed reservoir with a closed outer boundary. The transient rate decline type curves can also be divided into the following six flow periods shown in Figure 5.

3.2 Parameters Affecting Well Pressure and Production Performance

Various parameters have a significant influence on the shape of well testing type curves and on the well production performance, which is also the main objective of production engineers. Only by understanding the effect of these parameters on well performance can more accurate results be obtained from well test analysis.

Figures 7 and 8 show the effect of Langmuir volume (V_L) on well testing type curves and production performance with a closed outer boundary condition. The Langmuir volume is defined as the maximum amount of gas that adsorbed on the matrix particle at infinite pressure. So, the bigger the V_L, the more gas will diffuse from the micro-pores to the cleat fractures with a constant pressure drop.

With all other parameters being constant, V_L has a primary effect on the starting time of the transition flow and external boundaries-dominated flow period, and the depth of the trough (*Fig. 7*). The bigger the V_L, the earlier the transition, the later the pseudo-radial flow arrives, and the deeper the trough. For $V_L = 0$, the characteristics of derivative pressure curves become the same as the traditional homogeneous gas reservoir with a fractured well. As V_L represents the gas desorption capacity of the CBM reservoir, when the well produces at a constant bottom-hole pressure, the more gas will desorb and diffuse to the fractures, the bigger the V_L and the greater the production will be (*Fig. 9*). From the marked elliptical area in Figure 8, we can see that when $V_L = 40$ m^3/m^3, the duration of the plateau production is nearly 30 years, the average production rate is about 7.5 $\times$ 10^4 m^3/d, and the cumulative production at the end of this time period is up to 630 million m^3. However, for $V_L = 0$ m^3/m^3, which means no adsorbed gas stored on the matrix particle (as shown by the dashed line in *Fig. 8*), the production rate of the well declines quickly after 2-3 years. This phenomenon can explain why the CBM well and shale gas well have a long stable production period compared with the conventional reservoir.

Figures 9 and 10 show the effect of the SRV radial (r_m) on the well pressure and the production performance. The new concept of the "Fracture network" fracturing technique was put forward for the ultra-low permeability reservoir and the unconventional gas reservoir, such as CBM and shale, which is also called the SRV region. A large value of the SRV area is always expected by engineers and can lead to considerable economic profit. For the well testing type curves, r_m mainly affects the duration of the radial flow period of the SRV region. The larger the r_m, the longer the duration will be (*Fig. 9*). When a well produces at constant bottom-hole pressure, r_m has a significant effect on the well production rate. The production rate and cumulative production will coincide at the beginning but differ after the pressure wave propagation to the interface. A larger r_m will form a large seepage surface from the formation to the SRV region, which will result in a higher production rate. However, when a well produces at the end of its life, the same cumulative production will be obtained (according to the material balance theory, when other parameters are kept constant, no matter how the value of r_m changes, the control reserve of the well is the same) (*Fig. 10*).

Figures 11 and 12 present the pressure and production performance with different permeability of the SRV (k_{f1}). A large k_{f1} presents a development of the fracture network. When a well produces at a constant rate, the larger the k_{f1}, the smaller the production drop will be and the lower the location of the derivative curves will be on the log-log type curve (*Fig. 11*). For wells producing with a constant bottom-hole pressure, the bigger the k_{f1}, the higher the production rate. For example, after 100 days of production, the production rate will be nearly 10.2 $\times$ 10^4 m^3/d and the cumulative production is nearly 8 $\times$ 10^6 m^3 for $k_{f1} = 10k_{f2} = 2$ mD, but for a well without the SRV ($k_{f1} = k_{f2} = 0.2$ mD), the rate is only 7 $\times$ 10^4 m^3/d and cumulative production is 10.8 $\times$ 10^6 m^3. According to the data described above, the incremental production rate and cumulative production are both considerable.

CONCLUSIONS

In this paper, we investigated a comprehensive transport model for fractured CBM wells. With the assumption that a radial composite model with the inner region representing the SRV region induced by hydraulic fracturing is used, the solutions of this model at a constant production rate or bottom-hole pressure are obtained. Through the above analysis, the following conclusions can be summarized:

1. A mathematical model with multiple migration mechanisms to describe the fluid flow in the CBM reservoir is established, and the solutions of a fractured well with the SRV induced by hydraulic fracturing are obtained;
2. Well testing type curves and rate decline curves can be used to recognize the gas transport characteristics of CBM with different flow regimes. Compared with the conventional model, there is a trough on the pseudo-pressure derivative curves to reflect desorption and diffusion of CBM from the micro-pores to the cleat fractures;
3. The production rate and cumulative production for fractured wells with the SRV is much higher than the traditional single fracture well. And the larger the SRV radial and permeability, the higher the production rate will be;
4. Different parameters that govern the transport represent different physical characteristics, and also have different effects on the type of curves. As Langmuir Volume (V_L) represents the adsorption capacity of CBM, the bigger the V_L, and the deeper the trough will be;
5. The models and the results obtained in this paper give significant information on well tests and on the rate decline analysis of CBM wells.

ACKNOWLEDGMENTS

This work was supported by the National Science Fund for Distinguished Young Scholars of China (Grant

No. 51125019) and the Natural Science Foundation of China (Grant No. 51374181). The authors would also like to thank the reviewers and editors whose critical comments were very helpful in preparing this article.

REFERENCES

Agarwal G. (1979) Real Gas Pseudo-time a New Function for Pressure Buildup Analysis of MHF Gas Wells, *54th Technical Conference and Exhibition of the Society of Petroleum Engineers of AIME*, Las Vegas, NV, 23-26 Sept., *SPE Paper* 8279-MS.

Anbarci K., Ertekin T. (1990) A Comprehensive Study of Pressure Transient Analysis with Sorption Phenomena for Single-phase Gas Flow in Coal Seams, *65th annual Technical Conference and Exhibition of the Society of Petroleum Engineers*, New Orleans, LA, 23-26 Sept., *SPE Paper* 20568.

Anbarci K., Ertekin K. (1992) Pressure Transient Behavior of Fractured Wells in Coalbed Reservoirs, *SPE Annual Technical Conference and Exhibition*, Washington, DC, 4-7 Oct., *SPE Paper* 24703-MS.

Bayles G.A., Reznik A.A. (1986) Pressure Transient Analysis of Methane Production from Coal Beds: An Analytical Approach, *SPE Unconventional Gas Technology Symposium*, Louisville, Kentucky, 18-21 May, *SPE Paper* 15226-MS.

Blasingame T.A., McCray T.L., Lee W.J. (1991) Decline Curve Analysis for Variable Pressure Drop/Variable Flow Rate Systems, *SPE Gas Technology Symposium*, Houston, Texas, 22-24 Jan., *SPE Paper* 21513-MS.

Blasingame T.A., Lee W.J. (1994) The Variable-rate Reservoir Limits Testing of Gas Wells, *SPE Gas Technology Symposium*, Dallas, Texas, 13-15 June, *SPE Paper* 17708-MS.

Bumb A.C., Mckee C.R. (1988) Gas-well Testing in the Presence of Desorption for Coalbed Methane and Devonian Shale, *SPE Formation Evaluation* **3**, 1, 179-185.

Cai J.C., Yu B.M. (2011) A Discussion of the Effect of Tortuosity on the Capillary Imbibition in Porous Media, *Transp. Porous Media* **89**, 2, 251-263.

Clarkson C.R. (2009) Case Study: Production Data and Pressure Transient Analysis of Horseshoe Canyon CBM wells, *Journal of Canadian Petroleum Technology* **48**, 10, 27-38.

Clarkson C.R., Behmanesh H., Chorney L. (2013) Production-data and Pressure–transient Analysis of Horseshoe Canyon Coalbed-methane Wells, Part : Accounting for Dynamic Skin, *Journal of Canadian Petroleum Technology* **52**, 1, 41-53.

Claudia L.P., Joseph P. (2004) Production Analysis of Coalbed Wells Using Analytical Transient Solutions, *SPE Eastern Regional Meeting*, Charleston, West Virginia, 15-17 Sept., *SPE Paper* 91447-MS.

Engler T.W., Rajtar, J.M. (1992) Pressure Transient Testing of Horizontal Wells in Coalbed Reservoirs, *SPE Rocky Mountain Regional Meeting*, Casper, Wyoming, 18-21 May, *SPE Paper* 24374-MS.

Ertekin T., Sung W. (1989) Pressure Transient Analysis of Coal Seams in the Presence of Multi-mechanistic Flow and Sorption Phenomena, *SPE Gas Technology Symposium*, Dallas, Texas, 7-9 June, *SPE Paper* 19102.

Escobar F.H., Martinez J.A., Matilde M.M. (2013) Pressure Transient Analysis for a Reservoir with a Finite-conductivity Fault, *CT&F - Ciencia, Tecnología y Futuro* **5**, 2, 5-18.

Everdingen V., Hust A.F. (1949) The Application of the Laplace Transformation to Flow Problems in Reservoirs, *Journal of Petroleum Technology* **1**, 12, 305-324.

Gringarten A.C., Ramey H., Raghavan R. (1974) Unsteady-state Pressure Distributions Created by a Well with a Single Infinite-conductivity Vertical Fracture, *SPE Journal* **14**, 4, 347-360.

Hu G.Z., Wang H.T., Fan X.G., Yuan Z.G., Hong S. (2009) Mathematical Model of Coalbed Gas Flow with Klinkenberg Effects in Multi-physical Fields and its Analytic Solution, *Transp. Porous Med.* **76**, 3, 407-420.

Kamal M.M., Six J.L. (1993) Pressure Transient Testing of Methane Producing Coalbeds (includes associated papers 28013 and 28155), *SPE Advanced Technology Series* **1**, 1, 195-203.

King G.R., Ertekin T., Schwerer F.C. (1986) Numerical Simulation of the Transient Behavior of Coal-seam Degasification Wells, *SPE Formation Evaluation* **1**, 2, 165-183.

Langmuir L. (1916) The Constitution of Fundamental Properties of Solids and Liquids, *J. Am. Chem. Soc.* **38**, 11, 2221-2295.

Meng L., Luke D.C. (2007) A Dual-porosity Model for Gas Reservoir Flow Incorporating Adsorption Behaviour - Part 1. Theoretical Development and Asymptotic Analyses, *Transp. Porous Med.* **68**, 2, 153-173.

Nie R.S., Meng Y.F., Guo J.C., Jia Y.L. (2012) Modeling Transient Flow Behavior of a Horizontal Well in a Coal seam, *International Journal of Coal Geology* **92**, 1, 54-68.

Ozkan E., Raghavan R. (1991a) New Solutions for Well-test-analysis Problems: Part 1-Analytical Considerations (includes associated papers 28666 and 29213), *SPE Formation Evaluation* **6**, 3, 359-368.

Ozkan E., Raghavan R. (1991b) New Solutions for Well-test-analysis Problems: Part 2- Computational Considerations and Applications, *SPE Formation Evaluation* **6**, 3, 369-378.

Rushing J.A., Blasingame T.A., Johnston J.L., Lee W.J. (1991) Slug Testing in Multiple Coal Seams Intersected by a Single, Vertical Fracture, *SPE Annual Technical Conference and Exhibition*, Dallas, Texas, 6-9 Oct., *SPE Paper* 22945-MS.

Sarkar P.S., Rajtar J.M. (1994) Horizontal Well Transient Pressure Testing in Coalbed Reservoirs, *SPE Latin America/Caribbean Petroleum Engineering Conference*, Buenos Aires, Argentina, 27-29 April, *SPE Paper* 26995.

Stehfest H. (1970) Numerical Inversion of Laplace Transforms, *Communications of the ACM* **13**, 1, 47-49.

US Energy Information Administration, Washington, DC; http://www.eia.gov/dnav/ng/hist/rngr52nus_1a.htm.

Wang L., Wang X.D., Li J.Q., Wang J.H. (2013) Simulation of Pressure Transient Behavior for Asymmetrically Finite-conductivity Fractured Wells in Coal Reservoirs, *Transp. Porous Med.* **97**, 3, 353-372.

Zeng F.H., Zhao G. (2010) The Optimal Hydraulic Fracture Geometry Under Non-Darcy Flow Effects, *Journal of Petroleum Science and Engineering* **72**, 1-2, 143-157.

Zhang L.H., Zhao Y.L., Liu Q.G. (2014) Well-Test-Analysis and Applications of Source Functions in Bi-zonal Composite Gas Reservoir, *Petroleum Science and Technology* **32**, 8, 965-973.

Zhao Y.L., Zhang L.H., Wu F. (2012) Pressure Transient Analysis for Multi-Fractured Horizontal Well in Shale Gas Reservoirs, *Journal of Petroleum Science and Engineering* **90-91**, 31-38.

Zhao Y.L., Zhang L.H., Luo J.X., Zhang B.N. (2014) Performance of fractured horizontal well with stimulated reservoir volume in unconventional gas reservoir, *Journal of Hydrology* **512**, 447-456.

Zhao Y.L., Zhang L.H., Zhao J.Z., Luo J.X., Zhang B.N. (2013a) "Triple porosity" Modeling of Transient Well Test and Rate Decline Analysis for Multi-fractured Horizontal Well in Shale Gas Reservoirs, *Journal of Petroleum Science and Engineering* **110**, 253-262.

Zhao Y.L., Zhang L.H., Wu F., Zhang B.N., Liu Q.G. (2013b) Analysis of Horizontal Well Pressure Behavior in Fractured Low Permeability Reservoirs with Consideration of Threshold Pressure Gradient, *Journal of Geophysics and Engineering* **10**, 3, 1-10.

APPENDIX

Derivation of Instantaneous Linear Source Function when a Well Produces at a Constant Production Rate

Substituting the dimensionless groups defined in Equations (12-32) into the governing Equations (2) and (4), the mathematical model to describe an instantaneous linear source with a constant rate is (Zhao *et al.*, 2014):

1 For Transient Diffusion

Inner region

$$\frac{1}{r_D}\frac{\partial}{\partial r_D}\left(r_D\frac{\partial\psi_{1f}}{\partial r_D}\right)=\frac{\omega_1}{M_{12}}\frac{\partial\psi_{1f}}{\partial t_D}+\frac{(1-\omega_1-\omega_2)(1-\phi_{1f})}{M_{12}\lambda}\frac{\alpha q_{sc}p_{sc}T}{k_{f2}hT_{sc}}\left.\frac{\partial V}{\partial r_{Dm}}\right|_{r_{Dm}=1} \tag{A1}$$

Outer region

$$\frac{1}{r_D}\frac{\partial}{\partial r_D}\left(r_D\frac{\partial\psi_{2f}}{\partial r_D}\right)=\omega_2\frac{\partial\psi_{2f}}{\partial t_D}+\frac{(1-\omega_1-\omega_2)(1-\phi_{2f})}{\lambda}\frac{\alpha q_{sc}p_{sc}T}{k_{f2}hT_{sc}}\left.\frac{\partial V}{\partial r_{Dm}}\right|_{r_{Dm}=1} \tag{A2}$$

Using Fick's second law of diffusion, the partial differential equations describing transient state diffusive flow in the matrix are:

$$\frac{1}{r_{mD}^2}\frac{\partial}{\partial r_{mD}}\left(r_{mD}^2\frac{\partial V_D}{\partial r_{mD}}\right)=\lambda\frac{\partial V_D}{\partial t_D} \tag{A3}$$

The initial condition of the matrix system is:

$$V_D(t_D=0)=0 \tag{A4}$$

The inner and outer boundary conditions of the coal particle are:

$$V_D(r_{mD}=1)=V_D\left[\psi(p_f)\right] \tag{A5}$$

$$\left.\frac{\partial V_D}{\partial r_{mD}}\right|_{r_{mD}=0}=0 \tag{A6}$$

Combining Equation (23) with Equations (A3-A6), we have:

$$\overline{V}_D(r_{mD},t_D)=-\frac{\sigma\,\mathrm{sh}\left(r_{mD}\sqrt{\lambda s}\right)}{\mathrm{sh}\left(\sqrt{\lambda s}\right)r_{mD}}\Delta\overline{\psi}(p_f) \tag{A7}$$

So:

$$\left.\frac{\partial\overline{V}_D}{\partial r_{Dm}}\right|_{r_{Dm}=1}=-\sigma\left[\sqrt{\lambda s}\coth(\sqrt{\lambda s})-1\right]\frac{k_{f2}hT_{sc}\Delta\overline{\psi}(p_f)}{\alpha q_{sc}p_{sc}T} \tag{A8}$$

2 For Pseudo-Steady Flow

Inner region

$$\frac{1}{r_D}\frac{\partial}{\partial r_D}\left(r_D\frac{\partial \psi_{1f}}{\partial r_D}\right)=\frac{\omega_1}{M_{12}}\frac{\partial \psi_{1f}}{\partial t_D}+\frac{(1-\omega_1-\omega_2)(1-\phi_{1f})}{M_{12}\lambda}\frac{\alpha q_{sc}p_{sc}T}{k_{f2}hT_{sc}}(V_E-V) \tag{A9}$$

Outer region

$$\frac{1}{r_D}\frac{\partial}{\partial r_D}\left(r_D\frac{\partial \psi_{2f}}{\partial r_D}\right)=\omega_2\frac{\partial \psi_{2f}}{\partial t_D}+\frac{(1-\omega_1-\omega_2)(1-\phi_{2f})}{\lambda}\frac{\alpha q_{sc}p_{sc}T}{k_{f2}hT_{sc}}(V_E-V) \tag{A10}$$

Using Fick's first law of diffusion, the partial differential equation describing pseudo-steady diffusive flow in the matrix is:

$$\lambda\frac{\partial V_D}{\partial t_D}=(V_{ED}-V_D) \tag{A11}$$

Based in Equation (23) and using the Laplace transform method the solution of Equation (A-11) is:

$$\overline{V}_{ED}-\overline{V}_D=-\frac{\sigma s}{\lambda s+1}\frac{k_{f2}hT_{sc}\Delta\overline{\psi}(p_f)}{\alpha q_{sc}p_{sc}T} \tag{A12}$$

3 The Dimensionless Governing Equations

Inner region

$$\frac{1}{r_D}\frac{\partial}{\partial r_D}\left(r_D\frac{\partial \Delta\overline{\psi}_{1f}}{\partial r_D}\right)=f_1(s)\Delta\overline{\psi}_{1f} \tag{A13}$$

The parameter groups of $f_1(s)$ for different diffusion models are:

$$f_1(s)=\frac{\omega_1 s}{M_{12}}+\frac{\beta_1\sigma}{\lambda M_{12}}\left[\sqrt{\lambda s}\coth\left(\sqrt{\lambda s}\right)-1\right]\ \text{for transient flow} \tag{A14}$$

$$f_1(s)=\frac{\omega_1 s}{M_{12}}+\frac{\beta_1}{M_{12}}\frac{\sigma s}{\lambda s+1}\ \text{for pseudo-steady diffusion} \tag{A15}$$

where $\beta_1=(1-\omega_1-\omega_2)(1-\phi_{1f})$

Outer region

$$\frac{1}{r_D}\frac{\partial}{\partial r_D}\left(r_D\frac{\partial \Delta\overline{\psi}_{2f}}{\partial r_D}\right)=\omega_2 s\Delta\overline{\psi}_{2f}+\frac{\beta_2}{\lambda}\sigma\left[\sqrt{\lambda s}\coth\left(\sqrt{\lambda s}\right)-1\right]\Delta\overline{\psi}_{2f} \tag{A16}$$

The parameter groups of $f_2(s)$ for different diffusion models are:

$$f_2(s) = \omega_2 s + \frac{\beta_2 \sigma}{\lambda}\left[\sqrt{\lambda s}\coth\left(\sqrt{\lambda s}\right) - 1\right] \quad \text{for transient flow} \tag{A17}$$

$$f_2(s) = \omega_2 s + \beta_2 \frac{\sigma s}{\lambda s + 1} \quad \text{for pseudo-steady diffusion} \tag{A18}$$

where $\beta_2 = (1 - \omega_1 - \omega_2)(1 - \phi_{2f})$

4 Solving the Mathematical Model

Inner region

$$\frac{1}{r_D}\frac{\partial}{\partial r_D}\left(r_D \frac{\partial \Delta\overline{\psi}_{1f}}{\partial r_D}\right) = f_1(s)\Delta\overline{\psi}_{1f} \tag{A19}$$

Outer region

$$\frac{1}{r_D}\frac{\partial}{\partial r_D}\left(r_D \frac{\partial \Delta\overline{\psi}_{2f}}{\partial r_D}\right) = f_2(s)\Delta\overline{\psi}_{2f} \tag{A20}$$

In the case of an infinitely large reservoir, the outer boundary condition is:

$$\Delta\overline{\psi}_{2f}(r_D, t_D)\big|_{r_D \to \infty} = 0 \tag{A21}$$

According to characteristics of the Dirac delta function (Zhao *et al.*, 2012; Zhang *et al.*, 2014), the inner boundary condition becomes:

$$\lim_{\varepsilon \to 0^+} k_{1f} h \left(r_D \frac{\partial \overline{\psi}_{1f}}{\partial r_D}\right)_{r_D = \varepsilon} = -\alpha \hat{q} \frac{p_{sc} T}{T_{sc}} \frac{k_{2f}}{\Lambda L^2} \tag{A22}$$

The interface the conditions defining pressure and flow rate continuity in the reservoir are expressed as follows:

$$\Delta\overline{\psi}_{1f}(r_D, t_D)\big|_{r_D = r_{mD}} = \Delta\overline{\psi}_{2f}(r_D, t_D)\big|_{r_D = r_{mD}} \tag{A23}$$

and

$$\left.\frac{\partial \Delta\overline{\psi}_{1f}}{\partial r_D}\right|_{r_D = r_{mD}} = \frac{1}{M_{12}} \left.\frac{\partial \Delta\overline{\psi}_{2f}}{\partial r_D}\right|_{r_D = r_{mD}} \tag{A24}$$

The general solution of Equation (A19) and Equation (A20) is:

$$\Delta\overline{\psi}_{1f} = A_1 I_0\left(\sqrt{f_1(s)} r_D\right) + B_1 K_0\left(\sqrt{f_1(s)} r_D\right) \tag{A25}$$

$$\Delta\overline{\psi}_{2f} = A_2 I_0\left(\sqrt{f_2(s)} r_D\right) + B_2 K_0\left(\sqrt{f_2(s)} r_D\right) \tag{A26}$$

According to the outer boundary condition (Eq. A21), we have:

$$A_2 = 0 \tag{A27}$$

According to the inner boundary condition (Eq. A22), yields:

$$\lim_{\varepsilon\to 0^+} r_D\left[A_1 I_1\left(\sqrt{sf_1(s)}r_D\right) - B_1 K_1\left(\sqrt{sf_1(s)}r_D\right)\right]_{r_D=\varepsilon} = -\frac{\alpha}{k_{1f}h\sqrt{f_1(s)}}\hat{q}\frac{p_{sc}T}{T_{sc}}\frac{3.6k_{2f}}{\Lambda L^2} \tag{A28}$$

According to the characteristics of the Bessel function, $\lim_{x\to 0} xK_1(x) \to 1$, $\lim_{x\to 0} xI_1(x) \to 0$ gives:

$$B_1 = \alpha\hat{q}\frac{p_{sc}T}{T_{sc}}\frac{3.6k_{2f}}{\Lambda L^2}\frac{1}{k_{1f}h} \tag{A29}$$

Substituting Equations (A25) and (A26) into Equations (A23) and (A24):

$$A_1 I_0\left(\sqrt{f_1(s)}r_{mD}\right) + B_1 K_0\left(\sqrt{f_1(s)}r_{mD}\right) = B_2 K_0\left(\sqrt{f_2(s)}r_{mD}\right) \tag{A30}$$

$$A_1 I_1\left(\sqrt{f_1(s)}r_{mD}\right) - B_1 K_1\left(\sqrt{f_1(s)}r_{mD}\right) = -B_2\frac{\sqrt{f_2(s)}}{M_{12}\sqrt{f_1(s)}}K_1\left(\sqrt{f_2(s)}r_{mD}\right) \tag{A31}$$

Combining Equations (A30) and (A31), yields:

$$A_1 = B_1\frac{M_{12}\gamma_1 K_1(\gamma_1 r_{mD})K_0(\gamma_2 r_{mD}) - \gamma_2 K_0(\gamma_1 r_{mD})K_1(\gamma_2 r_{mD})}{M_{12}\gamma_1 I_1(\gamma_1 r_{mD})K_0(\gamma_2 r_{mD}) + \gamma_2 I_0(\gamma_1 r_{mD})K_1(\gamma_2 r_{mD})} \tag{A32}$$

where: $\gamma_1 = \sqrt{f_1(s)}, \gamma_2 = \sqrt{f_2(s)}$

So:

$$\Delta\bar{\psi}_{1f} = \alpha\hat{q}\frac{p_{sc}T}{T_{sc}}\frac{3.6k_{2f}}{\Lambda L^2}\frac{1}{\pi k_{1f}h}\left[K_0(\gamma_1 r_D) + A_C I_0(\gamma_1 r_D)\right] \tag{A33}$$

where: $A_C = \dfrac{M_{12}\gamma_1 K_1(\gamma_1 r_{mD})K_0(\gamma_2 r_{mD}) - \gamma_2 K_0(\gamma_1 r_{mD})K_1(\gamma_2 r_{mD})}{M_{12}\gamma_1 I_1(\gamma_1 r_{mD})K_0(\gamma_2 r_{mD}) + \gamma_2 I_0(\gamma_1 r_{mD})K_1(\gamma_2 r_{mD})}$

If the outer boundary is closed, the solution can be easily obtained by replacing Equation (A21) with the following equation (Zhang *et al.*, 2014):

$$\left.\frac{\partial\Delta\bar{\psi}_{2f}}{\partial r_D}\right|_{r_D=r_{eD}} = 0 \tag{A34}$$

3

Prediction of Mass Flow Rate in Supersonic Natural Gas Processing

Chuang Wen[1,3], Xuewen Cao[1]*, Yan Yang[2] and Yuqing Feng[3]

[1] College of Pipeline and Civil Engineering, China University of Petroleum, 66 Changjiang West Road, Huangdao District, Qingdao, 266580 - China

[2] School of Petroleum Engineering, Changzhou University, Zhonglou District, Changzhou, Jiangsu, 213016 - China

[3] CSIRO Computational Informatics, Clayton, Victoria 3169 - Australia

e-mail: wenchuang2008@gmail.com - caoxw@upc.edu.cn - flyloveyang@gmail.com - yuqing.feng@csiro.au

* Corresponding author

Résumé — Prédiction du débit massique dans les applications de traitement supersonique du gaz naturel — Le débit massique volumique de gaz naturel à travers un séparateur supersonique a été calculé numériquement en utilisant plusieurs équations d'état cubiques. Les résultats numériques indiquent que les coefficients de compressibilité et les coefficients de chaleur spécifique calculés avec les modèles de gaz réels diffèrent remarquablement des résultats issus de la loi des gaz parfaits pour une pression d'entrée élevée. Parallèlement, les calculs de débit massique effectués avec les modèles de gaz réels et avec la loi des gaz parfaits présentent une déviation de plus de 10 %. Cette déviation augmente lorsque la température d'entrée diminue, quelle que soit la pression en entrée. Une contre-pression plus élevée déplace l'onde de choc en amont. Le premier seuil à dépasser pour un fonctionnement normal du séparateur est obtenu pour un ratio de pression de 0,72. Un second seuil correspond à un ratio de pression de 0,95, le flux étant cependant dans ce cas subsonique, sans atteindre les conditions de saturation. Dans le cas des modèles de gaz réels, l'onde de choc est décalée vers l'amont par rapport à la loi des gaz parfaits dans la section de séparation cyclonique.

Abstract* — *Prediction of Mass Flow Rate in Supersonic Natural Gas Processing — *The mass flow rate of natural gas through the supersonic separator was numerically calculated by various cubic equations of state. The numerical results show that the compressibility factor and specific heat ratio for ideal gas law diverge remarkably from real gas models at a high inlet pressure. Simultaneously, the deviation of mass flow calculated by the ideal and real gas models reaches over 10%. The difference increases with the lower of the inlet temperature regardless of the inlet pressure. A higher back pressure results in an earlier location of the shock wave. The pressure ratio of 0.72 is the first threshold to get the separator work normally. The second threshold is 0.95, in which case the whole flow is subsonic and cannot reach the choked state. The shock position moves upstream with the real gas model compared to the ideal gas law in the cyclonic separation section.*

INTRODUCTION

A supersonic separator has been introduced for the natural gas processing, and its typical application is to remove the water vapor from natural gas for the control of the dew point [1, 2]. In this method, the natural gas liquids and heavier hydrocarbons may also be separated from natural gas depending on the low pressure and temperature resulted by the supersonic expansion. This revolutionary technique combines the theories of fluid mechanics, thermodynamics and phase transformation dynamics to emplace the processes of the gas expansion in supersonic velocities, cyclonic gas/liquids separation and re-compression in a compact tube. The supersonic separator enables high reliability and availability because there are no rotating parts in it. This new separation process prevents the hydrate formation and eliminates the needs for inhibitor and regeneration systems due to the short residence time in the device.

A supersonic separator mainly consists of four parts, namely, a swirling device, a Laval nozzle, a cyclonic separation section and a diffuser. The swirl device is used to transform part of the axial velocity to angular velocity to generate swirls. The Laval nozzle accelerates the natural gas flows, and the sonic speed is obtained at the nozzle throat. In the divergent part, the natural gas is further expanded to supersonic velocity resulting in a low pressure and temperature, which is the foundation condition for the condensation of water vapor. The cyclonic separation section is the extending of the Laval nozzle. It provides enough space and time conditions for the separation processing in supersonic velocities. In the diffuser, the supersonic velocity is changed to subsonic one when the natural gas flow meets the shock wave. Hence, the diffuser transforms the remaining kinetic energy to pressure for energy saving.

The supersonic separation technology has been extensively studied in the past. Alfyorov *et al.* [3] investigate the separation characteristics of the supersonic separator compared to the Joule-Thomson valve and turbo-expander for natural gas. Liu *et al.* [4] introduce an indoor supersonic separation rig and test its dehydration performance with moist air for the working fluids. Jassim *et al.* [5, 6] employe a computational fluid dynamics method to investigate the effects of the nozzle geometry and vorticity on the supersonic flow in Laval nozzle. Karimi and Abdi [7] analyze the influences of the gas dynamic parameters on the natural gas flow in the supersonic nozzle, such as the inlet pressure, the inlet temperature, and so on. Malyshkina [8, 9] obtaines the distribution of gas dynamic parameters of natural gas through a supersonic separator with a computational method, and a procedure is developed to predict the separation capability of water vapor and higher hydrocarbons from natural gas using a supersonic separator determined by the initial parameters. Zaporozhets *et al.* [10] introduce an expander-compressor unit namely the thermal gas-dynamic separator. In essence, it is a supersonic separator as it consists of a swirler, a nozzle, a separation part and a diffuser. Jiang *et al.* [11] employe the corrected Internally Consistent Classical Theory and Gyarmathy theory to modeling the nucleation and droplet growth of natural gas in the supersonic separation process. In our previous studies, we design a new supersonic separator incorporating a central body and analyze the effect of the nozzle geometry and swirls on the flow field [12, 13]. The Reynolds stress model is employed to predict the strong swirling flow of natural gas in the diffusers using the Navier-Stokes equations [14]. The discrete particle method is used to predict the particle separation characteristics in a supersonic separator and the separation efficiency is obtained [15].

As a novel technique for natural gas separation, the supersonic separator has a lot of advantages, such as impact design, low weight, no chemicals and unmanned operation. However, there is a non-negligible weakness, which the device is very sensitive to the inlet gas flow rate due to the presence of a Laval nozzle inside the device. The critical area at the nozzle throat determines the gas mass flow rate through this device. If the mass flow rate is less than the designed one, the natural gas is the subsonic flows in the whole supersonic separator, which cannot reach the required low pressure and temperature for the condensation of the water vapour, which means that it doesn't work in this condition. Therefore, it is important to study and evaluate the gas flow rate through a supersonic separator. The purpose of this study is to employ various cubic Equations Of State (EOS) to evaluate the gas properties and real mass flow rate of natural gas through the supersonic separator. The key pressure ratio will be discussed by analyzing its influence on the shock wave position.

1 COMPUTATIONAL APPROACH

1.1 Governing Equations

The high pressure natural gas through a supersonic separator reaches supersonic velocities where the Mach number is greater than unity. In the computational fluid dynamics simulation, the governing equations include the compressible conservation of mass (continuity equation), momentum, and energy equations. In this case, we didn't solve the partial differential equations for fully three-dimensional flows. Instead, the two-dimensional axisymmetric flows were solved in ANSYS FLUENT platform [16].

1.2 Real Gas Equations of State

An Equation Of State (EOS) is usually used to evaluate the fluid properties. The ideal gas law can be employed to describe the mathematical relationship of the Pressure-Volume-Temperature (PVT) of a gas at low pressure and high temperature. Errors in pressure and temperature evaluation lead to the poor prediction of the flow structures. Therefore, a real gas EOS is expected to calculate the thermodynamic parameters of a gas at higher pressures and lower temperatures.

In 1873, van der Waals presented an EOS with the valuable significance. Even though the equation is not accurate for most applications, it provides a perspective for researchers to represent the PVT behavior of gases. Since then, many EOS have been proposed to improve the accuracy in the predictions of thermodynamic properties by modifying the van der Waals EOS. In 1949, Redlich and Kwong [17] propose an Equation Of State (RK EOS) by correcting the attractive term. It has been recognized a most successful modification of the van der Waals EOS and proved its accuracy for many gaseous systems. The RK EOS can be expressed as the following form:

$$p = \frac{RT}{V-b} - \frac{a}{(T/T_c)^{0.5}V(V+b)} \tag{1}$$

$$a = 0.42747R^2T_c^2/p_c \tag{2}$$

$$b = 0.08664RT_c/p_c \tag{3}$$

where V is the gas volume; R is the gas constant; T_c and p_c are the critical temperature and pressure, respectively.

The success of RK EOS has inspired many researchers for further studies to improve the accuracy. Soave [18] suggests to redefine the term of $(T/T_c)^{-0.5}$ to $a(T)$ function depending on the temperature as follows:

$$a(T) = \left[1 + m\left(1 - (T/T_c)^{0.5}\right)\right]^2 \tag{4}$$

$$m = 0.480 + 1.574\omega - 0.176\omega^2 \tag{5}$$

where ω is the acentric factor.

After Soave's proposal, some other modifications were proposed for high accuracy. And the most famous of the studies is the Peng-Robinson EOS [19]. Based on the Soave's suggestion, they use the same $a(T)$ function with a different m and modified the volume dependency of the attractive term. PR EOS can be described in Equation (6):

$$p = \frac{RT}{V-b} - \frac{aa(T)}{V(V+b)+b(V-b)} \tag{6}$$

$$a = 0.457247R^2T_c^2/p_c \tag{7}$$

$$b = 0.07780RT_c/p_c \tag{8}$$

$$m = 0.37464 + 1.54226\omega - 0.2699\omega^2 \tag{9}$$

In 1995, Aungier [20] employes the critical point compressibility factor as the additional parameter and redefined the $a(T)$ function to modify the RK EOS. The Aungier EOS form differs from the original RK EOS by a non-zero parameter c which is added to improve the behavior of isotherms near the critical point, as well as setting the exponent n differently. Aungier EOS is expressed as the following form:

$$p = \frac{RT}{V-b+c} - \frac{aa(T)}{V(V+b)} \tag{10}$$

$$c = \frac{RT_c}{p_c + a/V_c(V_c+b)} + b - V_c \tag{11}$$

$$a(T) = (T/T_c)^{-m} \tag{12}$$

$$m = 0.4986 + 1.1735\omega + 0.4754\omega^2 \tag{13}$$

where V_c is the critical volume. a and b are given by Equation (2) and (3), respectively.

The *RK, SRK, ARK, PR* equations are widely used in oil and gas industry. The advantages of these equations are that they are easy to use and they often accurately represent the relation between temperature, pressure, and phase compositions in binary and multicomponent systems. These equations only require the critical properties and acentric factor for the generalized parameters. Little computer resources are required and those lead to good phase equilibrium correlation.

For the multi-component mixtures, such as natural gas, mixing laws are utilized to calculate the parameters a and b. The van der Waals mixing rules [21, 22] were applied to obtain the EOS parameters for the mixtures from those pure components. The mathematical expressions of this mixing rule can be written:

$$a = \sum_{i=1}^{n}\sum_{j=1}^{n} x_i x_j \sqrt{a_i a_j}(1 - k_{ij}) \tag{14}$$

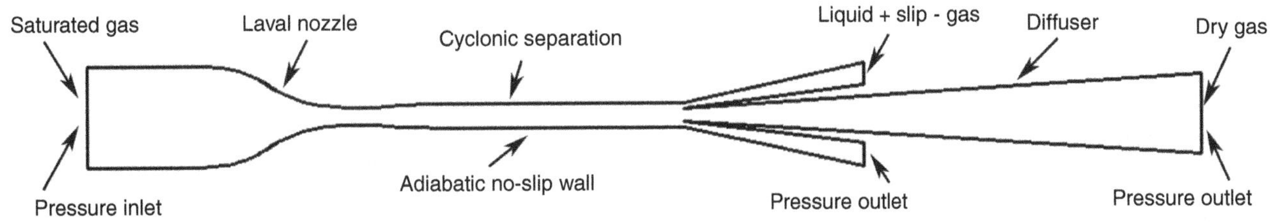

Figure 1

Schematic diagram of the supersonic separator.

$$b = \sum_{i=1}^{n} x_i b_i \tag{15}$$

where x is molar fraction; n is the total number of the gas components; k_{ij} is the binary interaction parameter between components i and j.

1.3 Computational Conditions

The supersonic separator employed in the computational study consisted of a Laval nozzle, a cyclonic separation section and a diffuser, as shown in Figure 1. The inlet and outlet diameters of the nozzle are 100.00 mm and 23.83 mm, respectively. The throat diameter is 17.40 mm. The profile of the convergent section of the nozzle is calculated by Equation (16), while the Foelsch's analytical calculation was employed to design the divergent section of the nozzle [23, 24]. The convergence and divergence half angle of the nozzle is 12.48° and 2.43°, respectively. The diffuser is designed as a straight conic one with the divergence angle of 6°. A constant area tube is inserted between the nozzle and diffuser representing as a cyclonic separation section. In addition, a straight tube with the length of 100 mm was connected to the upstream and downstream of the device for the computational calculation:

$$\begin{cases} \dfrac{D - D_{cr}}{D_1 - D_{cr}} = 1 - \dfrac{1}{X_m^2}\left(\dfrac{x}{L}\right)^3 & \left(\dfrac{x}{L} \le X_m\right) \\ \dfrac{D - D_{cr}}{D_1 - D_{cr}} = \dfrac{1}{(1 - X_m)^2}\left(1 - \dfrac{x}{L}\right)^3 & \left(\dfrac{x}{L} > X_m\right) \end{cases} \tag{16}$$

where D_1, D_{cr} and L are the inlet diameter, the throat diameter and the convergent length, respectively. $X_m = 0.3 - 0.7 \times x$ is the distance between arbitrary cross section and the inlet, and D is the convergent diameter at arbitrary cross section of x.

In this simulation case, the swirling and condensation flows were not considered in the prediction of the gas mass flow rates. Instead, we just considered the single gas flow in a supersonic separator. As a result of the high Reynolds number in the supersonic separator, the modeling of the turbulence plays a crucial role in the flow prediction [25]. For this sort of flow situation, the pressure gradient is not very large in the gas expansion, while the boundary layers are probably small, and the turbulent boundary layers act just as a correction. The k-ε model is a two equation model, and it is promisingly accurate for the flows with relatively small pressure gradients and for wall-bounded and internal flows where mean pressure gradients are very small. Hence, the standard k-ε model is used in this case because of its robustness, economy and reasonable accuracy.

The finite volume methods are employed to solve the governing equations with ANSYS FLUENT being the numerical platform. The semi-implicit method for pressure linked equations (SIMPLE) is adopted for the solution of the momentum equation by guessing the pressure gradient and getting a convergence [26, 27]. The boundary conditions employed in the present computation are shown as follows: pressure inlet for the separator entrance, pressure outlet condition for the exits of dry gas and liquids, respectively. No-slip and adiabatic boundary conditions are supposed for the solid walls. The convergence criterion is 10^{-6} for the energy equation and 10^{-3} for all other equations.

The natural gas used in the simulation is composed of methane, ethane, propane and water vapour. The mole composition of natural gas is as follows: 95.94% CH_4, 3.00% C_2H_6, 1.00% C_3H_8, 0.06% H_2O. The parameters are used in the simulation including the molecular weight, standard state entropy, critical temperature, critical pressure, critical specific volume, acentric factor, and so on.

2 RESULTS AND DISCUSSION

2.1 Validation of Computational Approach

The model validation work has been performed in a Laval nozzle and reported in our previous publication.

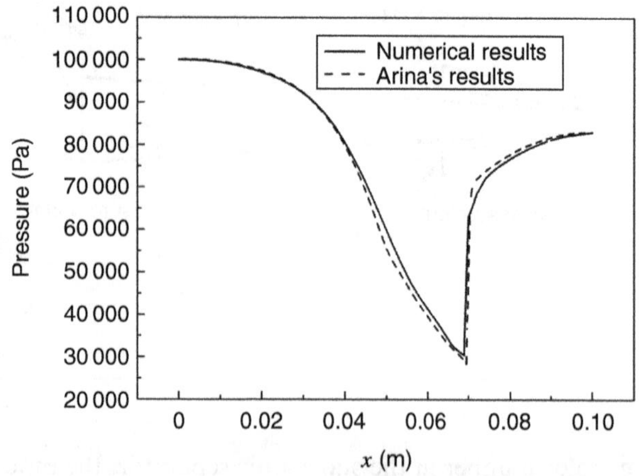

Figure 2

Pressure profiles for nozzle flow.

Detailed information can be found in [28]. Here, only one figure is re-plotted for general understanding. The comparisons of the static pressure in the central line of the Laval nozzle are shown in Figure 2. It can be seen that the same flow behavior is obtained and the shock-wave position is accurately captured by our simulation method. Therefore, our computational approach can predict the supersonic flow very well.

2.2 Ideal Mass Flow Rate of Natural Gas

The supersonic separator involves a Laval nozzle to expand the natural gas to supersonic velocities for lower pressure and temperature, which is the essential condition for the natural gas separation. Therefore, the mass flow rate to get the supersonic separator work is the choked mass flow rate, m_c:

$$m_c = \rho_c v_c A_c = \rho_c a_c A_c \tag{17}$$

where v_c and a_c are the gas velocity and sound speed at the nozzle throat, respectively. ρ_c is the gas density of the choked flow, and A_c is the section area at throat.

Assume that the natural gas follows the ideal gas law, we can obtain the mathematical relationship between the stagnation parameters and choked flow parameters as follows:

$$p_0 = \rho_0 R T_0 \tag{18}$$

$$T_0 = \left(1 + \frac{\gamma - 1}{2}\right) T_c \tag{19}$$

$$\rho_0 = \left(1 + \frac{\gamma - 1}{2}\right)^{\left(\frac{1}{\gamma - 1}\right)} \rho_c \tag{20}$$

$$a_c = \sqrt{\gamma R T_c} \tag{21}$$

where γ is the specific heat ratio. p_0, ρ_0 and T_0 are the stagnation parameters, respectively.

Substituting relationships (18-21) into Equation (17), and the choked mass flow rate can be expressed as:

$$m_c = \sqrt{\gamma}\left(\frac{2}{\gamma + 1}\right)^{\frac{\gamma+1}{2(\gamma-1)}} \frac{p_0}{\sqrt{RT_0}} A_c \tag{22}$$

To evaluate the real gas effects on the mass flow rate of natural gas through the supersonic separator, we define a dimensionless parameter deviation factor of the mass flow rate, m_e:

$$m_e = \frac{m_r - m_c}{m_c} \times 100\% \tag{23}$$

where m_r is the mass flow rate calculated by the real gas models.

2.3 Mass Flow Rate with Real Gas Models

The mass flow rate of natural gas through the supersonic separator is computationally simulated using various real gas models (RK, RKS, PR and RKA EOS in present simulations), respectively. The compressibility factor is the ratio of the molar volume of a gas to the molar volume of an ideal gas at the same temperature and pressure. It is a useful thermodynamic property for modifying the ideal gas law to account for the real gas behavior. The compressibility factor, Z, is defined as:

$$Z = pV/RT \tag{24}$$

where V is the molar volume, p is the pressure, T is the temperature, and R is the gas constant. The compressibility factor, Z, will be 1 for an ideal gas.

The compressibility factor was employed to assess the real gas effects with the inlet pressure of 5 atm and 60 atm, as shown in Figure 3. One can see that the natural gas has a compressibility factor of 0.99 in the whole supersonic separator with a lower inlet pressure (5 atm), which means the natural gas has a little difference from the ideal gas. On the contrary, when the inlet pressure is about 60 atm, the compressibility factor greatly increases in the Laval nozzle, and keeps nearly a constant in the cyclonic separation section. Then, the compressibility factor sharply declines as a result of the emergency of the shock. In the whole separation

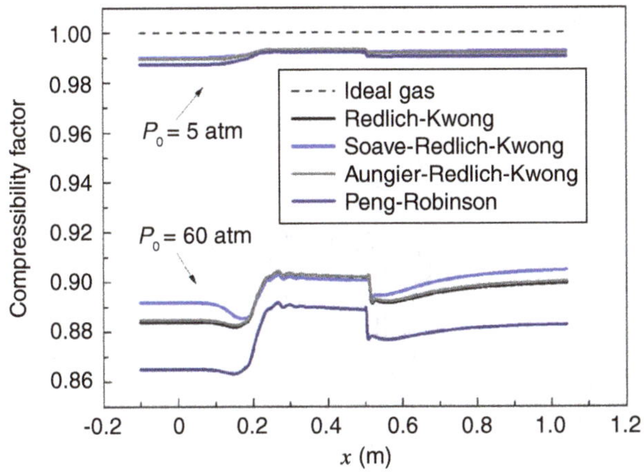

Figure 3

Compressibility factors of natural gas at the inlet pressure of 5 atm and 60 atm.

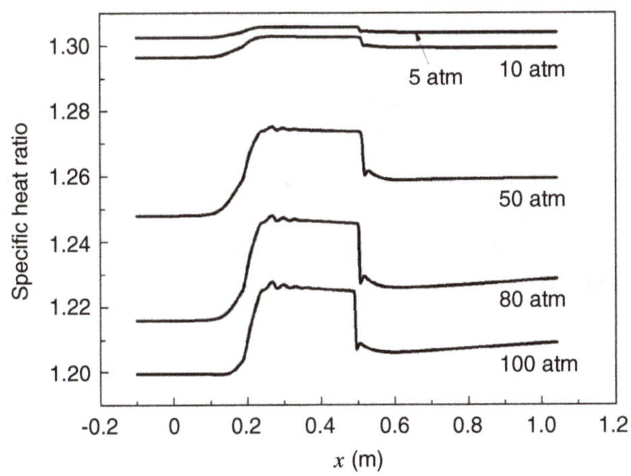

Figure 4

Specific heat ratio along the axis at various inlet pressures.

processing of natural gas in the supersonic separator, the compressibility factor diverges from the ideal value of unity, although the results calculated by the RK, RKS, PR and RKA EOS have a little difference from one to the other. Therefore, it can be concluded that the compressibility factor of natural gas affects the evaluation of the mass flow rate compared to the ideal gas through the supersonic separator at a high pressure.

Equation (22) indicates that the specific heat ratio plays an important role in the prediction of the mass flow rate of natural gas through a supersonic separator. Therefore, the specific heat ratio of natural gas is computationally calculated by PR EOS at various inlet pressures, and the axial profiles are depicted in Figure 4. The ideal ratio of specific heat of natural gas is constant

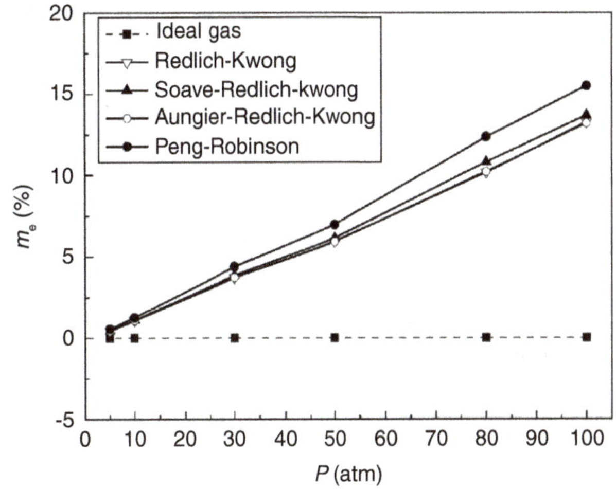

Figure 5

Deviation factor of mass flow of natural gas with real gas models.

at 1.304. One can see that the specific heat ratio rises suddenly in the Laval nozzle due to the acceleration of natural gas with an abrupt drop in the shock position, when the PR real gas model is performed. A similar tendency appears at various inlet pressures, but the amplitude of variation of the profiles of the specific heat ratio rises with higher inlet pressure. Moreover, the specific heat ratio decreases with the increase of the inlet pressure at a fixed position. That is, these variations at the nozzle throat affect significantly the calculation of the mass flow rate of natural gas through the supersonic separator.

Figure 5 shows the divergence of the mass flow rate of natural gas calculated by the ideal gas law and the real gas models. The inlet temperature was fixed at 288 K to evaluate the effect on the mass flow rate. It can be seen that the deviation factor, m_e increases with a higher inlet pressure for the real gas models. When the inlet pressure is less than 50 atm, the deviation factor is below 5%. If the inlet pressure becomes higher than 80 atm, the factor m_e will be more than 10%. In the operation of the supersonic separator, it usually works at a high inlet pressure, such as 100 atm or a little more. In this condition, the real gas model should be taken into account for the mass flow rate of natural gas through the supersonic separator.

2.4 Effects of Inlet Temperature on Mass Flow Rate

The ideal gas law and Peng-Robinson EOS are employed to predict the effects of inlet temperature on the mass flow rate of natural gas. Figure 6 depicts the profiles of the deviation factor at various inlet temperatures with

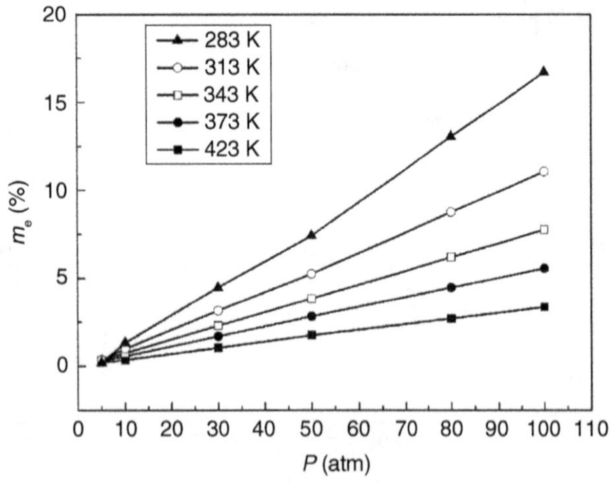

Figure 6

Effect of inlet temperature on the mass flow rate with PR EOS model.

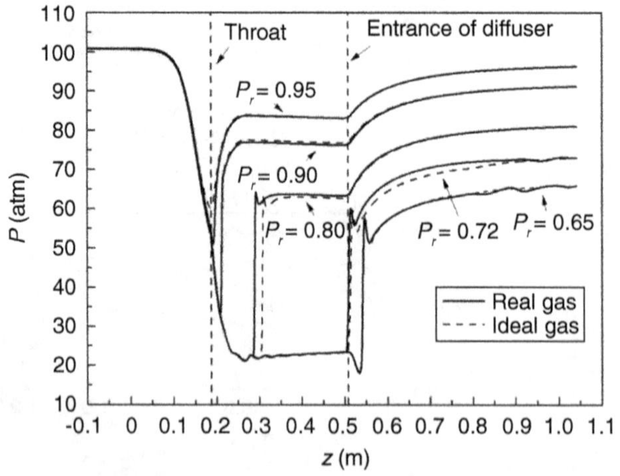

Figure 7

Distribution of the static pressure with different back pressures.

the inlet pressure from 5 atm to 100 atm. The deviation factor of natural gas rises with the decrease of the inlet temperature regardless of the inlet pressure. The growth rate of the deviation factor becomes larger at a higher inlet pressure. At a low inlet pressure, less than 30 atm, for example, the factor is below 5% even with a low temperature of 283 K. However, there would be a huge deviation of mass flow rate of natural gas if the ideal gas law was used to calculate the thermodynamic properties at the high pressure and low temperature. For instance, the calculation error reaches about 16.5% at the inlet pressure of 100 atm with the temperature of 283 K. Therefore, the ideal gas law will result in an error on the prediction of mass flow rate of natural gas through the supersonic separator. In the gas field, the typical operating condition for a supersonic separator is that the inlet temperature is normally less than 303 K with the inlet pressure of about 100 atm. In these conditions, the real gas model should be employed to predict the gas flow rate through a supersonic separator. Otherwise, the designed gas mass flow rate will be lower than 10% of the real one, if the ideal gas law is used. It causes that the natural gas flow can't be choked at the nozzle throat. That is, the gas flow is subsonic in the whole supersonic separator, which can't induce the low pressure and temperature for nucleation and condensation of a water vapour.

2.5 Pressure Ratio and Mass Flow Rate

The static pressure of natural gas along the supersonic separator axis is shown in Figure 7 at the different pressure ratio p_r, which is the ratio of the back pressure to the inlet one. The back pressure is the one at the separator outlet. The shock wave moves upstream with the increase of the pressure ratio from the diffuser to the nozzle. The shock wave can stay in the diffuser when the pressure ratio keeps below 0.72. If the pressure ratio becomes a little bigger, the shock wave will go into the cyclonic separation section, resulting in a sudden rise of the static pressure, which will destroy the supersonic separation processing. Thus, the pressure ratio of 0.72 is the first threshold to make the supersonic separator workable. When the pressure ratio reaches 0.95, the shock goes across the throat and the whole flow is subsonic through the supersonic separation. This is the second threshold of the pressure ratio, in which the mass flow rate cannot reach the choked flow. In the nozzle and the diffuser, both the ideal gas law and PR real gas predict the same distributions of the static pressure. In the cyclonic separation section, the real gas model predicts an earlier shock location compared to the ideal gas law. It means that the shock position is sensitive to the pressure ratio with the real gas model.

Moreover, the RK real gas model is employed to predict the vital pressure ratio representing the RK EOS, since they nearly predicted the similar results as shown in Figures 3 and 5. The data also show a similar result in the aspect of predicting the pressure ratio compared to the PR model. This indicates that the threshold of the pressure ratios predicted by the real gas model can be used in the design of a supersonic separator.

CONCLUSIONS

The mass flow rate of natural gas through the supersonic separator was predicted using the ideal gas law and cubic equations of state. At a high pressure, the compressibility factor calculated by the real gas models diverged significantly from the results of ideal gas, showing the real gas effects on the supersonic separation processing. In the PR real gas model, the specific heat ratio rised in the Laval nozzle and dropped sharply at the shock wave location. The variation of the specific heat ratio at the nozzle throat had an important effect on the evaluation of the mass flow rate of natural gas. The computational study indicated that the deviation of the mass flow rate of the ideal gas reached over 10% compared to the real gas models at higher pressure.

The deviation of mass flow rate of natural gas increased with the decline of the inlet temperature regardless of the inlet pressure, when the ideal and real gas models were used for the simulation. The calculation error of mass flow rate reached 16.5% in a higher inlet pressure and lower inlet temperature (100 atm, 15°C).

The position of the shock wave moved upstream from the diffuser to the nozzle with the increase of the back pressure. The first threshold of the back pressure of 0.72 times of the inlet one was found to get the supersonic separator work normally, where the real mass flow rate of natural gas was still the choked one. If the pressure ratio increased to 0.95, the shock went across the throat and the whole flow was the subsonic through the supersonic separator. This was the second threshold and the mass flow rate could not reach the choked flow.

ACKNOWLEDGMENTS

This work was supported in part by the National Natural Science Foundation of China (No. 51274232) and the Research Program for Excellent Doctoral Dissertation of China University of Petroleum. C. Wen acknowledges a fellowship from the China Scholarship Council (201206450035).

REFERENCES

1 Okimoto D., Brouwer J. (2002) Supersonic Gas Conditioning, *World Oil* **223**, 89-91.

2 Betting M., Epsom H. (2007) Supersonic Separator Gains Market Acceptance, *World Oil* **254**, 197-200.

3 Alfyorov V., Bagirov L.A., Dmitriev L., Feygin V., Imayev S., Lacey J.R. (2005) Supersonic Nozzle Efficiently Separates Natural Gas Components, *Oil Gas J.* **103**, 53-58.

4 Liu H., Liu Z., Feng Y., Gu K., Yan T. (2005) Characteristic of a Supersonic Swirling Dehydration System of Natural Gas, *Chin. J. Chem. Eng.* **1**, 9-12.

5 Jassim E., Abdi M.A., Muzychka Y. (2008) Computational Fluid Dynamics Study for Flow of Natural Gas through High-pressure Supersonic Nozzles: Part 1. Real Gas Effects and Shockwave, *Pet. Sci. Technol.* **26**, 1757-1772.

6 Jassim E., Abdi M.A., Muzychka Y. (2008) Computational Fluid Dynamics Study for Flow of Natural Gas through High-pressure Supersonic Nozzles: Part 2. Nozzle Geometry and Vorticity, *Pet. Sci. Technol.* **26**, 1773-1785.

7 Karimi A., Abdi M.A. (2009) Selective Dehydration of High-Pressure Natural Gas Using Supersonic Nozzles, *Chem. Eng. Process.* **48**, 560-568.

8 Malyshkina M.M. (2008) The Structure of Gas Dynamic Flow in a Supersonic Separator of Natural Gas, *High Temp.* **46**, 69-76.

9 Malyshkina M.M. (2010) The Procedure for Investigation of the Efficiency of Purification of Natural Gases in a Supersonic Separator, *High Temp.* **48**, 244-250.

10 Zaporozhets E.P., Zibert G.K., Zibert A.G. (2011) Thermal Gas Dynamic Separator, *Chem. Petrol. Eng.* **46**, 585-593.

11 Jiang D., Eri Q., Wang C., Tang L. (2011) A Fast and Efficient Numerical-Simulation Method for Supersonic Gas Processing, *SPE Projects, Facilities & Construction* **6**, 58-64.

12 Wen C., Cao X., Yang Y., Zhang J. (2011) Supersonic Swirling Characteristics of Natural Gas in Convergent-Divergent Nozzles, *Petrol. Sci.* **8**, 114-119.

13 Wen C., Cao X., Yang Y., Zhang J. (2011) Swirling Effects on the Performance of Supersonic Separators for Natural Gas Separation, *Chem. Eng. Technol.* **34**, 1575-1580.

14 Wen C., Cao X., Yang Y., Li W. (2012) Numerical Simulation of Natural Gas Flows in Diffusers for Supersonic Separators, *Energy* **37**, 195-200.

15 Wen C., Cao X., Yang Y., Zhang J. (2012) Evaluation of Natural Gas Dehydration in Supersonic Swirling Separators Applying the Discrete Particle Method, *Adv. Powder Technol.* **23**, 228-233.

16 ANSYS Fluent User Manual, 2011, ANSYS INC.

17 Redlich O., Kwong J.N.S. (1949) On the Thermodynamics of Solutions. V. An Equation of State. Fugacities of Gaseous Solutions, *Chem. Rev.* **44**, 233-244.

18 Soave G. (1972) Equilibrium Constants from a Modified Redlich-Kwong Equation of State, *Chem. Eng. Sci.* **27**, 1197-1203.

19 Peng D.Y., Robinson D.B. (1976) A New Two-Constant Equation of State, *Ind. Eng. Chem. Fundam.* **15**, 59-64.

20 Aungier R.H. (1995) A Fast, Accurate Real Gas Equation of State for Fluid Dynamic Analysis Applications, *J. Fluids Eng.* **117**, 277-281.

21 Kwak T.Y., Mansoori G.A. (1986) Van der Waals Mixing Rules for Cubic Equations of State. Applications for Supercritical Fluid Extraction Modeling, *Chem. Eng. Sci.* **41**, 1303-1309.

22 Benmekki E.H., Kwak T.Y., Mansoori G.A. (1987) *Supercritical Fluids*, American Chemical Society, Washington.

23 Wen C., Cao X., Yang Y. (2011) Swirling Flow of Natural Gas in Supersonic Separators, *Chem. Eng. Process.* **50**, 644-649.

24 Foelsch K. (1949) The Analytical Design of an Axially Symmetric Laval Nozzle for a Parallel and Uniform Jet, *J. Aero. Sci.* **16**, 161-166.

25 Pope S.B. (2000) *Turbulent flows,* Cambridge University Press, Cambridge.

26 Patankar S.V., Spalding D.B. (1972) A Calculation Procedure for Heat, Mass and Momentum Transfer in Three-Dimensional Parabolic Flows, *Int. J. Heat Mass Trans.* **15**, 1787-1806.

27 Patankar S.V. (1980) *Numerical Heat Transfer and Fluid Flow*, McGraw-Hill, New York.

28 Wen C., Cao X., Yang Y., Li W. (2011) An Unconventional Supersonic Liquefied Technology for Natural Gas, *Energy Educ. Sci. Technol.* **30**, 651-660.

4

Analytical Estimation of Water-Oil Relative Permeabilities through Fractures

Hadi Saboorian-Jooybari

Reservoir Studies Division, Department of Petroleum Engineering, Main Office Building, National Iranian South Oil Company (NISOC), P.O. Box 61735-1333, Ahvaz - Iran
e-mail: hadi.saboorian@gmail.com

Abstract — *Modeling multiphase flow through fractures is a key issue for understanding flow mechanism and performance prediction of fractured petroleum reservoirs, geothermal reservoirs, underground aquifers and carbon-dioxide sequestration. One of the most challenging subjects in modeling of fractured petroleum reservoirs is quantifying fluids competition for flow in fracture network (relative permeability curves). Unfortunately, there is no standard technique for experimental measurement of relative permeabilities through fractures and the existing methods are very expensive, time consuming and erroneous. Although, several formulations were presented to calculate fracture relative permeability curves in the form of linear and power functions of flowing fluids saturation, it is still unclear what form of relative permeability curves must be used for proper modeling of flow through fractures and consequently accurate reservoir simulation. Basically, the classic linear relative permeability (X-type) curves are used in almost all of reservoir simulators. In this work, basic fluid flow equations are combined to develop a new simple analytical model for water-oil two phase flow in a single fracture. The model gives rise to simple analytic formulations for fracture relative permeabilities. The model explicitly proves that water-oil relative permeabilities in fracture network are functions of fluids saturation, viscosity ratio, fluids density, inclination of fracture plane from horizon, pressure gradient along fracture and rock matrix wettability, however they were considered to be only functions of saturations in the classic X-type and power (Corey [35] and Honarpour et al. [28, 29]) models. Eventually, validity of the proposed formulations is checked against literature experimental data. The proposed fracture relative permeability functions have several advantages over the existing ones. Firstly, they are explicit functions of the parameters which are known for each of simulation computational cells or easy to measure in laboratory. It is also the first model that takes gravity effects and wettability of fracture walls into consideration and individually developed for water and oil-wet systems. Furthermore, the newly developed formulations are simple, efficient and accurate. Thus, they are recommended for implementation in dual and multiple continuum commercial reservoir simulators.*

Résumé — Estimation analytique des perméabilités relatives de l'eau et de l'huile d'un réseau de fractures — La modélisation de l'écoulement multiphasique à travers des fractures est un sujet-clé pour mieux modéliser les écoulements au sein des réservoirs pétroliers/géothermiques, des aquifères souterrains ou lors de la séquestration de dioxyde de carbone. Cette étape est essentielle pour évaluer la production d'un réservoir, la capacité de stockage ou prédire les productions futures. Un des sujets les plus difficiles dans la modélisation des réservoirs pétroliers fracturés est de quantifier l'interaction de différents fluides lorsqu'ils s'écoulent dans le réseau de fractures (courbes de la perméabilité relative). Malheureusement, il n'existe pas de technique expérimentale standard pour mesurer la perméabilité relative à travers des fractures et les méthodes existantes sont trop coûteuses, chronophages et incertaines. Toutefois, plusieurs formulations ont déjà été proposées pour calculer les courbes de perméabilité relative rendant compte de l'écoulement de fluides dans des fractures. Ces formulations sont des fonctions linéaires ou puissances de saturation des fluides. Cependant, la forme des courbes de perméabilité relative permettant de modéliser correctement l'écoulement à travers des fractures et par conséquent de simuler précisément le réservoir reste sujet à débat. En fait, les courbes classiques de perméabilité relative (type-X) sont utilisées dans presque tous les simulateurs de réservoir. Dans cette étude, les équations de l'écoulement des fluides sont combinées afin de développer un modèle simple et analytique pour le cas de l'écoulement diphasique eau/huile dans une seule fracture. Ce modèle fournit une formulation simple et analytique des perméabilités relatives de fractures. Ce modèle prouve explicitement que les perméabilités relatives eau/huile dans un réseau fracturé sont fonctions de la saturation des fluides, du rapport de viscosité, des densités des fluides, de l'inclination des fractures, du gradient de pression et de la mouillabilité de matrice (roche saine). Il est intéressant de noter que dans les précédents modèles, de type-X ou puissance (Corey [35] and Honarpour et al. [28, 29]), seule une dépendance aux saturations était considérée. Finalement, la validité des formulations proposées est vérifiée par comparaisons à des données expérimentales fournies dans la littérature. Les fonctions de perméabilité relative proposées offrent plusieurs avantages par rapport des celles déjà existantes. Premièrement, elles sont des fonctions explicites de paramètres qui sont soit connus pour chacune des cellules de simulation, ou soit simples à mesurer en laboratoire. Ce modèle est aussi le premier à considérer les effets gravitationnels et une mouillabilité variable de fracture et il a été mis au point pour des systèmes eau/huile. De plus, ces nouvelles formulations développées sont simples, efficaces et précises. Leurs utilisations sont donc recommandées dans les simulateurs de réservoirs commerciaux double ou multiples continuum.

NOMENCLATURE

A	Cross-sectional area to flow (m)
b	Fracture aperture (m)
b_o	Mean value of fracture half-aperture (m)
c_1, c_2	Constant of integration
g	Acceleration of gravity (9.81 m/s^2)
h	Height, thickness (m)
H_D	Ratio of gravity to capillary force (dimensionless)
k	Absolute permeability (m^2)
k_r	Relative permeability (fraction)
k_r*	End-point relative permeability (fraction)
n	Exponent
P	Pressure (Pa)
q	Volumetric flow rate (cm^3/s)
S	Saturation (fraction)
u	Velocity (m/s)
$\bar{u}$	Average velocity (m/s)
W	Fracture width (m)

GREEK

Δ	Difference
Φ	Flow potential (Pa)
σ	Interfacial tension (N/m)
μ	Viscosity (cp)
τ	Shear stress (Pa)
ρ	Density (kg/m^3)
γ	Specific weight (Pa/m)
θ	Inclination degree of fracture plates from horizon, contact angle
∂	Partial differential operator

SUBSCRIPTS AND SUPERSCRIPTS

*	Normalized
c	Connate
eff	Effective
g	Gas
i	Phase name
l	Liquid
m	Mean (average)
nw	Non-wetting phase
o	Oil
r	Residual
w	Water, wetting phase
x	x-direction (m)
z	z-direction (m)

INTRODUCTION

Naturally fractured reservoirs contain more than 50% of the known oil and gas reserve worldwide and supply a significant portion of daily oil and gas demands [1]. North Sea chalks, Texas carbonates and Asmari limestone in Iran and Iraq are examples of naturally fractured reservoirs. Despite their significant importance in supplying increasing oil demands, this type of reservoirs exhibits relatively low levels of oil recovery compared to conventional reservoirs. This is mainly attributed to the complex nature of these reservoirs due to presence of two distinct porosity systems (matrix blocks and fractures) that aggravates reservoir heterogeneity and makes reservoir characterization, modeling and simulation a complicated and erroneous task. Furthermore, the oil-wetness tendency of rocks is another cause of very low recovery efficiency in such reservoirs [2]. Multiphase flow in fractured formations is of particular importance for understanding the performance of groundwater hydrology, geothermal reservoirs and disposal of hazardous wastes into underground water. Hydrocarbon reservoirs produce fluids under a combination of mechanisms including capillary, gravity and viscous displacement as well as pore compaction and fluids expansion. The displacement process in a fractured reservoir takes place when gas or water moves rapidly in high permeable fractures and surrounds oil saturated matrix blocks partially or totally. On the other hand, invasion of matrix blocks by gas or water and expulsion of oil into the fracture network begins by the combination of capillary and gravity forces once the matrix block is besieged. This process is called imbibition when the displacing fluid is water, and gravity drainage when the displacing fluid is water. Depending on the flowing phases present in the fractured reservoir, either the capillary or the gravity force becomes dominant [3]. These forces can work in tandem or can oppose each other [4].

A number of researches [5-12] were conducted to shed additional light on production mechanisms of the fractured reservoirs. Reservoir simulation is the most effective and practical method for studying multiphase flow problems in heterogeneous porous media. Numerical modeling of fractured reservoirs can be conducted with several existing techniques [13-18]:

1) dual porosity,
2) dual permeability,
3) Multiple INteracting Continua (MINC),
4) subdomain,
5) Discrete Fracture Network (DFN).

The first four are referred to as multiple continuum techniques in which the reservoir is discretized into a series of high storage capacity matrix blocks and high permeability fracture systems. These approaches have very little differences with each other. In dual porosity models, fluid flow only takes place through fracture systems and matrix blocks act as source and sink. In dual porosity models, matrix blocks are assumed to be connected to each other only through a fracture network, whereas in dual permeability concept the communication can occur between the adjacent blocks, as well as within the fractures. Dual permeability models allow the wells to be completed in both matrix blocks and fractures, but all flow to the wells in dual porosity models takes place through fractures. MINC models generate secondary nested gridding of the matrix blocks based on the initial estimated solution. The subdomain allows user to refine matrix blocks in vertical direction for increasing the accuracy of calculations. The DFN approach generates the model of reservoir spatial distribution of fracture network and detailed knowledge of the fracture intrinsic properties. Although DFN is a very rigorous technique for modeling naturally fractured reservoirs, its application is limited due to lack of the detailed input data required. Therefore, multiple continuum techniques are the most frequently used methods for simulation of fractured porous media. In all of multiple continuum approaches including dual porosity models, the basic governing fluid flow equations are written separately for matrix blocks and fracture systems and then coupled together. Therefore, simulation of fractured reservoirs requires separate sets of relative permeability curves for the two comprising media [19-22].

Rock matrix relative permeabilities are easily measured using a standard experimental procedure on representative plug samples during Special Core Analysis (SCAL). Two techniques are commonly used in the industry for measurements of relative permeabilities: steady-state and unsteady-state methods. In the steady-state technique, two fluids are simultaneously injected at a fixed ratio until saturation and pressure equilibriums are established. Relative permeabilities are directly calculated from the pressure drop and flow rate data. Then, the fluids ratio is adjusted to obtain the relative permeabilities at other saturations. In the unsteady state method, the core plug is saturated with one fluid and the other fluid is injected into the sample either at constant rate or constant pressure while measuring the rate, pressure drop and the volume of the produced fluids. Relative permeabilities are then determined by one of the proposed mathematical methods [23-27]. Detailed descriptions of the experimental procedures of such measurements, mathematical calculation, their advantages and drawbacks and existing empirical correlations can be found in a number of works [28, 29]. Unfortunately, there is no standard technique for experimental measurement of relative permeabilities through fractures. Furthermore, the existing methods are very expensive, time consuming and sometimes fail to measure fracture properties correctly. For these reasons, development of analytical formulations for relative permeabilities through fractures is of great importance. Although, several formulations are presented to calculate fracture relative permeabilities in the form of linear and power functions of saturation, it is still unclear what form of relative permeability curves must be used for proper modeling of multiphase flow through fractures and consequently for accurate reservoir simulation. Upon this review, the necessity of the analytical development of relative permeability functions for fluid flow through fracture network is revealed.

1 LITERATURE REVIEW

The ability of a medium to conduct a single phase fluid is known as absolute permeability of that medium. A number of researchers [30-34] have studied absolute permeability of fractures with parallel planar plates and all have reached to the following expression for absolute permeability in single phase, isothermal, laminar and steady-state flow:

$$k = \frac{b^2}{12} \quad (1)$$

where b is the fracture aperture. In petroleum reservoirs, however, the rock pore spaces are usually occupied by two (or more) phases. In these cases, pores occupied by one phase are unavailable for flow by other phases and the concept of relative permeability is used to describe the competency of phases for flow:

$$k_{ri} = \frac{k_{eff,i}}{k} \quad (2)$$

In this relation, k_{ri} is the relative permeability to phase i, k is the absolute permeability of the porous medium and $k_{eff,i}$ is the effective permeability to phase i. When two or more fluids flow at the same time, the relative permeability is the measure of the conductance of the medium to each fluid phase. Indeed, the effective permeability of each fluid is the permeability of the porous medium to the fluid when the saturation of that fluid is less than 100% and the relative permeability of the fluid is the ratio of the effective permeability at a specific saturation of the fluid to the absolute permeability of the medium. On the other hand, when a medium is completely saturated with phase i, the effective and relative permeabilities of the phase are, respectively, equal to the absolute permeability of the medium and unity ($k_{eff,i} = k$ and $k_{ri} = 1$). For the phase i (considering immiscible flow), the generalized Darcy's law can be used to describe flow rates as:

$$q_i = -\frac{Akk_{ri}}{\mu_i} \cdot \frac{d\Phi_i}{dx} \quad (3)$$

where q_i is the volumetric flow rate of the phase, A is the cross-sectional area to flow, k is the absolute permeability, k_{ri} is the effective permeability, μ_i is viscosity of phase i, Φ_i is flow potential of phase i and x is direction of flow.

Since relative permeabilities are strongly saturation dependent, they are crucial to the modeling of multiphase flow in matrix blocks and fracture systems. Despite the importance, there is no consensus on experimental measurement and theoretical calculation of relative permeabilities through fractures. A few data can be found in literature for two phase flow in fractures. The earliest and simplest model was originated in experimental work of Romm [35]. He performed the experiment with water and kerosene through an artificial glass parallel-plate fracture with waxed surfaces. He found a linear relationship between relative permeability and saturation (*X*-curves), $k_{ri} = S_i$ with endpoints at 0 and 1. Since, in two phase flow:

$$S_w + S_{nw} = 1 \quad (4)$$

This will lead to unity value for the summation of the relative permeabilities (Fig. 1):

$$k_{r,w} + k_{r,nw} = 1 \quad (5)$$

where w and nw refer to wetting and non-wetting phases, respectively. This is equivalent to negligible phase interference at all saturations and is the most widely used model in commercial reservoir simulators for numerical modeling of fractured reservoirs. A lot of researchers in petroleum industry [36-39] took Romm's linear relative permeability curves in their works. After Romm [35], Grant [40] and Pruess *et al.* [41] concluded that the application of *X*-curve for relative permeabilities through fractures is valid in systems with significant phase changes such as steam-water system in geothermal reservoirs.

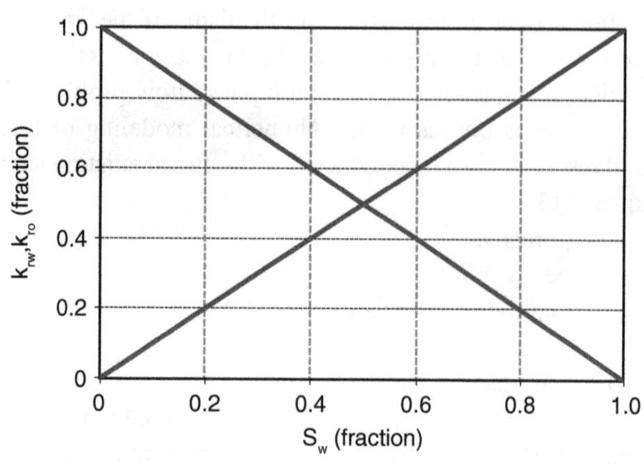

Figure 1

Romm's linear (*X*-type) relative permeability curves.

Among limited experimentally measured data available in the literature only a few works observed Romm's *X*-type model for relative permeabilities in fractures. Indeed, most of the data indicate considerable departure from straight line behavior. Romm's experiment with water and oil was duplicated by Merrill [42] in glass parallel-plates fracture and in natural porous media split along the length by sawing without using waxed paper strips as spacers. Since most of the data in his experiments scattered over a narrow range of water saturation (around a value of 71%), attempts failed to correlate relative permeability with saturation. However, the limited data indicated considerable deviation from the *X*-curve. A number of other analytical [43] and numerical [44] works for two phase flow in rough-walled and variable aperture fractures showed existence of considerable interference between phases and consequently $k_{r,w} + k_{r,nw} << 1$. Apart from oil-water system, other researchers [45-48] conducted experiments with air and water and observed failure of the *X*-curve to model fracture relative permeabilities.

A number of experimental studies found that two phase flow in fractures is better described using straight line relative permeabilities with slope less than one. Pan *et al.* [49] suggested that the slope of 0.6 for air and 0.7 for water relative permeability functions give rise to better match to experimental data. While, Rangel-German *et al.* [50] obtained the slope of 0.6 for both air and water relative permeabilities. Replacement of the unity slope of the *X*-curve model by smaller slopes in these works is an indication of occurrence of higher flow resistance in fractures because of strong interference between the flowing phases.

Some researchers [51-54] concluded that experimentally measured relative permeability curves are better matched with porous media approach (Corey model [55]) as:

$$k_{r,l} = S^{*4} \tag{6}$$

$$k_{r,g} = \left(1 - S^{*2}\right)\left(1 - S^{*}\right)^{2} \tag{7}$$

where S^* is normalized liquid saturation defined as:

$$S^* = \frac{S_l - S_{lr}}{1 - S_{lr} - S_{gr}} \tag{8}$$

and S_g and S_l are gas and liquid saturations, respectively. Meanwhile, the subscript r refers to residual saturations of the phases. In contrary to the above works, experimental studies of Persoff *et al.* [45] and Persoff and Pruess [46] concluded that fracture relative permeability cannot be well fitted neither by straight-line nor by Corey model.

The other model known as pipe flow model treated flow through fractures as flow through a system of pipes to construct fracture relative permeability curves [47, 48]. In this approach, a phase can be carried by the other phase as bubbles, mist, slug and other structures without need of establishing a continuous flow path. The models that were developed on the basis of pipe flow approach can be categorized in two classes:

- viscous coupling,
- equivalent homogeneous single-phase models.

The viscous coupling model was firstly introduced by Fourar *et al.* [56] for the simple case of lubricated film flow in a pipe equivalent to fracture as:

$$k_{r,l} = \frac{S_l^2}{2}(3 - S_l) \tag{9}$$

and

$$k_{r,g} = (1 - S_l)^3 + \frac{3}{2} \cdot \frac{\mu_g}{\mu_l} \cdot S_l(1 - S_l)(2 - S_l) \tag{10}$$

In 2010, Shad and Gates [57] extended the viscous coupling model to more than one phase systems. In another pipe flow model, which is known as equivalent homogeneous single-phase model, Fourar *et al.* [47] considered the two phase flow to be equivalent to single phase flow of a homogenous phase through a system of pipes. The average density and viscosity of the equivalent homogenous phase are calculated by weighting properties of each phases by the corresponding volumetric flow rates as:

$$\rho_m = \frac{\rho_1 q_1 + \rho_2 q_2}{q_1 + q_2} \tag{11}$$

$$\mu_m = \frac{\mu_1 q_1 + \mu_2 q_2}{q_1 + q_2} \tag{12}$$

where the subscript m refers to average value.

As it was here mentioned above, despite its essential importance, only a few experimental, numerical and analytical attempts have been made to clear the physics of multiphase flow and characteristic of relative permeabilities in fractures. Therefore, the extremely important question of proper form of relative permeability curves through fractures is still unanswered. This paper is an attempt to develop a new simple analytical model for description of characteristic behavior of water-oil relative permeabilities in fractures.

2 MATHEMATICAL FORMULATION

In two phase flow of immiscible fluids, the wetting-phase flows along the fracture wall and the non-wetting phase occupies the fracture interior [58, 59]. The potential gradient

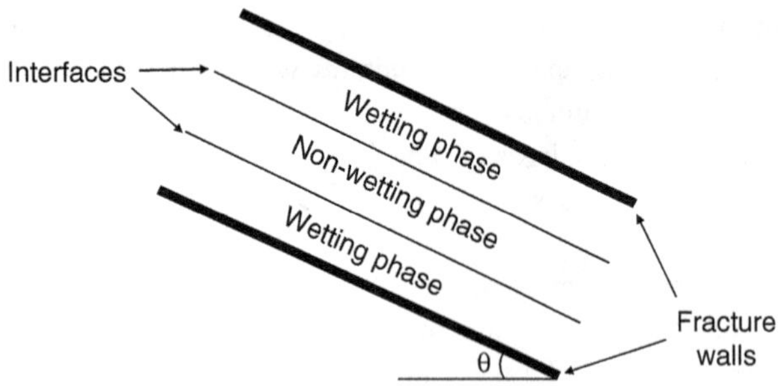

Figure 2

Illustration of wetting and non-wetting phase flow along the wall and interior of a fracture.

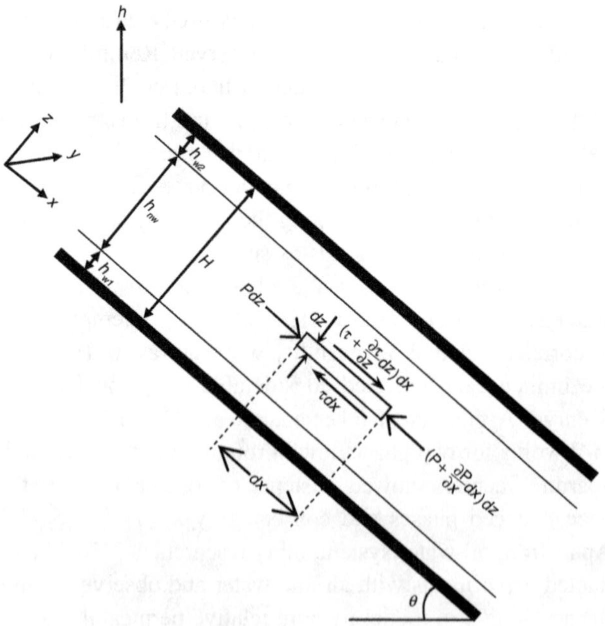

Figure 3

Momentum balance for two phase immiscible flow in an inclined fracture.

along fracture determines the thickness of each phase (Fig. 2). Water-oil two phase flow between the parallel fracture plates are considered to be laminar, incompressible and steady state. An infinitesimal thin lamina is taken as a control volume (free body) to analyze the flow (Fig. 3). Linear momentum of the element of a mass dm is a vector defined as $dm\,\boldsymbol{V}$. In this case, the mass dm is a part of velocity field:

$$\boldsymbol{V} = u\boldsymbol{i} + v\boldsymbol{j} + w\boldsymbol{k} \tag{13}$$

The fundamental statement of the second law of Newton in terms of linear momentum of the element is given as:

$$d\boldsymbol{F}_{ext} = \frac{D}{Dt}(dm\,\boldsymbol{V}) \tag{14}$$

where D is the essential derivative defined as:

$$\frac{D}{Dt} = \frac{\partial}{\partial t} + u\frac{\partial}{\partial x} + v\frac{\partial}{\partial y} + w\frac{\partial}{\partial z} \tag{15}$$

and $\boldsymbol{F}_{ext}$ is the external forces acting over the control volume. These forces can be classified as the body and surface forces. In absence of a magnetic/electric field, gravity is the only body force which can be expressed as $-\rho g dv \nabla h$. The surface forces on the particle are due to pressure and shear stress respectively in the form of $-dv\nabla P$ and $-dv\nabla\tau$. The shearing stress (τ) is the force that acts on the unit area of interface between the particles tangent to the direction of flow. As we know, substituting these forces in Equation (14) and dividing through by dv leads to the very famous Navier-Stokes equation:

$$\rho\left(\frac{\partial \boldsymbol{V}}{\partial t} + (\boldsymbol{V}\cdot\nabla)\boldsymbol{V}\right) = -\rho g\nabla h - \nabla P - \nabla\tau \tag{16}$$

where h is the vertical direction. Since for the laminar, incompressible and steady state condition shown in Figure 3, for flow in the x direction, we have $v = w = 0$. Thus, the only nonzero stress from τ_{ij} is τ_{xz}. With these in mind, the vector Equation (16) reduces to the following scalar equation of motion in the x, y and z directions:

$$\frac{\partial \tau_{xz}}{\partial z} = \frac{\partial P}{\partial x} + \rho g\frac{\partial h}{\partial x} \tag{17}$$

$$\frac{\partial P}{\partial y} = -\rho g\cos(\theta) \tag{18}$$

$$\frac{\partial P}{\partial z} = 0 \tag{19}$$

where θ is the inclination degree of the fracture from the horizontal line. The equations obtained in the y and z directions are trivial. Thus, Equation (17) is the only equation that should be employed for the further analysis as:

$$\frac{\partial \tau_{xz}}{\partial z} = \frac{\partial(P + \rho g h)}{\partial x} \tag{20}$$

Since the only nonzero component of the stress field is τ_{xz}, the subscript xz will be omitted hereafter for simplicity.

As we know, $P + \rho gh$ is called flow potential. Therefore, Equation (20) can be rewritten as:

$$\frac{\partial \tau}{\partial z} = \frac{\partial \Phi}{\partial x} \tag{21}$$

Since τ and Φ are only functions of z and x, respectively:

$$\frac{d\tau}{dz} = \frac{d\Phi}{dx} \tag{22}$$

Integration of this equation with respect to z gives:

$$\tau = \frac{d\Phi}{dx} \cdot z + c_1 \tag{23}$$

This equation is valid for the three streams in Figure 2:

$$\tau_{w_1} = \frac{d\Phi_{w_1}}{dx} \cdot z + c_1^{w_1} \tag{24}$$

$$\tau_{nw} = \frac{d\Phi_{nw}}{dx} \cdot z + c_1^{nw} \tag{25}$$

$$\tau_{w_2} = \frac{d\Phi_{w_2}}{dx} \cdot z + c_1^{w_2} \tag{26}$$

In which $c_1^{w_1}$, c_1^{nw} and $c_1^{w_2}$ are constants of integration. In order to make the formulations simpler, thickness of the two regions of wetting phase at the top and bottom walls of the fracture is assumed to be equal ($h_{w_1} = h_{w_2} = h_w$). Because of symmetry of the system in Figure 3:

$$\frac{d\Phi_{w_1}}{dx} = \frac{d\Phi_{w_2}}{dx} = \frac{d\Phi_w}{dx} \tag{27}$$

And flow potential gradient of the wetting and non-wetting phases are related as:

$$\frac{d\Phi_{nw}}{dx} = \frac{d\Phi_w}{dx} + (\rho_w - \rho_{nw})g \sin(\theta) \tag{28}$$

To evaluate Equations (24) through (26), the continuity of fluid mechanics properties at interface between three flow regions in Figures 2 and 3 necessitates:

$$z = 0, \tau_{w_1} = \tau_{nw} \tag{29}$$

$$z = h_{nw}, \tau_{nw} = \tau_{w_2} \tag{30}$$

Substituting these boundary conditions in Equations (24) through (26) yields:

$$c_1^{w_1} = c_1^{nw} = c_1^{w_2} = c_1 \tag{31}$$

Equations (24) through (26) can be combined with Newton's law of viscosity to get the following equations, respectively:

$$\mu_w \frac{du_{w_1}}{dz} = \frac{d\Phi_w}{dx} \cdot z + c_1 \tag{32}$$

$$\mu_{nw} \frac{du_{nw}}{dz} = \frac{d\Phi_{nw}}{dx} \cdot z + c_1 \tag{33}$$

$$\mu_w \frac{du_{w_2}}{dz} = \frac{d\Phi_w}{dx} \cdot z + c_1 \tag{34}$$

Integration of these equations with respect to z leads to velocity profiles in the three flow regions as follows:

$$u_{w_1} = \frac{1}{\mu_w}\left[\frac{d\Phi_w}{dx} \cdot \frac{z^2}{2} + c_1 \cdot z\right] + c_2^{w_1} \tag{35}$$

$$u_{nw} = \frac{1}{\mu_{nw}}\left[\frac{d\Phi_{nw}}{dx} \cdot \frac{z^2}{2} + c_1 \cdot z\right] + c_2^{nw} \tag{36}$$

$$u_{w_2} = \frac{1}{\mu_w}\left[\frac{d\Phi_w}{dx} \cdot \frac{z^2}{2} + c_1 \cdot z\right] + c_2^{w_2} \tag{37}$$

Determination of the integration constants (c_1, $c_2^{w_1}$, c_2^{nw} and $c_2^{w_2}$) needs the following boundary conditions (Streeter *et al.* [60]):

$$z = -h_w, u_{w_1} = 0 \tag{38}$$

$$z = 0, u_{w_1} = u_{nw} \tag{39}$$

$$z = h_{nw}, u_{nw} = u_{w_2} \tag{40}$$

$$z = h_{nw} + h_w, u_{w_2} = 0 \tag{41}$$

Substitution of the boundary conditions in Equations (35) through (37) yields:

$$c_2^{w_1} = -\frac{1}{\mu_w}\left[\frac{d\Phi_w}{dx} \cdot \frac{h_w^2}{2} - c_1 \cdot h_w\right] \tag{42}$$

$$c_2^{w_1} = c_2^{nw} \tag{43}$$

$$c_2^{w_2} = c_2^{nw} + \frac{1}{\mu_{nw}}\left[\frac{d\Phi_{nw}}{dx}\cdot\frac{h_{nw}^2}{2} + c_1\cdot h_{nw}\right] - \frac{1}{\mu_w}\left[\frac{d\Phi_w}{dx}\cdot\frac{h_{nw}^2}{2} + c_1\cdot h_{nw}\right] \quad (44)$$

$$c_2^{w_2} = -\frac{1}{\mu_w}\left[\frac{d\Phi_w}{dx}\cdot\frac{(h_{nw}+h_w)^2}{2} + c_1\cdot(h_{nw}+h_w)\right] \quad (45)$$

Solving the system of four equations (Eqs. 42 through 45) and four unknowns (c_1, $c_2^{w_1}$, c_2^{nw} and $c_2^{w_2}$) results in the following solutions:

$$c_1 = -\frac{\frac{h_w h_{nw}}{\mu_w}\cdot\frac{d\Phi_w}{dx} + \frac{h_{nw}^2}{2\mu_{nw}}\cdot\frac{d\Phi_{nw}}{dx}}{\frac{2h_w}{\mu_w} + \frac{h_{nw}}{\mu_{nw}}} \quad (46)$$

$$c_2^{w_1} = c_2^{nw} = -\frac{1}{\mu_w} \times \left[\frac{h_w^2}{2}\cdot\frac{d\Phi_w}{dx} + \frac{\frac{h_w^2 h_{nw}}{\mu_w}\cdot\frac{d\Phi_w}{dx} + \frac{h_{nw}^2 h_w}{2\mu_{nw}}\cdot\frac{d\Phi_{nw}}{dx}}{\frac{2h_w}{\mu_w} + \frac{h_{nw}}{\mu_{nw}}}\right] \quad (47)$$

$$c_2^{w_2} = -\frac{1}{\mu_w}\cdot\left[\frac{(h_w+h_{nw})^2}{2}\cdot\frac{d\Phi_w}{dx} - (h_w+h_{nw})\frac{\frac{h_w h_{nw}}{\mu_w}\cdot\frac{d\Phi_w}{dx} + \frac{h_{nw}^2}{2\mu_{nw}}\cdot\frac{d\Phi_{nw}}{dx}}{\frac{2h_w}{\mu_w} + \frac{h_{nw}}{\mu_{nw}}}\right] \quad (48)$$

Substituting the integration constants in Equations (35) to (37) yields the velocity profiles in the three flow regions. Knowing the velocity profiles, it is simple to calculate average velocity for each of the flowing streams:

$$\overline{u_{w_1}} = \frac{\int_{-h_w}^{0} u_w dz}{\int_{-h_w}^{0} dz} = \frac{1}{\mu_w}\left[\frac{d\Phi_w}{dx}\cdot\frac{h_w^2}{6} - c_1\cdot\frac{h_w}{2}\right] + c_2^{w_1} \quad (49)$$

$$\overline{u_{nw}} = \frac{\int_0^{h_{nw}} u_w dz}{\int_0^{h_{nw}} dz} = \frac{1}{\mu_{nw}}\left[\frac{d\Phi_{nw}}{dx}\cdot\frac{h_{nw}^2}{6} + c_1\cdot\frac{h_{nw}}{2}\right] + c_2^{nw} \quad (50)$$

$$\overline{u_{w_2}} = \frac{\int_{h_{nw}}^{h_{nw}+h_w} u_{w_2} dz}{\int_{h_{nw}}^{h_{nw}+h_w} dz} = \frac{1}{\mu_w h_w}\left[\frac{d\Phi_w}{dx}\cdot\left(\frac{(h_{nw}+h_w)^3}{6} - \frac{h_{nw}^3}{6}\right) + c_1\cdot\left(\frac{(h_{nw}+h_w)^2}{2} - \frac{h_{nw}^2}{2}\right)\right] + c_2^{w_2} h_w \quad (51)$$

Substitution of the integration constants in above equations we obtain:

$$\overline{u_{w_1}} = \frac{\int_{-h_w}^{0} u_w dz}{\int_{-h_w}^{0} dz} = -\frac{1}{\mu_w}\cdot\frac{d\Phi_w}{dx}\cdot\left(\frac{h_w^2}{3} + \frac{h_w h_{nw}}{4}\right) - \frac{h_w h_{nw}^2(\rho_w - \rho_{nw})g\sin(\theta)}{4\mu_w\mu_{nw}\left(\frac{2h_w}{\mu_w} + \frac{h_{nw}}{\mu_{nw}}\right)} \quad (52)$$

$$\overline{u_{nw}} = \frac{\int_0^{h_{nw}} u_w dz}{\int_0^{h_{nw}} dz} = -\frac{1}{2}\cdot\frac{d\Phi_{nw}}{dx}\cdot\left(\frac{h_{nw}^2}{6\mu_{nw}} + \frac{h_w^2 + h_w h_{nw}}{\mu_w}\right) + \frac{h_w}{2\mu_w}(h_w + h_{nw})(\rho_w - \rho_{nw})g\sin(\theta) \quad (53)$$

$$\overline{u_{w_2}} = \frac{\int_{h_{nw}}^{h_{nw}+h_w} u_{w_2} dz}{\int_{h_{nw}}^{h_{nw}+h_w} dz} = -\frac{1}{\mu_w}\cdot\frac{d\Phi_w}{dx}\cdot\left(\frac{h_w^2}{3} + \frac{h_w h_{nw}}{4}\right) - \frac{h_w h_{nw}^2(\rho_w - \rho_{nw})g\sin(\theta)}{4\mu_w\mu_{nw}\left(\frac{2h_w}{\mu_w} + \frac{h_{nw}}{\mu_{nw}}\right)} \quad (54)$$

Comparison of Equations (52) and (54) indicates that average velocity of the wetting phase regions are equal to each other. The volumetric flow rate of each of the three regions is the product of average velocity and area of the stream:

$$q_{w_1} = q_{w_2} = Wh_w\left[-\frac{1}{\mu_w}\cdot\frac{d\Phi_w}{dx}\cdot\left(\frac{h_w^2}{3} + \frac{h_w h_{nw}}{4}\right) - \frac{h_w h_{nw}^2(\rho_w - \rho_{nw})g\sin(\theta)}{4\mu_w\mu_{nw}\left(\frac{2h_w}{\mu_w} + \frac{h_{nw}}{\mu_{nw}}\right)}\right] \quad (55)$$

$$q_{nw} = Wh_{nw}\left[-\frac{1}{2}\cdot\frac{d\Phi_{nw}}{dx}\cdot\left(\frac{h_{nw}^2}{6\mu_{nw}} + \frac{h_w^2 + h_w h_{nw}}{\mu_w}\right) + \frac{h_w}{2\mu_w}(h_w + h_{nw})(\rho_w - \rho_{nw})g\sin(\theta)\right] \quad (56)$$

Therefore, total flow rate of the wetting phase would be:

$$q_w = 2Wh_w\left[-\frac{1}{\mu_w}\cdot\frac{d\Phi_w}{dx}\cdot\left(\frac{h_w^2}{3}+\frac{h_wh_{nw}}{4}\right)\right.$$
$$\left.-\frac{h_wh_{nw}^2(\rho_w-\rho_{nw})g\sin(\theta)}{4\mu_w\mu_{nw}\left(\frac{2h_w}{\mu_w}+\frac{h_{nw}}{\mu_{nw}}\right)}\right] \tag{57}$$

For flow through the fracture, the volumetric flow rate also can be obtained from Darcy's law:

$$q_w = -\frac{kk_{rw}W(2h_w+h_{nw})}{\mu_w}\cdot\frac{d\Phi_w}{dx} \tag{58}$$

$$q_{nw} = -\frac{kk_{rnw}W(2h_w+h_{nw})}{\mu_{nw}}\cdot\frac{d\Phi_{nw}}{dx} \tag{59}$$

where k is absolute permeability of the fracture that is calculated using Equation (1) as:

$$k = \frac{(2h_w+h_{nw})^2}{12} \tag{60}$$

Substituting Equations (58) and (59) in Equations (57) and (56) yields the following expressions for wetting and non-wetting phase relative permeabilities, respectively:

$$k_{rw} = \frac{24h_w}{(2h_w+h_{nw})^3}$$
$$\times\left[\left(\frac{h_w^2}{3}+\frac{h_wh_{nw}}{4}\right)+\frac{h_wh_{nw}^2(\rho_w-\rho_{nw})g\sin(\theta)}{4\mu_{nw}\left(\frac{2h_w}{\mu_w}+\frac{h_{nw}}{\mu_{nw}}\right)\frac{d\Phi_w}{dx}}\right] \tag{61}$$

$$k_{rnw} = \frac{12\mu_{nw}h_{nw}}{(2h_w+h_{nw})^3}\left[\frac{1}{2}\left(\frac{h_{nw}^2}{6\mu_{nw}}+\frac{h_w^2+h_wh_{nw}}{\mu_w}\right)\right.$$
$$\left.-\frac{h_w}{2\mu_w}(h_w+h_{nw})\frac{(\rho_w-\rho_{nw})g\sin(\theta)}{\frac{d\Phi_{nw}}{dx}}\right] \tag{62}$$

In order to define relative permeabilities in terms of saturations, wetting and non-wetting phase saturations are defined as:

$$S_w = \frac{2h_w}{2h_w+h_{nw}} \tag{63}$$

$$S_{nw} = \frac{h_{nw}}{2h_w+h_{nw}} \tag{64}$$

Eventually, simple rearrangement of Equations (61) and (62) followed by substitution of Equations (63) and (64) gives rise to the following analytical expressions to estimate relative permeabilities in an inclined fracture:

$$k_{rw} = 12S_w\left[\left(\frac{S_w^2}{12}+\frac{S_wS_{nw}}{8}\right)\right.$$
$$\left.+\frac{S_{nw}^2(\rho_w-\rho_{nw})g\sin(\theta)}{8\left(\frac{1}{\mu_r}+\frac{S_{nw}}{S_w}\right)\left(\frac{dP}{dx}-\rho_wg\sin(\theta)\right)}\right] \tag{65}$$

$$k_{rnw} = 12S_{nw}\left[\frac{1}{2}\left(\frac{S_{nw}^2}{6}+\frac{S_w^2+2S_wS_{nw}}{4\mu_r}\right)\right.$$
$$\left.-\frac{S_w}{4\mu_r}\left(\frac{S_w}{2}+S_{nw}\right)\frac{(\rho_w-\rho_{nw})g\sin(\theta)}{\frac{dP}{dx}-\rho_{nw}g\sin(\theta)}\right] \tag{66}$$

where dP/dx is the pressure gradient along the fracture and μ_r is viscosity ratio which is the wetting to non-wetting phase viscosity ratio (μ_w/μ_{nw}). The newly developed relative permeability formulas in Equations (65) and (66) are explicit functions of phase saturations, viscosity ratio, fluids density, inclination of fracture from horizon, pressure gradient along fracture and rock matrix wettability. On the other hand, the newly developed relative permeability curves in Equations (65) and (66) are explicit functions of the parameters which are known for each of simulation computational cells or easy to measure in laboratory. Moreover, it is also the first relative permeability model that takes the gravity and wettability effects into account.

3 EFFECTS OF WETTABILITY AND VISCOSITY RATIO

3.1 Fracture Surface Wettability

In the previous section, general formulations of relative permeability curves were proposed for two-phase flow through fractures. In this section, it is intended to apply them to water-oil systems with strong water- and oil-wetness of the fracture walls and investigate the effects of wettability on the relative permeabilities. In the case of strongly water wet reservoir rock, Equations (65) and (66) take the following forms:

$$k_{rw} = 12S_w\left[\left(\frac{S_w^2}{12}+\frac{S_w(1-S_w)}{8}\right)\right.$$
$$\left.+\frac{(1-S_w)^2(\rho_w-\rho_o)g\sin(\theta)}{8\left(\frac{1}{\mu_r}+\frac{1-S_w}{S_w}\right)\left(\frac{dP}{dx}-\rho_wg\sin(\theta)\right)}\right] \tag{67}$$

$$k_{ro} = 12(1 - S_w)\left[\frac{1}{2}\left(\frac{(1 - S_w)^2}{6} + \frac{S_w^2 + 2S_w(1 - S_w)}{4\mu_r}\right) - \frac{S_w}{4\mu_r}\left(1 - \frac{S_w}{2}\right)\frac{(\rho_w - \rho_o)g\sin(\theta)}{\frac{dP}{dx} - \rho_o g\sin(\theta)}\right] \quad (68)$$

and for strong oil-wetness of the fracture walls:

$$k_{rw} = 12S_w\left[\frac{1}{2}\left(\frac{S_w^2}{6} + \frac{(1 - S_w)^2 + 2S_w(1 - S_w)}{4\mu_r}\right) - \frac{(1 - S_w)}{4\mu_r}\left(\frac{1 + S_w}{2}\right)\frac{(\rho_o - \rho_w)g\sin(\theta)}{\frac{dP}{dx} - \rho_w g\sin(\theta)}\right] \quad (69)$$

$$k_{ro} = 12(1 - S_w)\left[\left(\frac{(1 - S_w)^2}{12} + \frac{S_w(1 - S_w)}{8}\right) + \frac{S_w^2(\rho_o - \rho_w)g\sin(\theta)}{8\left(\frac{1}{\mu_r} + \frac{S_w}{1 - S_w}\right)\left(\frac{dP}{dx} - \rho_o g\sin(\theta)\right)}\right] \quad (70)$$

where the subscripts w and o stand for water and oil phases, respectively. Figure 4 shows a comparison between the relative permeabilities for strongly water and oil-wet fractures using Equations (67) to (70) for the rock and fluids properties of Table 1. This figure clearly indicates strong dependence of fracture relative permeability curves to the wettability of the walls. The other important point that can be inferred from this figure is that the value of oil relative permeability in the water-wet system exceeds unity over a limited range of saturation for the data of Table 1 ($\mu_r = 1$). The reason for observing such behavior is described in the following section.

Equations (67) through (70) can be used for analytical estimation of oil and water relative permeability curves through an inclined fracture as long as the flowing phases configuration is as depicted in Figures 2 and 3. In two-phase flow of water and heavy oils, the aqueous phase is underridden by the more viscous oleic phase regardless of wettability of the fracture walls (Fig. 5). Following the same procedure as here presented above leads to the following expressions for this kind of stratified flow pattern:

$$k_{rw} = \frac{S_w^2}{S_w(1 - \mu_r) + \mu_r} \cdot \left[S_w^2 + 4\mu_r S_w(1 - S_w) + 3\mu_r(1 - S_w)^2 \times \left(1 + \frac{(\rho_o - \rho_w)g\sin(\theta)}{\frac{dP}{dx} - \rho_w g\sin(\theta)}\right)\right] \quad (71)$$

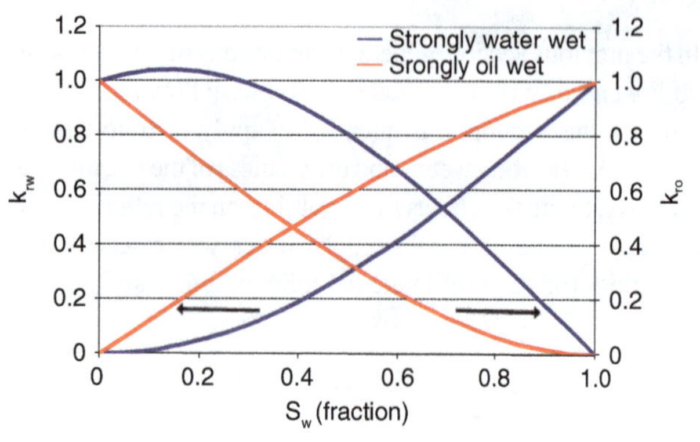

Figure 4

Comparison between relative permeabilities for strongly water and oil wet fractures for the rock and fluids properties of Table 1 ($\mu_r = 1$).

TABLE 1

Summary of the rock and fluids properties used to plot Figure 4

Parameters	Value
Inclination (deg)	45
dP/dx (kPa/m)	−2.26
Oil density (kg/m^3)	862
Water density (kg/m^3)	1 115
Oil viscosity (cp)	1.5
Water viscosity (cp)	1.5

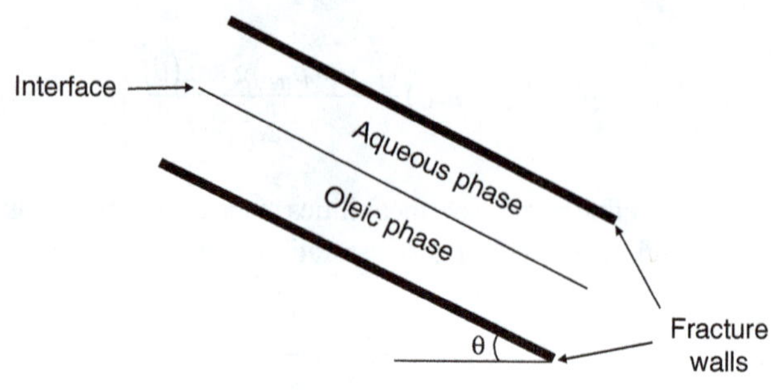

Figure 5

Illustration of wetting and non-wetting phase flow in stratified flow of oil along the bottom and water along the top walls of a fracture.

$$k_{ro} = \frac{(1-S_w)^2}{(1-S_w)(\mu_r-1)+1} \cdot \left[\mu_r(1-S_w)^2 + 4S_w(1-S_w) + 3S_w^2\left(1-\frac{(\rho_o-\rho_w)g\sin(\theta)}{\frac{dP}{dx}-\rho_o g\sin(\theta)}\right)\right] \tag{72}$$

where $\mu_r = \mu_w/\mu_o$ in Equations (71) and (72). It should be overemphasized here that the wetting tendency of the fracture walls plays no rule in two phase flow of water and heavy oil. In other words, the more viscous oil phase flows along the bottom wall, while water wets the top wall (Fig. 5). Therefore, these expressions are recommended for estimation of water and heavy oil relative permeabilities when the viscosity ratio is very small (*e.g.* $\mu_r < 0.1$). Otherwise, Equations (67) to (70) should be used in water and typical oil two phase flow depending on different rock matrix wetting preferences.

3.2 Viscosity Ratio

Series of sensitivities were performed to investigate dependence of the relative permeability curves on the viscosity ratio in strongly water- and oil-wet systems (Figs. 6, 7). The data of Table 1 with different viscosity ratios are used in the analysis. According to these figures and functional form of Equations (67) to (70), the viscosity ratio of the flowing phases has substantial effect on the relative permeability curve of non-wetting phase and can lead to values much larger than unity, while it almost plays negligible rule in calculation of wetting phase relative permeability. In other words, values of the viscosity ratio less than unity give rise to water and oil relative permeabilities greater than unity in oil and water-wet systems, respectively (Figs. 6, 7).

Furthermore, the less the viscosity ratio, the more the non-wetting phase relative permeability exceeds unity. This unusual behavior respect to what is practically accepted, was previously observed by Pan [61] in experimental measurement of water-oil relative permeabilities in water-wet Hele-Shaw flow cell. They reported significant increase of oil (the non-wetting phase) relative permeability with its viscosity, but relative permeability curve of water (the wetting phase) wasn't affected by viscosity ratio very much. Additionally, they found that the relative permeability of oil (the non-wetting phase) could be as high as four at low water saturation in the water wet cell, while water (the wetting phase) relative permeability didn't exceed unity. This finding is in line with what is observed in this work. Pan [61] justified this behavior by lubrication effects because of flow of the water film(s) between the oil and the fracture wall(s) and its impacts on the flowing streams velocities. Indeed, when the less viscous phase is flowing along the fracture wall, the oil flow rate might be greater than that would result for single phase flow under the same condition.

Physically speaking, the meaning of the viscous coupling is that the velocity of each of the flowing streams is affected by the pressure gradient and viscosity of the adjacent streams, as well as its own pressure gradient and viscosity which would give rise to momentum transfer across the interfaces. This justification was adopted from the analyses of Dullien [62] about possibility of occurrence of lubrication phenomenon during multiphase flow in real porous media. It should be noted here that because of dominance of viscous forces for flow in fractures compared to matrix, much more significant effects of viscous coupling are expected for relative permeabilities through fractures.

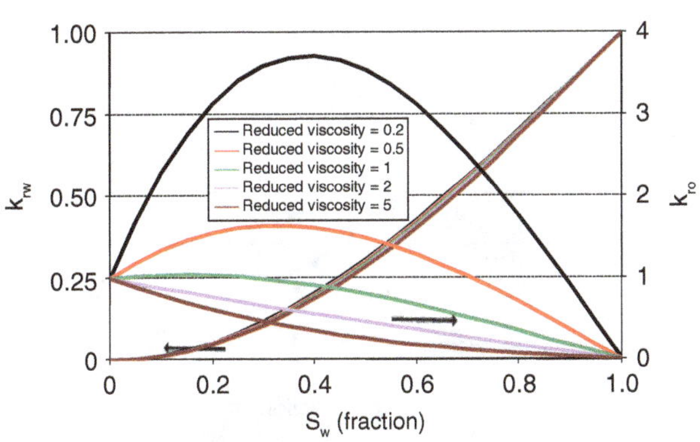

Figure 6

Effect of viscosity ratio on relative permeability curves for water-wet fracture walls.

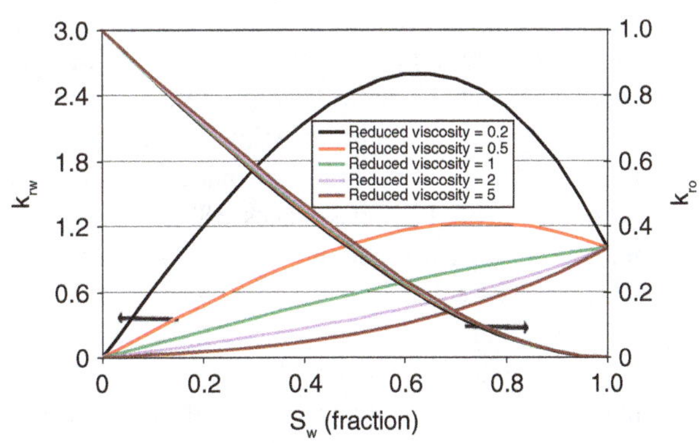

Figure 7

Effect of viscosity ratio on relative permeability curves for oil-wet fracture walls.

In summary, the proposed analytical formulations demonstrate that relative permeabilities through fractures are functions of several rock and fluids properties including phase saturations, viscosity ratio, fluids density, inclination of fracture plate from the horizontal position, pressure gradient along fracture and rock matrix wettability. On the other hand, the newly developed relative permeability expressions are explicit functions of the parameters that are known for each of simulation computational cells or easy to measure in laboratory. It is also the first relative permeability model that takes the gravity terms into consideration. The other advantage of these formulations for applying in reservoir simulators is that it wouldn't require import of fracture relative permeabilities in tabular form. Therefore, these formulations are simple, efficient and accurate for implementation in dual-, multiple continuum and DFN commercial reservoir simulators.

4 VALIDATION WITH EXPERIMENT

Our extensive survey in the literature indicates that limited number of experimental works have been done to examine two phase flow in a single fracture. But, the majority of them used gas (nitrogen) and water in their experiments. On the other hand, among experimentally measured data of relative permeability curves in fractures only a few works have been done with oil and water as the flowing phases to be used for confirming the validity of the proposed formulations. Unfortunately, fracture orientation was horizontal in almost all of these experiments. Therefore, none of the existing relative permeabilities in the literature can be used to evaluate the validity of the gravity effects in the newly proposed model. Anyway, the validity of the proposed models is checked against the experimental data of Pan [61]. He measured oil and water relative permeabilities in a horizontally oriented cell of impermeable parallel fracture plates (the so-called Hele-Shaw cell). He developed three parallel-plates models which differ from each other in the injection method and experimental design. It was analyzed that the design of their third model can better mimic the basic physics of fluid flow in fractured reservoirs. It is because the fluids entered the fracture through a porous medium (sintered metal) instead of individual tubes in their first two models. This model was constructed with Plexiglas plates 38 mm thick. The adjustable aperture of the fracture could be changed from zero to 0.5 mm. Three pairs of differential pressure transducers were installed on the bottom plate to measure the pressure gradient along the fracture. The plates were placed horizontally with a width and length of 100 and 550 mm. The cells developed above analytical formulations are used to calculate the measured fracture relative permeabilities for water of viscosity 1.012 cp and different oils (viscosities of 10.2, 66 and 106 cp) which correspond to viscosity ratios of 0.1, 0.015, 0.0096, respectively. The fracture average apertures in these tests were 106, 112 and 125 μm. Meanwhile, the range of the pressure gradients along the fractures were 16-130, 5-932 and 25-462 kPa/m for the viscosity ratio of 0.1, 0.015, 0.0096, respectively.

Figures 8, 9 and 10 show comparison between the calculated and measured relative permeabilities. Clearly, good agreement is found between them. Therefore, using the newly developed functions is sufficient to match the experimental data. In order to further validate the developed analytical models, the useful intermediate data of the above experiments are used to numerically extract the relative permeabilities. A nonlinear history-matching is performed for numerical derivation of the relative permeabilities. A commercial reservoir simulator is coupled with a

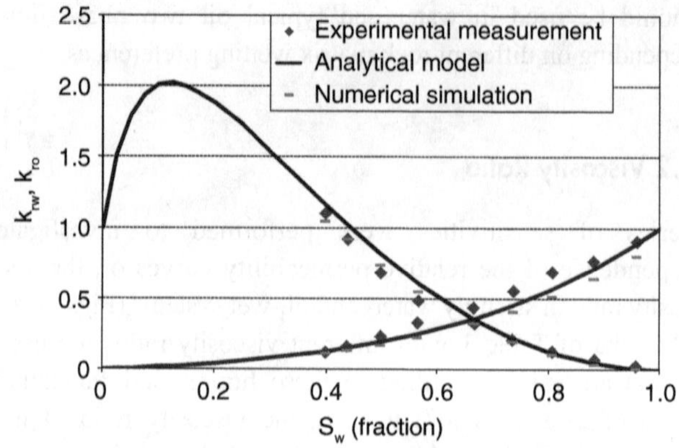

Figure 8

Comparison of calculation of the analytical model with experimental measurement of Pan (1999) [61] for $\mu_r = 0.1$.

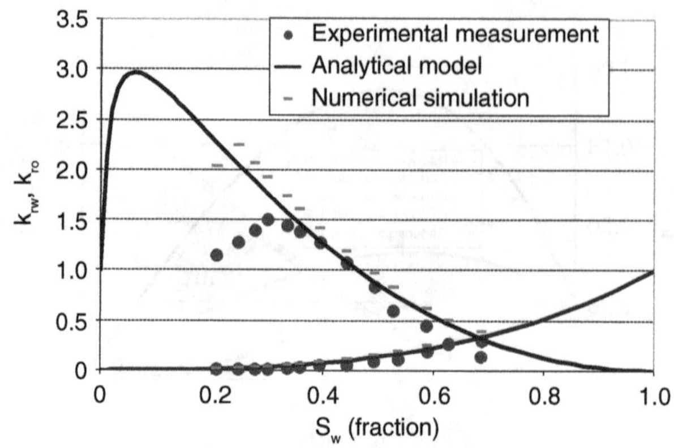

Figure 9

Comparison of calculation of the analytical model with experimental measurement of Pan (1999) [61] for $\mu_r = 0.015$.

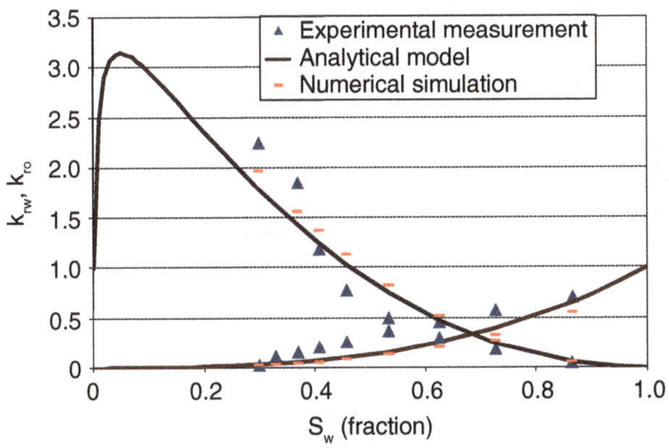

Figure 10

Comparison of calculation of the analytical model with experimental measurement of Pan (1999) [61] for $\mu_r = 0.0096$.

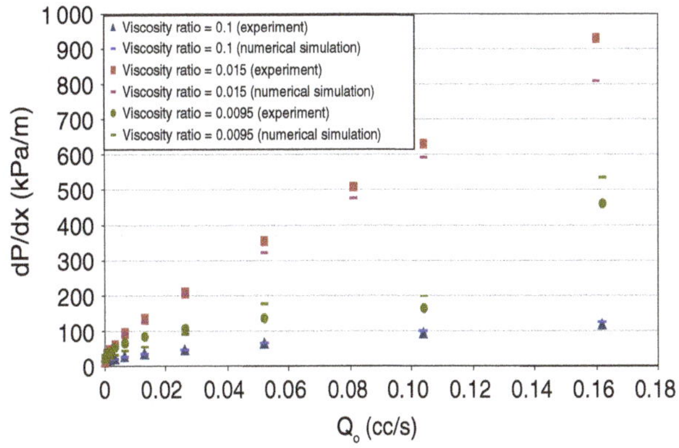

Figure 12

Experimental pressure gradient data compared with results of the numerical simulator.

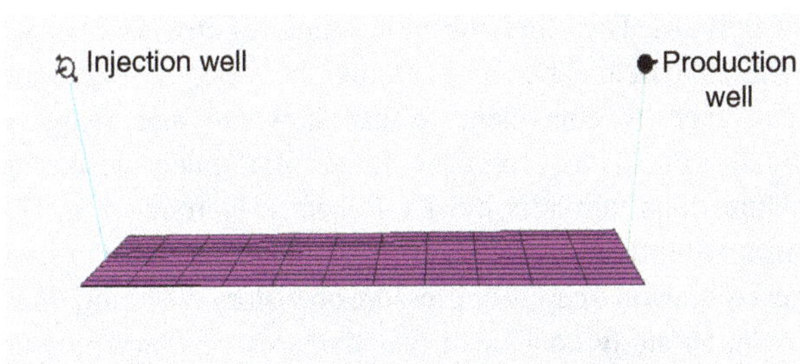

Figure 11

Gridding of the fracture for numerical validation of the analytical relative permeabilities.

MATLAB-based optimization code to minimize the following objective function:

$$\sum_{i=1}^{N}\left(\left(\frac{dP}{dx}\right)_i^{experiment} - \left(\frac{dP}{dx}\right)_i^{simulation\ model}\right)^2 \tag{73}$$

where $\left(\frac{dP}{dx}\right)_i$ are pressure gradients for injection of the fluids with different flow rate ratios and N is the number of experiments with different injection rates for a fixed set of fracture and fluid properties. In order to prevent numerical dispersion, the fractures are subdivided into many grid cells and single porosity fine grid simulations are performed (Fig. 11). Grid cells porosity was set equal to unity and absolute permeabilities are calculated from Equation (1). The linear relative permeabilities are used as the initial guess and capillary forces are neglected for flow in the fractures. The model is initiated with water filled fracture grid cells. Figure 12 compares the experimental measurements of pressure gradient with numerically estimated values for the three experiments described above. The slight deviation of the numerical calculation of pressure gradient from the experimental measurement, particularly at the high flow rates can be attributed to the possible deflection of the fracture walls at high pressure gradients. The satisfactory matches between the analytical, numerical and experimentally-derived relative permeabilities in Figures 8, 9 and 10 guarantee validity of the newly developed formulations for accurate calculation of relative permeabilities of water and oil through fractures.

Unfortunately, due to lack of the experimental measurement of relative permeabilities in non-horizontal fractures it wasn't possible to evaluate the validity of the gravity effects in the newly proposed model. Therefore, it is highly recommended that the future experimental researches are conducted in inclined fractures. This will surely provide valuable experimental data to validate the analytical prediction of the gravity effects contribution in multiphase flow through fractures.

5 EVALUATION OF THE ASSUMPTIONS

Two main assumptions that were made in the development of the mathematical model of this article include laminar and stratified flow in fractures. These assumptions are examined here.

Laminar flow in fractures. It is evident that laminar flow assumption is reasonable for the ranges of the velocity usually occur in the bulk of reservoirs (1 ft/day ~ 31 cm/day).

Segregated stratified flow pattern: this assumption is equivalent to large viscous and gravity forces compared to capillary force in fractures. Capillary pressure for flow

TABLE 2

Fracture capillary pressure as a function of aperture for water-oil system

Fracture aperture (μm)	Capillary pressure (kPa)
1	80
10	8
50	1.6
100	0.8
200	0.4
300	0.27
400	0.2
500	0.16
600	0.13
700	0.11
800	0.099
900	0.089
1 000	0.08
6 000	0.013

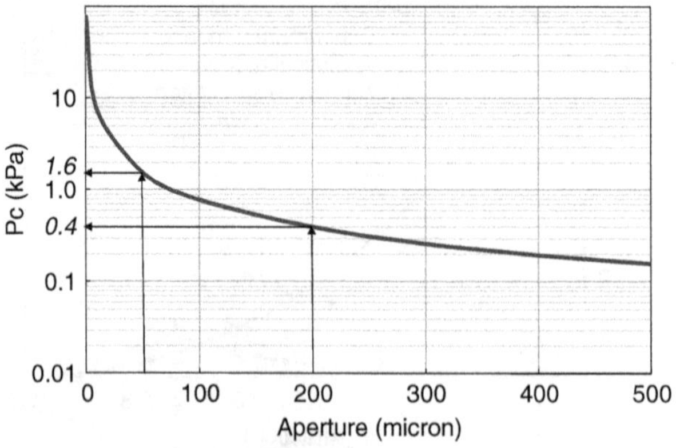

Figure 13

Fracture capillary pressure as a function of aperture for water-oil system.

between two flat parallel plates can be calculated by the following equation:

$$P_c = \frac{2\sigma \cos(\theta)}{b} \tag{74}$$

where σ is interfacial tension between fluids, θ is contact angle and b is fracture aperture. This equation was used by Wang and Narasimhan [63] to study fluid flow in partially saturated fractured porous media. The usefulness of Equation (74) can be utilized to show the dependence of fracture capillary pressure on its aperture. According to this equation, capillary forces are weaker in wider fractures and stronger in narrower ones. For typical water-oil systems, we assume that $\sigma = 0.04\,\mathrm{N/m}$ and $\theta = 0$. For these values, fractures capillary pressure as a function of aperture is given in Table 2 and shown in Figure 13. In order to judge about the assumption of dominance of viscous and gravity forces to capillary force in a single fracture using Figure 13, it is essential to answer one important question: how wide are natural fractures? Several researchers [64-71] reported observation of fracture apertures ranging from about 30 to over 6 000 μm. These researchers stated that the apertures in the range of 50 to 200 μm seem to be very common, while 6 000 μm fracture apertures were rarely observed. However, Weber and Bakker [72] stated that the size of fracture aperture can be enlarged by leaching as much as 5 000 μm. Therefore, the typical range of observed fracture apertures in field experiences around the world is 50-200 μm. This range is equivalent to capillary pressure range of 1.6-0.4 kPa. Thus, capillary forces are much weaker in fractures than in matrix blocks. Recently, Noroozi *et al.* [73] performed full field simulation on an Iranian reservoir to show effect of fractures capillary pressure on history matching. Their simulation study concluded that fractures capillary pressure can be considered zero in water-oil system.

It is worth noting that fracture walls were assumed to be strongly water wet by considering contact angle equal to zero in the above computations. In carbonate reservoirs with oil wetness tendency, fractures capillary pressure is several times less than the values of Table 2 and Figure 13. Since fractures are usually observed in dense and brittle reservoirs with high capillary pressures of tens of pounds per square inch, it can be neglected in fractures compared to matrix blocks. Furthermore, if the assumption was unrealistic, the good agreements between the calculated relative permeabilities using the new formulations and measured data wouldn't be observed.

CONCLUSIONS

The following conclusions can be drawn from this work:

- although, several formulations are available to calculate fracture relative permeability curves in the form of linear and power functions of saturations, the classic Romm's *X*-type curves are still used in almost all of the reservoir simulators;
- the basic fluid flow equations were combined the above formulations to propose new simple analytical formulations for two phase relative permeabilities through

fractures. The newly developed relative permeability expressions are explicit functions of phase saturations, viscosity ratio, fluids density, inclination of fracture from the horizon, pressure gradient along fracture and rock matrix wettability. It is also the first relative permeability model that takes the gravity effects and fracture wall wettability into consideration;

- good agreements of calculations of the model with literature experimental data verify the validity of the analytical formulations;
- it was explicitly shown that viscosity the ratio has substantial effect on the relative permeability curve of non-wetting phase and can lead to values greater than unity, while it almost plays no rule in calculation of wetting phase relative permeability. This unusual behavior respect to what is practically accepted was observed in experimental measurement of water-oil relative permeabilities. We justified such observation by the lubrication and viscous coupling effects for multiphase flow in fractures;
- the main advantage of these formulations for applying in commercial reservoir simulators is that it would not require the import of fracture relative permeabilities in tabular form. Therefore, these formulations are simple, efficient and accurate for implementation in dual-, multiple continuum and DFN reservoir simulators.

ACKNOWLEDGMENTS

The author would like to thank National Iranian Oil Company (NIOC) and National Iranian South Oil Company (NISOC) for their help.

REFERENCES

1 Ahmed T. (2010) *Reservoir Engineering Handbook*, 4th edn, Gulf Professional Publishing.

2 Bourbiaux B., Fourno A., Nguyen Q.L., Norrant F., Robin M., Rosenberg E., Argillier J.F. (2014) Experimental and Numerical Assessment of Chemical EOR in Oil-wet Naturally-Fractured Reservoirs, *SPE Improved Oil Recovery Symposium*, Tulsa, Oklahoma, USA, April 12-16, *SPE Paper* 169140.

3 Ramirez B., Kazemi H., Al-Kobaisi M., Ozkan E., Atan S. (2009) A Critical Review for Proper Use of Water/Oil/Gas Transfer Functions in Dual-Porosity Naturally Fractured Reservoirs: Part I, *SPE Res. Eval. Eng.* **12**, 2, 200-210.

4 Gilman J.R. (2003) Practical aspects of simulation of fractured reservoirs, *International forum on reservoir simulation*, Buhl, Baden-Baden, Germany.

5 Saidi A.M. (1983) Simulation of Naturally fractured Reservoirs, *SPE Reservoir Simulation Symposium*, San Francisco, CA, 15-18 Nov, SPE 12270.

6 Fung L.S.K. (1991) Simulation of Block-to-Block Processes in Naturally fractured Reservoirs, *SPE Res. Eval. Eng.* **6**, 4, 477-484.

7 Uleberg K., Kleppe J. (1996) Dual Porosity, Dual Permeability Formulation for Fractured Reservoir Simulation, Norwegian University of Science and Technology, *Trondheim RUTH Seminar*, Stavanger.

8 Kazemi H., Gilman J.R. (1993) Multiphase Flow in Fractured Petroleum Reservoirs, in *Flow and Contaminant Transport in Fractured Rock*, Bear J., Tsang C.F., de Marsily G. (eds), San Diego, California, Academic Press, pp. 267-323.

9 Al-Kandari H.A., Kazemi H., Van Kirk C.W. (2002) Gas Injection Enhanced Oil Recovery in High Relief Naturally Fractured Reservoirs, *Kuwait University Technical Conference and Exhibition*, November.

10 Liu H., Di Donato G., Blunt M.J. (2006) General Transfer Functions for Multiphase Flow, *SPE Annual Technical Conference and Exhibition*, San Antonio, TX, 24-27 Sept., *Paper SPE* 102542.

11 Saboorian-Jooybari H., Ashoori S., Mowazi G. (2012) Development of an Analytical Time-Dependent Matrix/Fracture Shape Factor for Countercurrent Imbibition in Simulation of Fractured Reservoirs, *Transp. Porous Media* **92**, 3, 687-708.

12 Saboorian-Jooybari H., Khademi N. (2014) Traveling Wave Analysis of Cocurrent Imbibition in Porous Media, *Journal of Porous Media* **17**, 3, 185-195.

13 Barenblatt G.I., Zheltov I.P., Kochina I.N. (1960) Basic concepts in the theory of seepage of homogeneous liquids in fissured rocks, *PMM. Sov. Appl. Math. Mech.* **24**, 5, 852-864.

14 Warren J.E., Root P.J. (1963) The behavior of naturally fractured reservoirs, *SPE J.* 245-255.

15 Kazemi H. (1990) Naturally fractured reservoirs, *3rd International Forum on Reservoir Simulation*, Baden, Austria.

16 Pruess K., Narasimhan T.N. (1985) A practical method for modeling fluid and heat flow in fractured porous media, *SPE J.* **25**, 14-26.

17 Stothoff S., Or D. (2000) A discrete-fracture boundary integral approach to simulating coupled energy and moisture transport in a fractured porous medium, in *Dynamics of Fluids in Fractured Rocks, Concepts and Recent Advances*, Faybishenko B., Witherspoon P.A., Benson S.M. (eds), AGU Geophysical Monograph 122, American Geophysical Union, Washington, DC, pp. 269-279.

18 Collins F. (1988) Evaluation of the improved dual-porosity model for the simulation of gravity effects in naturally fractured reservoirs, *39th Ann. Tech. Meeting of the Petroleum Society of Canadian Institute of Mining*, Calgary, Canada, CIM paper No. 88-39-05.

19 Bourbiaux B. (2010) Fractured Reservoir Simulation: a Challenging and Rewarding Issue, *Oil & Gas Science and Technology – Rev. IFP* **65**, 2, 227-238.

20 Lemonnier P., Bourbiaux B. (2010) Simulation of Naturally Fractured Reservoirs State of the Art - Part 1 – Physical Mechanisms and Simulator Formulation, *Oil & Gas Science and Technology – Rev. IFP* **65**, 2, 239-262.

21 Lemonnier P., Bourbiaux B. (2010) Simulation of Naturally Fractured Reservoirs. State of the Art - Part 2 – Matrix-Fracture Transfers and Typical Features of Numerical Studies, *Oil & Gas Science and Technology – Rev. IFP* **65**, 2, 263-286.

22 Henn N., Quintard M., Bourbiaux B., Sakthikumar S. (2004) Modelling of Conductive Faults with a Multiscale Approach, *Oil & Gas Science and Technology – Rev. IFP* **59**, 2, 197-214.

23 Johnson E.F., Bossler D.P., Naumann V.O. (1959) Calculation of relative permeability from displacement experiments, *Trans. AIME* **216**, 370.

24 Saraf D.N., McCaffery F.G. (1982) Two- and Three-Phase Relative Permeabilities: a Review, Petroleum Recovery Institute Report #81-8, Calgary, Alberta, Canada.

25 Jones S.C., Roszelle W.O. (1978) Graphical techniques for determining relative permeability from displacement experiments, *J. Pet. Technol.* **5**, 807.

26 Sarem A.M. (1959) Significance of water-oil relative permeability data calculated from displacement tests, *Conference of Theory of Fluid Flow in Porous Media*, University of Oklahoma, Norman, 189.

27 Archer J.S., Wong S.W. (1973) Use of a reservoir simulator to interpret laboratory waterflood data, *SPE Journal* **13**, 343-347.

28 Honarpour M., Koederitz L., Harvey H. (1986) *Relative Permeability of Petroleum Reservoirs*, Boca Raton, FL, CRC Press.

29 Honarpour M., Mahmood S.M. (1988) Relative-permeability measurements-an overview, *Journal of Petroleum Technology* **40**, 963-966.

30 Lomize G.M. (1951) *Flow in Fractured Rocks* (in Russian), Gosenergoizdat, Moscow 127 pp.

31 Snow D.T. (1965) Anisotropic permeability of fractured media, *Water Resour. Res.* **5**, 6, 1273-1289.

32 Louis C.A. (1969) A study of groundwater flow in jointed rock and its influence on the stability of rock masses, *Rock Mech. Res. Rep.* **10**, Imperial College, London, 90 pp.

33 Bear J. (1972) *Dynamics of Fluids in Porous Media*, Elsevier, New York, p. 764.

34 Witherspoon P.A., Wang J.S.Y., Iwai K., Gale J.E. (1980) Validity of cubic law for fluid flow in deformable rock fracture, *Water Resour. Res.* **16**, 6, 1016-1024.

35 Romm E.S. (1966) *Fluid Flow in Fractured Rocks* (in Russian), Nedra Publishing House, Moscow.

36 Van Golf-Racht T. (1982) *Fundamentals of Fractured Reservoir Engineering*, Elsevier Scientific, New York.

37 Thomas L.K., Dixon T.N., Pierson R.G. (1983) Fractured reservoir simulation, *SPE J.* **23**, 1, 42-54.

38 Gilman J.R., Kazemi H. (1983) Improvements in simulation of naturally fractured reservoirs, *SPE J.* **23**, 4, 695-707.

39 Kazemi H., Gilman J.R. (1989) Multiphase flow in fractured petroleum reservoirs, *Advanced Workshop on Heat and Mass Transport in Fractured Rocks*, Lab. Nac. Eng. Civ. (LNEC), Lisbon, June.

40 Grant M.A. (1977) Permeability reduction factors at Wairakei, *AIChE-ASME Heat Transfer Conference*, Am. Inst. of Chem. Eng. and Am. Soc. of Mech. Eng., Salt Lake City, Utah.

41 Pruess K., Bodvarsson G.S., Stefansson V., Eliasson E.T. (1984) The Krafla geothermal field, Iceland, 4, History match and prediction of individual well performance, *Water Resour. Res.* **20**, 11, 1561-1584.

42 Merrill L.S. (1975) Two phase flow in fractures, *PhD Thesis*, University of Denver.

43 Pruess K., Tsang Y.W. (1990) On two-phase relative permeability and capillary pressure of rough-walled rock fractures, *Water Resour. Res.* **26**, 9, 1915-1926.

44 Rossen W.R., Kumar A.T.A. (1992) Single- and two-phase flow in natural fractures, *67th Annual Technical Conference of the Society of Petroleum Engineers*, Washington, DC, 4-7 Oct., *SPE Paper* 24195.

45 Persoff P.K., Pruess K., Myer L. (1991) Two-Phase Flow Visualization and Relative Permeability Measurement in Transparent Replicas of Rough-Walled Rock Fractures, *16th Workshop on Geothermal Reservoir Engineering*, Stanford, California, 23-25 Jan.

46 Persoff P., Pruess K. (1995) Two-Phase Flow Visualization and Relative Permeability Measurement in Natural Rough-Walled Rock Fractures, *Water Resources Research* **31**, 5, 1175-1186.

47 Fourar M., Bories S., Lenormand R., Persoff P. (1993) Two-Phase Flow in Smooth and Rough Fractures: Measurement and Correlation by Porous-Medium and Pipe Flow Models, *Water Resources Research* **29**, 11, 3699-3708.

48 Fourar M., Bories S. (1995) Experimental Study of Air-Water Two-Phase Flow through a Fracture (Narrow Channel), *Int. J. Multiphase Flow* **21**, 4, 621-637.

49 Pan X., Wong R.C., Maini B.B. (1996) Steady State Two-Phase Flow in a Smooth Parallel Fracture, Presented at the *47th Annual Technical Meeting of the Petroleum Society in Calgary*, Alberta, Canada, June 10-12.

50 Rangel-German E., Akin S., Castanier L. (1999) Multiphase-Flow Properties of Fractured Porous Media, *SPE Western Regional Meeting*, Anchorage, AK, *SPE Paper* 54591.

51 Pieters D.A., Graves R.M. (1994) Fracture Relative Permeability: Linear or Non-Linear Function of Saturation, *SPE International Petroleum Conference & Exhibition of Mexico*, Veracruz, Mexico, 10-13 Oct, *SPE Paper* 28701.

52 Horne R.H., Satik C., Mahiya G., Li K., Ambusso W., Tovar R., Wang C., Nassori H. (2000) Steam-Water Relative Permeability, *World Geothermal Congress*, Kyushu-Tohoku, Japan, 28 May - 10 June.

53 Diomampo G.P. (2001) Relative Permeability Through Fractures, Report SGP-TR-170 prepared under DOE contract No. DE-FG07-95ID13370, August.

54 Chen A. (2005) Liquid-Gas Relative Pemeabilities in Fractures: Effects of Flow Structures, Phase Transformation and Surface Roughness, *PhD Thesis*, Stanford University.

55 Corey A.T. (1954) The interrelationship between gas and oil relative permeabilities, *Prod. Mon.* **19**, 38-41.

56 Fourar M., de Nancy E., Lenormand R. (1998) A Viscous Coupling Model for Relative Permeabilities in Fractures, *SPE Annual Technical Conference and Exhibition*, New Orleans, LA, USA, *SPE Paper* 49006.

57 Shad S., Gates I.D. (2010) Multiphase Flow in Fractures: Co-current and Counter-Current Flow in a Fracture, *Journal of Canadian Petroleum Technology* **49**, 2, 48-55.

58 Brown S.R., Scholz C.H. (1985) Broad Bandwidth Study of the Topography of Natural Rock Surfaces, *J. Geophys. Res.* **90**, 12575-12582.

59 Toledo P.G., Nvoy A.R., Davis H.T. (1990) Hydraulic Conductivity of Porous Media at Low Water Content, *Soil Science of America J.* **54**, 3, 673.

60 Streeter V.L., Wylie E.B., Bedford K.W. (1997) *Fluid Mechanics*, 9th edn, McGraw-Hill publishing company.

61 Pan X. (1999) Immiscible Two-Phase Flow in a Fracture, *PhD Thesis*, University of Calgary, Calgary, Alberta, Canada.

62 Dullien A.L. (1993) Physical interpretation of hydrodynamic coupling in a steady-state two phase flow, *13th Annual Hydrology Meeting of American Geophysical Union*, Colorado State University, Fort Collins, Colorado, pp. 363-377.

63 Wang J.S.Y., Narasimhan T.N. (1985) Hydrologic Mechanisms Governing Fluid Flow in a Partially Saturated, Fractured, Porous Medium, *Water Resource Research* **21**, 12, 1861-1874.

64 Banward S.E., Gustafsson M., Laaksoharju E.L.T., Wallin B. (1994) Large Scale intrusion of shallow water into a vertical fracture zone in crystalline bedrock: Initial hydrochemical perturbation during tunnel construction at the Äspö Hard Rock Laboratory, southeastern Sweden, *Water Resources Research* **30**, 6, 1747-1763.

65 Folger P.F., Poeter E., Wanty R.B., Frishman D., Day W. (1996) Controls on 222Rn variations in a fractured crystalline rock aquifer evaluated using aquifer tests and geophysical logging, *Groundwater* **34**, 2, 250-261.

66 Himmelsbach T., Hbtzl H., Maloszewski P. (1998) Solute Transport Processes in a Highly Permeable Fault Zone of Lindau Fractured Rock Test Site (Germany), *Groundwater* **36**, 5, 792-800.

67 Kumar A.T.A., Majors P.D., Rossen W.R. (1997) Measurement of Aperture and Multiphase Flow in Fractures using NMR Imaging, *SPE Formation Evaluation* **12**, 2, 101-108.

68 Schrauf T.W., Evans D.D. (1986) Laboratory Studies of Gas Flow through a Single Natural Fracture, *Water Resources Research* **22**, 7, 1038-1050.

69 Shapiro A.M., Nicholas J.R. (1989) Assessing the validity of the channel model of fracture aperture under field conditions, *Water Resources Research* **25**, 5, 817-828.

70 Novakowski K.S., Lapcevic P.A. (1994) Field measurement of radial solute transport in fractured rock, *Water Resources Research* **30**, 1, 37-44.

71 Hensel W. (1987) A Perspective Look at Fracture Porosity, *62nd Annual Technical Conference and Exhibition of the Society of Petroleum Engineers*, Dallas, TX, 27-30 Sept., *SPE Paper* 16806.

72 Weber K.J., Bakker M. (1981) Fracture and Vuggy Porosity, *SPE Annual Technical Conference and Exhibition*, 4-7 Oct., San Antonio, Texas, *Paper SPE* 10332

73 Noroozi M.M., Moradi B., Bashiri G. (2010) Effects of Fracture Properties on Numerical Simulation of a Naturally Fractured Reservoir, *Trinidad and Tobago Energy Resources Conference*, Port of Spain, Trinidad, 27-30 June, *SPE Paper* 132838.

5

Hollow Silica: A Novel Material for Methane Storage

Vangala Dhanunjana Chari[1,2], Pinnelli S.R. Prasad[1]* and Sarabu Ramana Murthy[2]

[1] Gas Hydrate Group, National Geophysical Research Institute (Council of Scientific & Industrial Research), Hyderabad – 500 007 - India
[2] Department of Physics, Osmania University, Hyderabad – 500 007 - India
e-mail: psrprasad@ngri.res.in

* Corresponding author

Abstract — *Methane gas storage in the form of Methane Hydrate (MH) in hollow silica was studied and compared with solid silica and pure water systems. The gas hydrate growth/dissociation was monitored by following the pressure (gas intake) – temperature variations in a classical isochoric process. The effect of stirring on the hydrate formation kinetics and yield was clearly evidenced in the case of solid and pure water systems, whereas it did not show any influence in hollow silica; and in fact, the yields remained identical in both stirring and non-stirring experiments. Approximately 3.6 m.mol of methane per gram of water was consumed as MH in the hollow silica matrix and the formation kinetics was extremely fast (~180 min). However, the methane gas conversion into MH in solid silica and pure water systems was ~ 10 times higher in a stirred reactor when compared with a non-stirred system.*

Résumé — **La silice creuse : un nouveau matériau pour le stockage de méthane** — Le stockage de gaz méthane sous la forme d'hydrate de méthane (MH, *Methane Hydrate*) dans de la silice creuse a été étudié et comparé à un système de silice pleine avec de l'eau pure. La croissance/dissociation de l'hydrate gazeux a été contrôlée en suivant les variations de pression (admission du gaz) et de température dans un processus isochore classique. L'effet de l'agitation sur la cinétique de formation des hydrates et le rendement a été clairement mis en évidence, dans le cas de systèmes d'eau pure et de silice pleine, tandis qu'il ne présente aucune influence dans la silice creuse ; et en fait, les rendements restent identiques à la fois dans des expériences avec agitation et sans agitation. Environ 3,6 m.mol de méthane par gramme d'eau ont été consommés sous forme de MH dans la matrice de silice creuse et la cinétique de formation a été extrêmement rapide (~180 min). Toutefois, la conversion du méthane en MH dans le système de silice pleine avec de l'eau pure s'est avérée environ 10 fois supérieure dans un réacteur agité par rapport à un système non-agité.

INTRODUCTION

Natural gas hydrate, ice like crystalline compound, forms at elevated pressure and in low temperature conditions, in which guest molecules occupy the cages constructed by water molecules through hydrogen bonding. Gas Hydrates (GH) are classified into three different structures, sI, sII and sH, depending on the size of the guest molecules and the shape of the polyhedral water cages [1-3]. The vast occurrences of GH in the ocean sediments and certain permafrost regions and the present estimations show that their resources are almost double that of all types of available carbon fuels in nature. The interest in the GH is because of the fact that the amount of gas stored in this form is extremely high ($\sim$ 180 times STP equivalent). Further, the main constituent in the GH is methane, which is a less polluting and more eco-friendly clean fuel.

This highly dense natural gas in the hydrate form has generated interest in making use of hydrate technology for natural gas storage and transportation [4-6]. However, gas transportation through hydrate technology has to overcome some challenging constraints; among them, a few are the slow hydrate formation rate, the hydrate stability and the low conversion ratio of gas to solid hydrate, resulting in insufficient storage capacity, and the cost involved at various stages of the process. Further, converting the water and gas mixtures into gas hydrates, under suitable thermodynamic conditions, is also an attractive way to store large quantities of gas, such as hydrogen [7, 8] and carbon dioxide [9]. Extensive efforts have been made to develop efficient storage techniques in both scientific and industrial fields. Thus, the research on gas hydrates gained momentum for separating CO_2 from flue gases and storing it under ocean floors [10, 11].

GH formation is possible at sub-ambient temperature and high pressure, but these can be altered by adding some additives to the hydrate forming system. In general, it is known that gas hydrate forms on a gas-liquid interface, and thus mainly the gas-liquid interfacial area can affect hydrate formation. Therefore, an efficient way to increase these factors is necessary for forming gas hydrate solid in a gas-liquid system. For example, several processes to increase the gas hydrate formation have been demonstrated. To increase the effective temperature at a specific pressure for hydrate formation, thermodynamic promoters such as THF (Tetra Hydro Furan) were used [12]. However, the disadvantage with such system is a significant decrease in the storage capacity of the fuel gas (guest) molecules due to the fact that part of the hydrate cages are occupied by the promoter guest molecules. The addition of surfactants such as Sodium Dodecyl Sulfate (SDS) can promote the gas hydrate formation without affecting the thermodynamics of hydrate crystallization. SDS accelerates the methane hydrate growth by increasing the total surface area / morphology of hydrate particles and the gas liquid interfacial area. Nevertheless, as the hydrate formation is exothermic, therefore a certain amount of heat is liberated during the hydrate formation [13-14]. A higher amount of such heat could lead to substantial rise in the system temperature and could effectively decrease the growth kinetics. The effect of the heat of hydrate formation on the hydrate growth was lowered by the addition of a small amount of aluminium foam, but at high pressures it has its own limitations [15]. Recently, the use of Dry Water (DW) was reported by Wang *et al.* [16] to increase the effective interacting area of methane gas molecules with water molecules to achieve high hydrate yield, but the reusability and rigorous preparative conditions associated with the system are the main constraints. Further use of dry gel along with DW offered a limited solution for reusability as the hydrate conversion capability was highly reduced in subsequent freezing and thawing cycles [17].

The literature shows that continuous stirring conditions have a predominant effect on the hydrate nucleation and growth, since the water-gas interfacial area is continuously renewed [18-20]. The induction time depends on the degree of agitation, the presence of foreign particles, and the rate of heat and mass transfer. In the case of hydrate formation in a porous media it is necessary to account for additional forces such as capillary ones during the growth of the hydrate, though in case of stirring these forces are neglected due to continuous agitation in the system. In our previous paper we reported the effect of silica particles on methane hydrate stability and the kinetic promotion behaviour on hydrate formation in a stirred vessel [18]. Even though hydrate yield was lower in silica + water + CH_4, the addition of a small amount of silica acts as a kinetic promoter for the formation of MH. Most of the experiments on methane hydrate formation kinetics in sediments are reported in isobaric conditions. The studies on the isochoric cooling process with temperature and pressure both decreasing are scarce. Experimentally the constant gas pressure is maintained in isobaric conditions when the system temperature is within the Gas Hydrate Stability Zone (GHSZ), while in the isochoric process a fixed amount of gas is added to a reactor and the temperature is slowly brought into the GHSZ. In the present paper methane hydrate formation, growth kinetics and hydrate yield in two types of silica systems are reported and compared with a pure water system. The significance of stirring on the methane hydrate formation with different silica systems in an isochoric process in similar *P-T* conditions is examined.

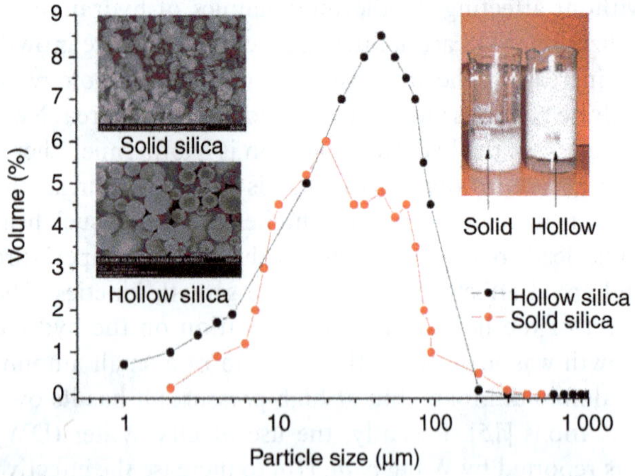

Figure 1

FESEM images and grain size distribution of silica samples used for methane hydrate synthesis.

1 EXPERIMENTAL SECTION

1.1 Materials and Characterization

The silica powders were obtained from Nanoshel (*Intelligent Materials Pvt. Ltd.*). Deionized ultra-pure water (*Millipore* – type 1) was used, and dissolved gases were removed by evacuation. The methane gas used was 99.95% pure. In Figure 1 we show the grain size distribution and the morphology of the silica. The morphology of the silica, analyzed by a Field Emission Scanning Electron Microscope (FESEM - *Carl Zeiss* Ultra 55), shows that the silica particles are spherical with a mean diameter of 30-80 μm (inset figures). The grain size distribution was measured by the Dynamic Light-Scattering (DLS) method in a water suspension using a particle size analyzer (*Malvern* Mastersizer 2000), and the maximum volume of particles was in a particle size range of 20-90 μm. The bulk density of hollow silica was measured as 0.10 $g.cm^{-3}$, while the same for solid silica having similar particle size distributions was 1.5 $g.cm^{-3}$. The BET specific surface area of the hollow silica was 2.4 $m^2.g^{-1}$. For the experimental studies, we used the same amounts of silica (12.5 gm) and water (100 mL) in the solid and hollow systems. In the inset of Figure 1 we show that the solid silica was settled at the bottom of the water column, whereas the hollow silica was floating on the water with some amount of water trapped (we used ~ 10 g of silica and the water was 50 mL) [21] in the pore spaces and between the voids of hollow silica particles.

1.2 Apparatus and Procedure

The main part of the apparatus is a SS-316 cylindrical vessel, which can withstand pressures up to 10 MPa, and the volume of the vessel is 250 mL. The silica powders were used as such and the required amounts of silica and degassed ultra-pure water were measured by using a *Metler* Toledo (AB104-S) accurate analytical balance. Cold fluid (water + glycol mixture) circulated around the vessel with the help of a circulator (coolant temperature was kept at 275 K) to maintain the temperature inside the cell (~ 277 K) at the desired level. A platinum resistance thermometer (Pt100) was inserted into the vessel to measure temperatures with an accuracy of ±0.2 K. The pressure in the vessel was measured with a pressure transducer (*WIKA*, type A-10 for pressure range 0-25 MPa with ±0.5 accuracy). A stirrer with variable speed was installed in the vessel to agitate the fluids. The effect of the stirring rate on methane hydrate formation was studied by Vysniauskas and Bishnoi [22], and it was observed that at lower rpm (less than 300) the hydrate crystals tended to remain on the surface, and at least 400 rpm was required to remove the hydrate crystals from the surface. Also, the studies of Golombok *et al.* [23], revealed that the induction time was dependent on the stirrer rpm and induction times can be brought down by increasing the stirring rates. However, higher rpm (~ 800) decreases the induction time but at the cost of a decrease in the hydrate yield [20]. Therefore, all the stirred experiments were conducted with a fixed speed of 500 rpm. The methane gas was supplied from a cylinder to the desired level (7.5-10 MPa) using a Teledyne *ISCO* Syringe pump. After obtaining the temperature and pressure stability far above the hydrate formation region, the valve inline connecting the vessel and the *ISCO* pump/cylinder was closed. Subsequently, the cold fluid from the chiller was circulated and the hydrate formation in the vessel at a particular temperature was detected by a sharp pressure drop. The insignificant head-pressure drop in the reactor over longer duration indicates the saturation of hydrate conversion. After the completion of hydrate conversion, if the temperature is increased stepwise to bring the system out of the hydrate forming region, hydrate crystals partially dissociate, thereby substantially increasing the pressure. The temperature and pressure were logged at 60-second time intervals. In this way, a pressure temperature diagram (cooling and warming) was obtained for each experimental run, from which we determined the hydrate formation and dissociation pattern. In the case of the non-stirred system the water + silica were mixed and agitated with a glass rod by hand only; no vigorous mixing was applied as in the case

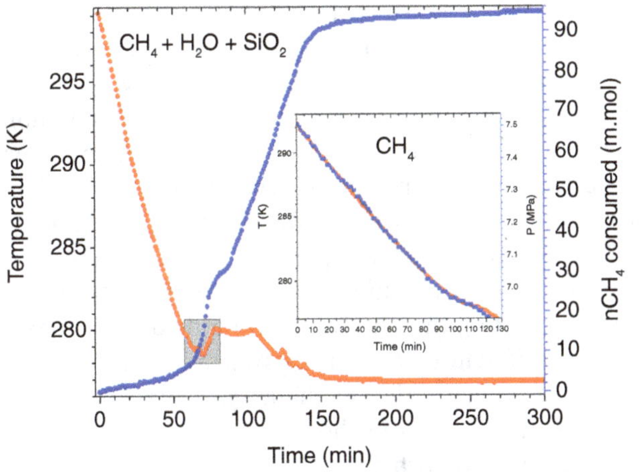

Figure 2

Observed temporal variations of temperature and methane gas consumption (calculated from *Eq. 1*) in the silica + water + CH_4 system during the cooling cycle. In the inset, we show time variations of gas pressure and temperature for CH_4 (methane gas without water).

of the DW system. The molar concentration of methane gas (ΔnH, t) in the hydrate phase during the experiment is defined by the following equation [21]:

$$\Delta n\mathrm{H}, t = ng, 0 - ng, t = (P_0 V / Z_0 R\mathrm{T}_0) - (P_t V / ZR\mathrm{T}_t) \tag{1}$$

where Z is the compressibility factor calculated by the Peng-Robinson equation of state. The gas volume (V) was assumed to be constant during the experiments, *i.e.*, the volume changes due to phase transitions were neglected. $ng,0$ and ng,t represent the number of moles of feed (methane) gas taken at zero time and in the gas phase at time t, respectively (n represents the number of moles of feed gas). The hydrate yield was computed from the observed methane gas consumption to the expected values with ideal stoichiometry compositions ($8CH_4{\cdot}46H_2O$)[1].

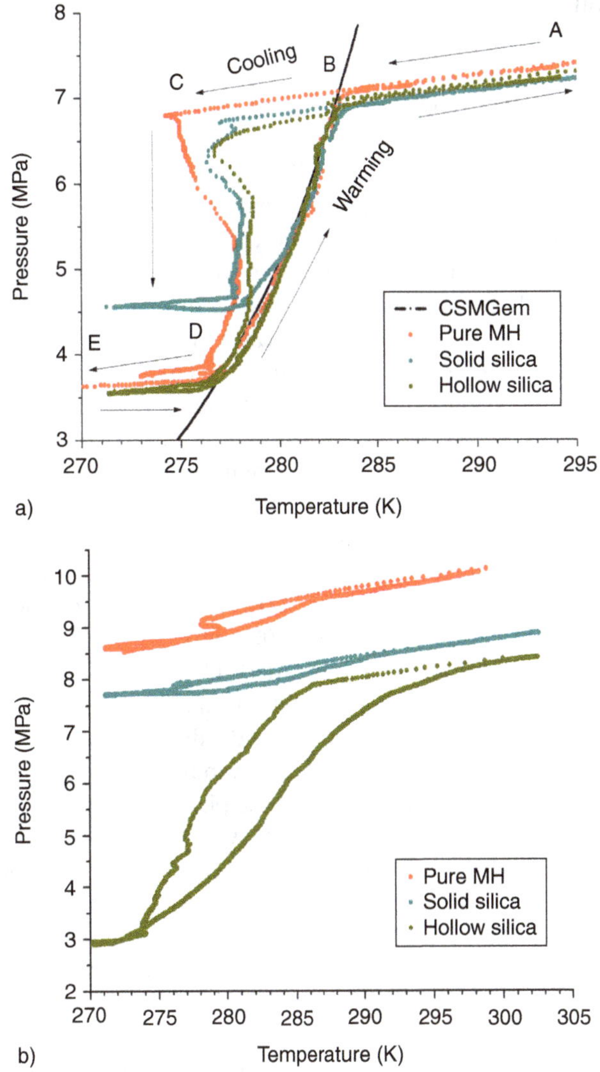

Figure 3

The pressure-temperature trajectories of formation and dissociation of methane hydrate synthesized in a stirred (a) and non-stirred (b) reactor.

2 RESULTS AND DISCUSSION

In Figure 2, we show the temporal variations in the recorded temperature and gas consumption for: CH_4 + H_2O + SiO_2 (hollow) in a continuously stirred reactor vessel. In the inset, we also show a plot for temporal variations in the temperature and pressure of CH_4 gas alone as a control experiment. From the inset it is clearly seen that the pressure and temperature decrease progressively with time during the cooling process, indicating real gas behaviour, whereas when water (an essential constituent for gas hydrates) is also present along with the methane, a much more rapid decrease in the vapor state pressure (*Fig. 3*) or much more rapid consumption of gas (*Fig. 2*), indicates the hydrate formation. Further, a smaller pressure drop (0.8 MPa in the temperature range of 298-277 K) is due to a temperature decrease, which is due to methane gas dissolution in

[1] Yield = (observed mass ratio)/(expected mass ratio). The mass of methane gas consumed (mCH_4) is calculated from the observed pressure drop during MH formation. Observed:

$$\text{weight ratio} = \frac{(mCH_4)}{(mCH_4 + mH_2O)};$$

expected:

$$\text{weight ratio} = \frac{(8 \times \text{molar mass of } CH_4)}{(8 \times \text{molar mass of } CH_4 + 46 \times \text{molar mass of } H_2O)}.$$

water and the compressibility of methane gas in the reactor. However, a much sharper pressure drop in the narrow temperature zone, as shown in the *p-T* trajectories (*Fig. 3*), clearly indicates the hydrate formation. During the first 50 min in the case of $CH_4 + H_2O$ present with SiO_2, the system entered into the GHSZ and as hydrate formation is an exothermic phenomenon, a small rise in temperature (283 K at 7MPa, *Fig. 3*) indicates the hydrate nucleation in the system. We also noted that a considerable rise in the temperature (278 to 280 K) indicates the growth of the hydrate (shown as a shaded square). The pressure drop was around 1.5 MPa in the temperature range of 282-279 K. Further, it took about 70 min to complete the hydrate formation and to reach equilibrium temperature. Thereafter there is no measurable pressure decrease for a longer time, indicating the saturation of hydrate formation.

In Figure 3, we show the MH formation and dissociation with constant stirring (500 rpm) [20, 22, 23] and without stirring. The points A, B, C and D marked in the *p-T* trajectory (*Fig. 3a*) show observed variations during hydrate formation and dissociation. The point A (starting point) is clearly outside the GHSZ, whereas points B and D are on the phase stability curve, and point C is inside the GHSZ. A slow decrease in the pressure with temperature in the A-B-C region indicates the dissolution of methane gas in water. The portion BC signifies the super-cooled state of the reaction mixture without nucleation of hydrate particles even though the pressure and temperature are conducive for hydrate formation. The system requires some amount of sub-cooling for hydrate nucleation and growth. The temperature difference between the temperature at point B called equilibrium temperature (T_{eqm}) and that at point C is called the sub-cooling (T_{sub}), where $T_{sub} = T_{eqm}(B) - T(C)$, and the time required to reach C from B is called induction time [2]. By retaining the system there for some more time, a small but noticeable temperature rise in the hydrate forming system indicates the nucleation of hydrates. Afterward, a small decrease in the temperature (C to D) results in a rapid decrease in the gas pressure, which indicates the growth of hydrates. This process continues for some time, after which the pressure-temperature variations are marginal (D to E), signifying the saturation of hydrate conversion. A similar pattern is also observed in the warming cycle. The phase change of methane gas and liquid water into the solid gas hydrate is typically a first-order transition, and thus, occurs with a finite hysteresis loop. Even though the hysteresis loop area depends on other factors such as the rate of temperature variations, typically this is cor related with the volume fraction of the hydrates at constant temperature variations [24]. It is well known that undercooling significantly depends on the sediment matrix and also depends on the particle size [25]. It is observed from Figure 3a, b that in stirred and non-stirred systems, pure methane hydrates require more undercooling to form hydrates when compared with a silica + water + CH_4 system. Moreover, the experimental *p-T* trace shows that the pressure drop in the case of hollow silica is maximum in stirred and non-stirred systems, indicating the gas conversion to hydrate differs in the two different types of silica systems. Even though in a stirred system, the methane gas conversion into hydrates in the case of solid silica is significant, it is low when compared with the other two systems. Approximately the same amount of methane is consumed in pure and hollow silica. In the case of the non-stirred system, during the cooling process, the pressure drop is very small for water + CH_4, solid silica + water + CH_4 and the drop in pressure is approximated to the real gas law with very low methane gas to methane hydrate conversion (20-30 m.moles). Though a high driving force favours more hydrate yield [27], we did not observe any increase in the pressure drop even at high pressures in case of water + CH_4, solid silica + water + CH_4. By contrast, as shown in figure 3b, the MH formation occurred and there is a substantial MH growth in the case of the hollow silica system. A dramatic pressure drop upon cooling and a rapid pressure rise upon heating are noted.

From some previous reports the role of stirring in improved kinetics for hydrate formation was clearly evidenced [18-20], as the stirring is very useful to renew the contact surface between the liquid and gas phases, thus preventing the formation of a solid crust at the interface. The heat and mass transfer scenarios were different in both stirred and non-stirred systems and play a significant role in the hydrate nucleation, the induction time and the growth of the hydrate after nucleation [28]. Therefore, in the case of the non-stirred system, the sub-cooling ($\sim \Delta t = 5°\text{-}8°$) required was found to be different when compared with stirring, for which undercooling was approximately the same with a marginal difference of 2-3°. It was interesting to note that in the case of non- stirred hollow silica system, the hydrate growth started as the system just entered into the hydrate forming region (GHSZ, *Fig. 3b*), showing no requirement of sub-cooling.

In order to accelerate the formation of methane hydrate, the enhancement of gas-water contacts and hydration heat removal are two crucial approaches. As reported in the literature [13, 14], the addition of SDS to water increases the gas hydrate formation rate and storage capacity. However, fast hydrate formation is

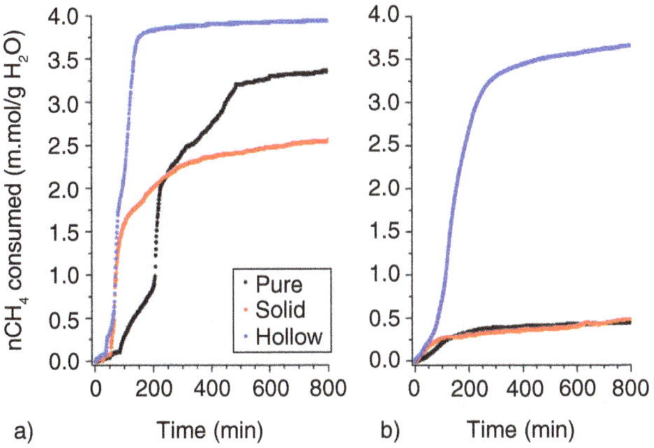

Figure 4

Methane hydrate formation kinetics in pure, solid and hollow silica systems (a- stirred and b- non-stirred).

always associated with the generation of great hydration heat, that weakens the acceleration effect of SDS on hydrate formation. According to Liang *et al.* [15], aluminum foam (AF) is an excellent thermal conductor possessing good thermal conductivity and a large rough surface. When AF combined with SDS is used as the additive, they observed faster formation and growth of methane hydrate, reducing the effect of the heat of hydrate formation; however, this system has some limitations. As mentioned above from the literature it is very clear that the methane gas uptake and the induction time were considerably reduced in the stirred tank reactor. At different locations in the system, once the critical hydrate nucleus size was reached then rapid hydrate growth occurred. The method followed to calculate the amount of methane gas consumed in the form of hydrate and the hydrate yield calculations are given elsewhere [18].

The reports of Linga *et al.* [28] show that during the initial stage of hydrate growth in sand systems, the stirring dominated, but later on it had less effect on hydrate conversion when compared with a non-stirred system. Over longer time scales, the methane gas convertion to methane hydrate (hydrate yield) was found to be greater in the case of non-stirred system. When it comes to fuel gas transportation in the form of hydrates, hydrate yield as well as the formation kinetics are both important. From Figures 4a, b, the formation kinetics was very much quicker in hollow silica in the case of stirred and non-stirred systems. The conversion kinetics of CH_4 gas to MH was also faster than that of the aforementioned reference [27] with less hydrate yield. For the hollow silica system, the time required to bring the system into the GHSZ in both the cases before the nucleation was about ($\sim$ 40 to 45 min). After the nucleation the average hydrate growth rate was around: 0.0542 m.mol/g.H_2O/min (50-120 min) in stirred and 0.0216 m.mol/g.H_2O/min (50-200 min) in non-stirred systems. The time required for 90% conversion of methane gas to hydrate, from the start of the experiment, was found to be around 250 minutes (non-stirred) and only 120 minutes in the case of the stirred system. The experiments were repeated at least five times and we found there was an infinitesimal change in the hydrate conversion factor.

The effect of stirring on methane hydrate formed in hollow silica was completely absent, as the kinetics and hydrate yield were the same in both cases. From Figures 4a, b, the slope variations during the hydrate growth in the case of hollow silica (stirred) were assumed to be due to the heat and mass transfers; it was due to gas diffusion and capillary force in the case of the non-stirred system. In Table 1, we show the details of the number of moles of CH_4 gas converted to hydrate in stirred and non-stirred reactors. The kinetics and hydrate yield (the methane gas storage capacity) in different silica systems confirm that the hollow silica has an anomalous behavior in stirred and non-stirred systems. With advantages such as an increase in the hydrate growth kinetics without any stirring and high hydrate yield, the hollow silica system could definitely

TABLE 1

The methane gas consumed in hydrate formation in the case of pure and different types of silica systems in non-stirred and stirred systems

S. No	System	nCH_4 consumed (m.mol/g.H_2O) (stirred)	Hydrate yield (%) (stirred)	nCH_4 consumed (m.mol/g.H_2O) (non-stirred)	Hydrate yield (%) (non-stirred)
1	Pure MH	3.48 ($\pm$0.11)	39.21 ($\pm$1.07)	0.30 ($\pm$0.07)	3.60 ($\pm$0.51)
2	Solid silica	2.80 ($\pm$0.12)	31.34 ($\pm$0.95)	0.32 ($\pm$0.05)	3.80 ($\pm$0.62)
3	Hollow silica	4.09 ($\pm$0.21)	43.22 ($\pm$1.90)	3.89 ($\pm$0.23)	43.69 ($\pm$1.7)

be considered as a useful system for methane gas storage and transport applications in the hydrate form.

CONCLUSION

We studied methane hydrate formation and the effect of stirring in two different types of silica systems and compared them with the pure MH system. Significant methane hydrate formation was observed in all systems when stirring was employed. The methane gas conversion to MH was found to be minimum in solid silica and maximum in hollow silica. However, in the case of the non-stirred reactor, the hydrate yield was almost negligible in pure/solid silica when compared with hollow silica. In hollow silica, the methane hydrate formation kinetics and the methane gas to MH conversion were the same as those of the stirred system, showing no effect of stirring. The type of silica (solid or hollow) plays an important role in hydrate conversion. The fast hydrate formation kinetics and high hydrate yield without stirring in the hollow silica system effectively reduce the cost of hydrate synthesis. Therefore, hollow silica is a viable platform for the storage and transportation of methane gas in the form of MH.

ACKNOWLEDGMENTS

The authors acknowledge the funding from the Council for Scientific and Industrial Research, Department of Science & Technology, Ministry of Earth Sciences (Gas Hydrates program) and the Directorate of Hydrocarbons (NGHP), Government of India, through sponsored projects. The authors sincerely thank the Director of the National Geophysical Research Institute, Hyderabad, for his encouragement, and permission to publish this paper.

REFERENCES

1 Englezos P. (1993) Clathrate Hydrates, *Ind. Eng. Chem. Res.* **32**, 1251-1274.

2 Sloan E.D., Koh C.A. (2008) *Clathrate Hydrates of Natural Gases*, 3rd ed., CRC Press, Taylor & Francis Group, Boca Raton.

3 Sloan E.D. (2003) Fundamental principles and applications of natural gas hydrates, *Nature* **426**, 353-359.

4 Makogon Y.F., Holditch S.A., Makogon T.Y. (2007) Natural gas-hydrates — A potential energy source for the 21st Century, *J. Pet. Sci. Eng.* **56**, 14-31.

5 Gudmundson J.S., Parlaktuna M., Levik O.I., Andersson V. (2000) Laboratory for Continuous Production of Natural Gas Hydrates, *Ann. New York Aca. Sci.* **912**, 851-858.

6 Daimaru T., Fujii M., Yamasaki A., Yanagisawa Y. (2004) Energy saving potential for natural gas hydrate transportation, *Prepr. Pap.-Am. Chem. Soc., Div. Fuel Chem.* **49**, 190-191.

7 Prasad P.S.R., Sugahara T., Sum A.K., Sloan E.D., Koh C. A. (2009) Hydrogen Storage in double clathrates with tert-butylamine, *J. Phys. Chem. A.* **113**, 6540-6453.

8 Sugahara T., Haag J.C., Prasad P.S.R, Warntjes A.A., Sloan E.D., Sum A.K., Koh C.A. (2009) Increasing hydrogen storage capacity using tetrahydrofuran, *J. Am. Chem. Soc.* **131**, 14616-14617.

9 Linga P., Kumar R., Englezos P. (2007) The clathrate hydrate process for post and pre-combustion capture of carbon dioxide, *J. Hazard. Mater.* **149**, 625-629.

10 Kang S.P., Lee J., Seo Y. (2013) Pre-combustion capture of CO_2 by gas hydrate formation in silica gel pore structure, *Chem. Eng. J.* **218**, 126-132.

11 Babu P., Kumar R., Linga P. (2013) A New Porous Material to Enhance the Kinetics of Clathrate Process: Application to Precombustion Carbon Dioxide Capture, *Env. Sci. Tech.* **47**, 13191-13198.

12 Chari V.D., Sharma D.V.S.G.K., Prasad P.S.R. (2011) Methane hydrate phase stability with lower mole fractions of tetrahydrofuran (THF) and tert-butylamine (t-$BuNH_2$), *Fluid Phase Equilib.* **315**, 126-130.

13 Zhong Y., Rogers R.E. (2000) Surfactant effects on gas hydrate formation, *Chem. Eng. Sci.* **55**, 4175-4187.

14 Ganji H., Manteghian M., Zadeh K.S., Omidkhah M.R., Mofrad H.R. (2007) Effect of different surfactants on methane hydrate formation rate, stability and storage capacity, *Fuel* **86**, 434-441.

15 Liang Y., Shuanshi F., Wang Y., Xuemei L., Xie D. (2011) Accelerated Formation of Methane Hydrate in Aluminum Foam, *Ind. Eng. Chem. Res.* **50**, 11563-11569.

16 Wang W., Bray C.L., Adams D.J., Cooper A.I. (2008) Methane Storage in Dry Water Gas Hydrates, *J. Am. Chem. Soc.* **130**, 11608-11609.

17 Carter B.O., Wang W., Adams D.J., Cooper A.I. (2010) Gas Storage in "Dry Water" and "Dry Gel" Clathrates, *Langmuir* **26**, 3186-3193.

18 Prasad P.S.R., Chari V.D., Sharma D.V.S.G.K., Murthy S.R. (2012) Effect of silica particles on the stability of methane hydrates, *Fluid Phase Equilib.* **318**, 110-114.

19 Mork M., Gudmundsson J.S. (2002) Hydrate Formation Rate in a Continuous Stirred Tank Reactor: Experimental Results and Bubble-to-Crystal Model, *4th International Conference on Gas Hydrates.*

20 Wenfeng H., Wang J., Shuanshi F., Wenbin H. (2007) Study on methane hydration process in a semi-continuous stirred tank reactor, *Energy Convers. Manage* **48**, 954-960.

21 Chari V.D., Raju B., Prasad P.S.R., Rao D.N. (2013) Methane hydrates in spherical silica matrix: Optimization of capillary water, *Energy Fuels* **27**, 3679-3684.

22 Vysniauskas A., Bishnoi P.R. (1983) A kinetic study of methane hydrate formation, *Chem. Eng. Sci.* **38**, 1061-1972.

23 Golombok M., Ineke E., Luzardo J.C.R., He Y.Y., Zitha P. (2009) Resolving CO_2 and methane hydrate formation kinetics, *Envi. Chem. Lett.* **7**, 325-330.

24 Anderson R., Tohidi B., Webber J.B.W. (2009) Gas hydrate growth and dissociation in narrow pore networks: capillary inhibition and hysteresis phenomena, *The Geological Society, London, Special Publications* **319**, 145-159.

25 Guanli J., Qingbai W., Jing Z. (2010) Effect of cooling rate on methane hydrate formation in media, *Fluid Phase Equilib.* **298**, 225-230.

26 Chari V.D., Sharma D.V.S.G.K., Prasad P.S.R., Murthy S. R. (2013) Methane hydrates formation and dissociation in nano silica suspension, *J. Nat. Gas Sci. Eng.* **11**, 7-11.

27 Linga P., Daraboina N., Ripmeester J.A., Englezos P. (2012) Enhanced rate of gas hydrate formation in a fixed bed column filled with sand compared to a stirred vessel, *Chem. Eng. Sci.* **68**, 617-623.

6

Production of Hydroprocessed Esters and Fatty Acids (HEFA) – Optimisation of Process Yield

Laurie Starck[1]*, Ludivine Pidol[1], Nicolas Jeuland[1], Thierry Chapus[2], Paul Bogers[3] and Joanna Bauldreay[3]

[1] IFP Energies nouvelles, 1-4 avenue de Bois-Préau, 92852 Rueil-Malmaison - France
[2] IFP Energies Nouvelles, Rond-point de l'échangeur de Solaize, BP 3, 69360 Solaize - France
[3] Shell Global Solutions Downstream, Shell Technology Centre Thornton, P.O. Box 1, Chester CH1 3SH - United Kingdom
e-mail: laurie.starck@ifpen.fr - ludivine.pidol@ifpen.fr - nicolas.jeuland@ifpen.fr - thierry.chapus@ifpen.fr
paul.bogers@shell.com - joanna.bauldreay@shell.com

* Corresponding author

Abstract — *Both Fischer-Tropsch (FT) and Hydroprocessed Esters and Fatty Acids (HEFA) Synthetic Paraffinic Kerosine (SPK) fuels are considered as leading alternative replacements for conventional jet fuel. To satisfy the requirements of Civil Aviation Authorities (CAA), their drop-in incorporations have been subjected to a rigorous certification process. To reach the ambitious incorporation targets, new routes for biofuels incorporation may need to emerge, involving optimizing the production processes and the blending strategies. This paper focuses on a new strategy for incorporating HEFA, allowing the process yield to be optimised.*

One of the major steps limiting the process yield for HEFA remains the isomerisation that allows production of a biofuel with very good cold flow properties. But this step introduces a substantial decrease of the overall yield (fuel component per kg of starting material) due to the production of light compounds, unsuitable for conventional jet fuel. In this work relaxing the freezing point requirement for the neat HEFA component (by decreasing the severity of the isomerisation step) is proposed in order to minimize the production of less valuable light compounds. This strategy could lead to a significant additional biofuel yield with respect to the oil compared to a process making a better freezing point component. This allows the land surface area necessary for HEFA feedstock cultivation to be reduced for a given amount of bio-jet fuel produced.

Résumé — **Production d'huiles végétales hydrotraitées (*Hydroprocessed Esters and Fatty Acids, HEFA*) – Optimisation du rendement** — Le développement des carburants alternatifs est en plein essor, notamment dans le domaine aéronautique. Cela se concrétise par la possibilité, d'incorporer jusqu'à 50 % de carburants de synthèse de type Fischer-Tropsch (FT) ou *Hydroprocessed Esters and Fatty Acids* (HEFA) dans du carburéacteur. Ces cibles d'incorporation sont ambitieuses. C'est pourquoi, l'objectif de cet article est d'étudier une stratégie innovante pour l'incorporation des carburants alternatifs, et plus précisément des carburants de type HEFA, dans le domaine aéronautique en optimisant les stratégies de mélanges c'est-à-dire en cherchant à optimiser les rendements des procédés.

En effet, l'un des moyens d'action permettant d'améliorer les rendements des procédés HEFA est d'agir sur l'étape d'hydrotraitement. Cette étape permet d'améliorer les propriétés à froid. Cependant la contre partie est l'impact que cela peut avoir sur le rendement : améliorer les propriétés à froid est synonyme de perte en rendement (carburant produit par kg par rapport à la masse de matière première) à cause de la production de produits légers qui ne sont pas utilisables dans la coupe jet. Ce travail propose donc de relâcher la contrainte sur le point de disparition des cristaux, exigée pour les carburants de type HEFA, en jouant sur la sévérité de l'étape d'hydrotraitement et donc en minimisant la production de produits légers afin de maximiser les rendements. Cette stratégie pourrait amener à avoir un meilleur rendement en biojet par rapport à l'huile de départ comparée à une stratégie dans laquelle on recherche un meilleur point de disparition des cristaux. Ainsi, il serait envisageable de réduire les surfaces agricoles utilisées pour la culture des plantes pour un même volume de biojet produit.

DEFINITIONS / ABBREVIATIONS

ASTM	American Society for Testing and Materials International
CAA	Civil Aviation Authorities
DCO	Decarboxylation
EU-ETS	European Union Greenhouse Gas Emission Trading Scheme
FT	Fischer-Tropsch
HDO	Hydrodeoxygenation
HEFA	Hydroprocessed Esters and Fatty Acids
HIS	Hydroisomerization
HRJ	Hydroprocessed Renewable Jet
HVO	Hydrotreated Vegetable Oil
*n*P	Normal (Linear) Paraffins
*i*P	Iso Paraffins
SPK	Synthetic Paraffinic Kerosene
SWAFEA	Sustainable Way for Alternative Fuel and Energy in Aviation

INTRODUCTION

World wide air traffic has been steadily increasing for many years and is predicted to grow at a rate of close to 4-5% per year, with even higher growth rates in the Middle East and Asia [1, 2].

Moreover, the increased focus on climate change over the last decade has created pressure to reduce greenhouse gases emissions. It has been estimated that the aeronautics sector represents 2 to 3% of the global CO_2 emissions [3]. Such a contribution may seem to be minor but the air traffic is expected to strongly increase in the next years while other industries move to lower carbon options. This is one reason it has been decided to include the aeronautics sector in the EU-ETS (European Union Greenhouse Gas Emission Trading Scheme) from 2012.

International Air Transport Association (IATA) has adopted a voluntary ambitious fuel efficiency goal: reducing CO_2 emissions by 50% by 2050, compared to the 2005 level, as illustrated in Figure 1. Consequently, the search for new alternative fuels for aircraft seems to be a promising and necessary solution from an energy security and environmental perspective. If aviation wants to reduce its greenhouse gas emissions, it has to turn to biofuels, which are the only fuels having the potential to achieve significant greenhouse emissions savings. Some of the steps towards a reduced CO_2 goal can be achieved with the efficiency increases anticipated through aircraft improvement, operational measures or infrastructure changes. But these measures are not enough and additional reductions are required. It is why biofuel appears to be the main candidate to achieve these reductions. Significant emissions reduction can be achieved with biofuel, provided that low emissions are achieved in the cultivation step of the biomass and if there is a rigorous control of land use change. Without these controls, some biofuels can have very poor CO_2 footprints.

The effort to develop alternative aviation fuels has already begun, focused on Fischer-Tropsch (FT) fuels.

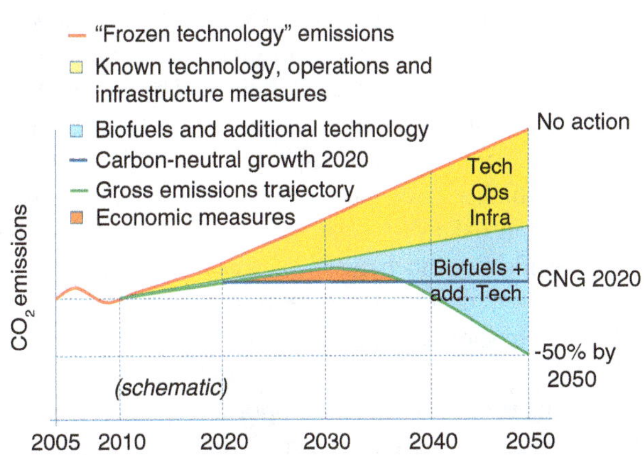

Figure 1

CO_2 emissions reduction targeted by IATA (Source IATA).

Generic FT Synthetic Paraffinic Kerosene (SPK) was approved for use in blends, at up to 50% volume, with Jet A-1 in ASTM D7566 in August 2009 [4, 5]. The second class of alternative fuels approved for certification is Hydroprocessed Esters and Fatty Acids (HEFA), also called Hydrogenated Vegetable Oil (HVO) or Hydroprocessed Renewable Jet (HRJ); it covers hydrocarbon aviation fuel produced from animal oils or vegetable oils (triglycerides) by hydroprocessing. This fuel has also been called BioSPK, although one should remember that the FT SPK include biomass to liquid fuels that can equally be called BioSPK. The ASTM D7566 specification is structured to support various classes of alternative fuels in its appendices and HEFA was approved for use at up to 50% volume in blends with conventional kerosene in ASTM D7566 in July 2011 [6]. It can now be used in ASTM D1655 fuels and, following an update to DEF STAN 91-91, HEFA has also been approved under "Check List". The results of a study by the EU-funded Sustainable Way for Alternative Fuels and Energy in Aviation (SWAFEA) [7] have shown that the incorporation of biofuels in aviation fuel is necessary to reach the ambitious IATA targets in terms of CO_2 reduction. These conclusions also show that a massive incorporation ratio will be needed, so that huge investments need to be made. Moreover, the question of biofuels deployment has been raised, linked with a need for an optimised production yield to meet those targets.

There is competition for biofuel components which affects the design and economics of facilities that convert bio-oils into fuel. Specifically, there is competition between gasoil (*e.g.* automotive Diesel, heating and industrial Diesel) and kerosene users. Most bio-oils naturally yield product in the gasoil range so further processing steps are needed to make a product that is technically better suited to kerosene (jet fuel) production, in particular having better low temperature features than required for Diesel applications. The objective of this work is to study product quality trade-offs that could affect the economics of biojet production in the initial periods when biofuel availability will be limited. Specifically, it considers incorporating low levels (considerably less than the 50% volume now approved) of SPK with freezing points higher than those currently approved (−40°C maximum) into Jet A-1, while keeping the blend freezing point specification unchanged. Increasing (worsening) the freezing point decreases the level of processing of the SPK, which should produce higher biofuel component yields [8] and a potentially better overall profitability than a route that produces a higher proportion of better freezing point products that can be incorporated at higher percentages into the final Jet A-1. The recommendation for low or medium HVO (HEFA) concentrations is also studied for Diesel engines [9].

This paper does not deal with the economics themselves but provides technical data that could assist in understanding the relationship between process costs (usually related to process complexity), process yield, HEFA quality, and blending rate. IFPEN and *Shell* Global Solutions have joined their expertise to produce two different qualities of HEFA from the same original oil for the purposes of the SWAFEA project. The level of hydroisomerisation applied to a hydroprocessed oil to improve the low temperature properties was varied between the two HEFA:

- if, as expected, important constraints are put on oil availability and HEFA production capacity, in the short term the incorporation process will have to be optimised to incorporate this product into conventional Jet A-1. In this case, the global optimisation of the production yield will be of major importance. A major factor limiting the process yield remains the isomerisation step, which allows production of a biofuel with very good cold flow properties. It also leads to a substantial decrease in the overall yield because of the co-production of light compounds that cannot be incorporated in conventional jet fuel. To evaluate the impact of this production step, an extreme view has been chosen: a specific HEFA has been produced to have limited cold flow properties (target: freezing point around −20°C) and the resulting product has been blended into conventional jet fuel at different blending ratios. The overall yield has been calculated, to evaluate if a substantial gain can be obtained with such a strategy, while keeping the final blend in the limits of the Jet A-1 specification (freezing point ≤ −47°C);
- the current specification limits the incorporation rate of HEFA to a maximum of 50% volume. To meet the IATA targets, higher blending rates may eventually be needed. Indeed, during the ramp-up period, this product may not be available everywhere. The capacity to blend higher ratios in some locations can therefore help to meet the incorporation targets, taking into account the local availability of the product. A specific HEFA meeting the ASTM D7566 specifications has been made to evaluate the potential of such a product being blended at a high ratio (75%) in Jet A-1.

To summarize the different strategies:

- Case 1: aim for a reduced HEFA incorporation ratio for HEFA with poor cold flow properties (target: freezing point around −20°C). Blends of this HEFA production (called HEFA1) with a conventional Jet A-1 have been prepared with 10%, 20% and 30% volume of HEFA1;
- Case 2: aim for a larger HEFA incorporation ratio than permitted by ASTM D7566: one blend of HEFA

production (called HEFA2) with the same conventional Jet A-1, has been prepared with an incorporation ratio of 75% volume of HEFA2.

1 GENERAL INFORMATION ON THE HYDROTREATMENT PROCESS OF VEGETABLE OILS

1.1 Vegetable Oil Composition

Renewable sources such as vegetable oils are mainly composed of triglycerides, which are tri-esters comprising 3 fatty acid hydrocarbon chains and a glycerol backbone, as shown in Figure 2. The degree of unsaturation of the fatty acid chains (*i.e.* the number of double bonds), as well as the carbon distribution, varies according to the nature of the oil.

Vegetable oils contain 6 oxygen atoms per mole, which leads to oxygen contents of about 10 to 12 wt%. According to the nature of the starting vegetable oils, fatty acid hydrocarbon chains can exhibit various carbon atom distributions, which are generally in the range C8 to C24. The nature of the feedstock also impacts the degree of unsaturation of the hydrocarbon chain; there can be between 3 and 6 unsaturated bonds. Table 1 gives some characteristics typically seen for rapeseed oils [10].

1.2 Main Process Steps

The present work has been achieved using a rapeseed oil sent to hydrotreatment, followed by a hydroisomerization step; rapeseed oil was chosen for reasons of feedstock availability. It is important to note that it would have been possible to use any other type of oil to demonstrate the two strategies for HEFA incorporation, though the absolute results obtained could vary according to the fatty acid distribution of the source oil.

The hydrogenation process generally comprises two steps [11, 12]:
- hydrotreatment, which is a treatment involving hydrogen, using an hydrotreatment catalyst in appropriate operating conditions;

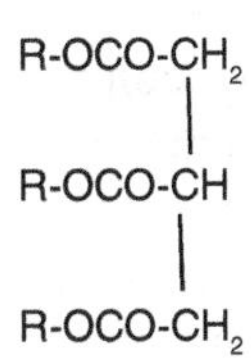

Figure 2

Typical structure of triglycerides, major components of vegetable oils.

TABLE 1

Composition of a typical rapeseed oil

Origin	Method	Unit	Rapeseed oil
Density 15/4	NF EN ISO 12185	kg/m^3	920.0
Sulphur	ASTM D5453	wt ppm	4.2
Nitrogen	ASTM D4629	wt ppm	23.3
Carbon	ASTM D5291	wt%	77.7
Hydrogen	ASTM D5291	wt%	11.75
Oxygen	ASTM D5622	wt%	11.17
Acid value	EN14104	mg KOH/g	0.03
Water	EN ISO 12937	wt ppm	485

- hydroisomerization.

The hydrotreatment step consists of oxygen removal, which leads to the production of a paraffinic middle distillate (boiling at 150°C+). Oxygen removal can be achieved *via* 2 pathways:
- hydrogenation of the fatty acid chain, to produced water and a paraffinic product, maintaining the hydrocarbon chain length (HDO pathway for hydrodeoxygenation, *Eq. 1*):

$$\begin{gathered}\text{Fatty acid chain in}\\ \text{C18} + 16^{i}\,\text{H}_2 \rightarrow 3\,\text{C18} + 6\,\text{H}_2\text{O} + \text{C}_3\text{H}_8\\ 100\,\text{kg of oil} + 3.6\,\text{kg of H}_2 \rightarrow 86.3\,\text{kg of } n\text{-paraffins}\end{gathered} \quad (1)$$

- hydrogenation of the fatty acid chain with production of carbon oxides (such as CO_2 and CO), and a paraffinic product with a loss of one carbon atom in the chain length (DCO pathway for decarboxylation, *Eq. 2*):

$$\begin{gathered}\text{Fatty acid chain in}\\ \text{C18} + 7^{i}\,\text{H}_2 \rightarrow 3\,\text{C17} + \text{CO}_2 + \text{C}_3\text{H}_8\\ 100\,\text{kg of oil} + 1.6\,\text{kg of H}_2 \rightarrow 81\,\text{kg of } n\text{-paraffins}\end{gathered} \quad (2)$$

This is all a rather theoretical depiction; CO and CH are also made. It is also important to note that, despite the increased carbon yield in Equation (1), more H is required. In practice the mechanisms of Equation (1) and Equation (2) both occur but their relative contributions can vary

[i] This number depends on the number of unsaturation of triglycerides. This equation has been established for triglycerides with 4 unsaturated bonds.

TABLE 2
Boiling and melting points of *n*C16, *n*C17 and *n*C18

	*n*C16	*n*C17	*n*C18
Boiling point (°C)	286.9	302.2	316.7
Melting point (°C)	18	22	28.6

depending on conditions, catalyst, etc. However, the objective is to favour the HDO pathway, to maximize the yield of valuable fuel products.

After this first hydrotreatment section, oxygen is completely removed, and a pure paraffinic product is obtained, with typically the same carbon atom distribution as in the starting feedstock. For instance, starting from rapeseed oil, the paraffinic product after hydrotreatment is mainly composed of linear (normal) paraffins (*n*P) such as *n*C16, *n*C17 and *n*C18. This product exhibits a very high cetane, but very poor cold flow properties. This product is located in the Diesel boiling range, and the yield typically obtained is in the range 80-85 wt% relative to the feed.

Table 2 below gives the boiling and melting points of some linear paraffins, which are the major hydrocarbon components of the paraffinic product after hydrotreatment. Note: for pure products, there is a discrete melting point; for fuels, the melting of wax crystals occurs over a temperature range and the freezing point for jet fuels is the temperature at which, on warming up a cold fuel, the last wax crystals disappear. Therefore for "melting point", think freezing point.

After hydrotreatment, the liquid hydrocarbon product is a 100% paraffinic Diesel product, composed mainly of *n*C16 – *n*C18 linear paraffins. The boiling range of the product is consequently 280°C+, well above the range seen for typical jet fuels.

To improve the cold flow properties of this paraffinic Diesel, it is necessary to subject the product to a hydroisomerization (HIS) treatment, which will convert linear paraffins into iso-paraffins. This improves low temperature performance for kerosene applications but has a detrimental effect on cetane quality in Diesel engine applications as shown in Figure 3 [13].

The HIS step is generally achieved using an appropriate catalyst in hydrogenation conditions. The catalyst is a bifunctional one, comprising both a hydrogenation function and an acid function. The objective, ideally, is to convert linear paraffins (*n*P) into iso-paraffins (*i*P), while avoiding cracking reactions which lead to yield loss by formation of gases and naphtha or gasoline cuts (that boil below 150°C). To be as close as possible to this ideal situation, an appropriate choice of the catalyst and operating conditions has to be made.

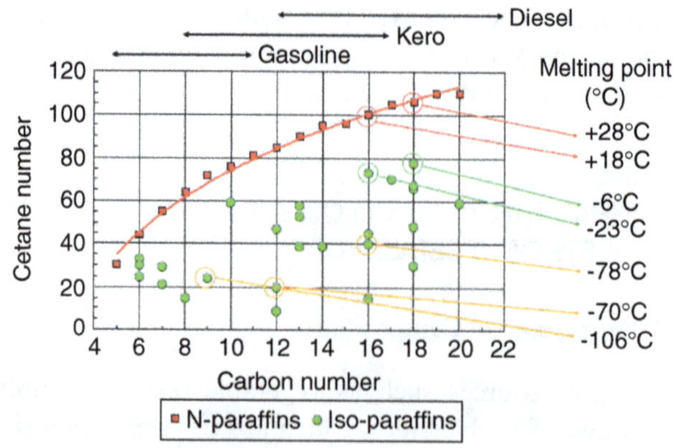

Figure 3
Melting point and cetane number of some normal and iso-paraffins present in the jet fuel and Diesel range [13].

Here is a general description of the reactions which take place in the HIS section:

- after hydrotreatment, the hydrocarbon product is a 100% paraffinic Diesel product, with a boiling range 280°C+. It is therefore necessary to convert *n*P into *i*P, to improve cold flow properties, and convert the Diesel boiling range product into a jet fuel boiling range product (150-300°C);
- when the severity of the HIS section is increased, it leads to a higher conversion of *n*P into *i*P, which leads to lower boiling and melting points (improved cold flow) of the hydrocarbons. Consequently Diesel is progressively converted into jet fuel (kerosene), and then into gasoline and gases at higher HIS severities. These lighter cuts (gasoline and gases) may be of lower value, for example in the EU where there is already a: surplus of gasoline. From biofuel producers' perspectives, the discussion would need to be: a) Diesel *versus*; b) Diesel + low amounts high freeze jet component *versus*; c) not much Diesel + larger amounts of low freeze jet component.

At the end of the HIS step, there is a final distillation step to remove gasoline range material.

To summarize, the challenge of HIS operation for HEFA production is to convert a paraffinic Diesel into a paraffinic jet fuel. As jet fuel is an intermediate product between Diesel and gasoline, there is an optimum degree of conversion in terms of jet fuel yield. This optimum can be attained by selecting the more appropriate catalyst as well as carefully tuning the operating conditions. This is the condition to reach the maximum jet fuel yield, together with jet fuel quality meeting the specifications (distillation range and cold flow properties, *i.e.* freezing point when referring to jet fuel).

Figure 4 below shows the typical evolution of yields during HIS operation. There is a maximum jet fuel yield, which can be reached by carefully tuning the severity of HIS operation. The severity has to be sufficient to convert the hydrocarbons from the Diesel range (280°C+) into the jet fuel range (150-300°C), but not too high, to avoid excessive production of less economically attractive gases and gasoline (150°C) components.

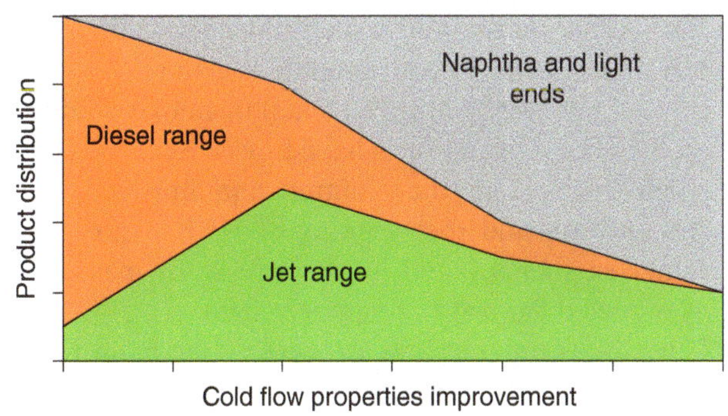

Figure 4

Impact of cold flow properties on kerosene yield of HEFA (source: IFP Energies nouvelles).

1.3 Final Product Properties

The specifications requirements included in Table 3 have to be met in the final jet fuel after blending. It should be noted that it is necessary to blend the paraffinic jet fuel with a conventional jet fuel from petroleum, since an 8% volume minimum aromatics content is specified, to comply with material compatibility requirements. The current certification for HEFA allows no more than 50% volume HEFA in the final jet fuel, in part to help achieve this minimum aromatics level.

2 PRODUCTION OF HEFA JET FUELS WITH VARIOUS QUALITY REQUIREMENTS

2.1 Neat Materials (HEFA and Jet Fuels)

The previous section showed that there is an optimum conversion in terms of jet fuel yield and quality.

TABLE 3

Characteristics of the Jet A-1 fuels used in this work

Analysis	Unit	Jet A-1 specifications	Jet fuel A	Jet fuel B
Density 15°C	kg/m^3	775-840	803.3	821.0
Sulfur	wt ppm	3 000 maximum	697	< 5
Total aromatics	vol%	25.0 max (FIA)	16.1	20.2
		26.5 max (hplc)	18.9	20.4
Distillation Initial point	°C		147.6	152.1
T10 D86	°C	205 max	167.0	170.2
T20 D86	°C		175.5	176.6
T50 D86	°C		199.0	201.5
T90 D86	°C		245.3	240.2
Final point D86	°C	300 max	270.6	258.9
T90-T10 D86	°C	22 min	78.3	70.0
Freezing point	°C	−47 max	−49.6	−64.9
Smoke point	mm	25 min or 19 min (*)	21	19
Viscosity at −20°C	mm^2/s	8.0 max	4.04	4.19
Flash point	°C	38 min	40.5	42.5

(*) and Naphthalenes < 3.0 wt% (D1840).

TABLE 4
Cold flow properties of HEFA1 and HEFA2

Analysis	Unit	HEFA1	HEFA2
Freezing point	°C	–27°C	–57.5°C
Viscosity at –20°C	mm^2/s	11.72	7.52

The target which is set for HEFA jet fuel quality (freezing point in particular) consequently has an impact on the overall jet fuel yield.

The objective of this work was to examine the possibility of relaxing the quality requirement for the freezing point of the HEFA jet fuel component, and to estimate the corresponding gain of yield. It was decided to set two distinct freezing point targets for the jet fuel candidate HEFA:

- one HEFA to have a freezing point close to –20°C;
- one HEFA to have a freezing point below –47°C.

Table 4 presents the cold flow properties, freezing point and viscosity at –20°C, for the HEFA that were produced from rapeseed oil for this study.

Two different jet fuels were used to prepare further blends with HEFA1 and HEFA2. A first Jet A-1 (Jet fuel A) with conventional characteristics and a second Jet A-1 (Jet fuel B), with some more extreme but still "on spec" properties (in particular, exceptionally good freezing point). The main characteristics of the conventional Jet A-1 used for this work are gathered in Table 3.

2.2 Strategy of Blending and Characterization

Two blending strategies were adopted to optimise the amount of each HEFA which could be incorporated into Jet fuel A and Jet fuel B based on freezing point and viscosity at –20°C.

Jet fuel A: this fuel offered a typical freezing point of –49.6°C and a low blending ratio strategy of 10%, 20% and 30% volume was adopted for HEFA1 due to the –27°C freezing point of this product. HEFA2 however, offered an excellent –57.5°C freezing point was therefore blended at one high concentration, 75% volume.

Jet fuel B: this fuel offered an excellent freezing point of –64.9°C allowing greater scope for blending HEFA1 and a concentration range of 20%, 30%, 40%, 50%, 60%, 70% volume was selected. While no freezing point issues were expected for blending HEFA2, a broad range of concentrations was also chosen 25%, 40%, 50%, 60% and 75% volume to investigate if any other specification constraints might limit incorporation.

Tables 5 and 6 present the characteristics of blends of Jet fuel A with HEFA1 and HEFA2, respectively.

The main conclusions from the analysis are as follows:

- all blends have densities that match predicted densities, within method precision – the blending is reliably linear by volume and the mixing steps have produced homogeneous blends;
- all the blends met the current ASTM specifications except for freezing point (for some blends with HEFA1) and aromatic content (blend with HEFA2);
- blending HEFA in conventional fuels produces substantial increases (improvements) in smoke point;
- all the HEFA blend fuels have Lower Heating Values (LHV) higher than the minimum limit imposed by the specification and the blending of HEFA in conventional jet fuel resulted in an increase of mass LHV as expected from the component data.

If we focus on cold flow properties, the main results obtained for freezing points are presented in Figure 5.

The main conclusions are:

- HEFA1 blending ratio is limited by the freezing point. Nevertheless, significant blending ratios can be obtained before reaching this limit. With Jet fuel A this maximum blending ratio has been estimated to be 16%volume. Nevertheless, this point has to be further checked and does not represent a fixed value; there can be measurement uncertainties with freezing points of highly hydroprocessed fuels and the maximum ratio will also be very dependent on the freezing point of the crude oil based component;
- HEFA2 can be blended at any concentration without any limitation on the freezing point, due to the very good freezing point of the pure product;
- with Jet fuel A, the blend freezing point behaviour for both HEFA seems to be close to linearity, which could mean that the maximum blending ratio could be easily calculated according to pure product properties. The blending ratio of these HEFA could consequently be adjusted, taking into account the initial freezing point of the conventional Jet A-1. Nevertheless, this point still has to be confirmed using other Jet A-1, to check that this linearity is not dependent on Jet A-1 composition.

As far as viscosity is concerned, the best fit to the results is not always linear, particularly for HEFA1, a fuel whose freezing point is only slightly lower than the –20°C test condition. Nevertheless, starting from a viscosity of 4 mm^2/s for the pure Jet A-1 (which is representative of typical Jet A-1 available on the market), a blend of 75% HEFA2 still stays below the specification viscosity limit. Moreover, all the blends made with HEFA2 are expected to pass the ASTM D7566 viscosity limit (8 mm^2/s at –20°C). These results are presented in Figure 6.

TABLE 5
Characteristics of blends of Jet fuel A with HEFA1

Analysis	Unit	Jet A-1 specifications	Jet fuel	HEFA1	Jet + 10% HEFA1	Jet + 20% HEFA1	Jet + 30% HEFA1
Density 15°C	kg/m^3	775-840	803.3	773.5	800.0	797.0	794.0
Total aromatics (hplc)	vol%	26.5 max	18.9	0.0	16.9	15.0	13.3
Distillation Initial	°C		147.6	174.2	148.7	153.6	152.4
T10	°C	205 max	167.0	201.2	170.2	174.5	174.5
T20	°C		175.5	218.4	179.6	183.3	184.0
T50	°C		199.0	267.5	204.7	210.7	214.5
T90	°C		245.3	288.9	260.8	271.1	277.4
Final	°C	300 max	270.6	293.2	282.8	288.5	289.5
T90-T10	°C	22 min	78.3	87.7	90.6	96.6	102.9
Freezing point	°C	−47 max	−49.6	−27	−49.0	−46.5	−44.5
Smoke point	mm	25 min or 19 min (*)	21	> 50	25.0	26.0	28.0
Viscosity at −20°C	mm^2/s	8.000 max	4.04	11.72	4.426	4.859	5.363
Flash point	°C	38 min	40.5	67	43	43.5	45
Specific energy, net	MJ/kg	42.8 min	43.25	44.07	43.31	43.41	43.49

(*) and Naphthalenes < 3.0 wt% (D1840).

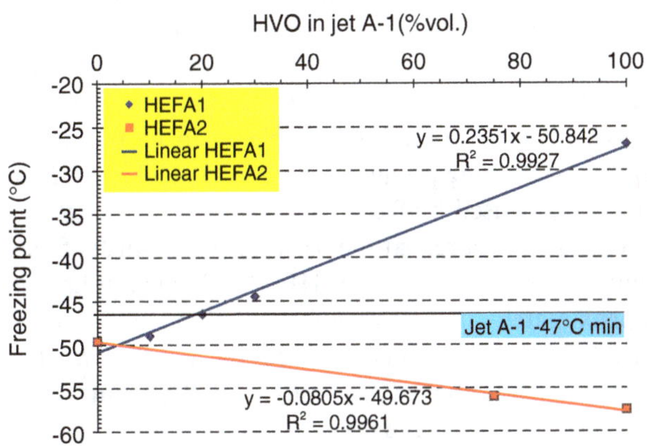

Figure 5

Impact of HEFA content on final blend freezing point for blends in Jet fuel A.

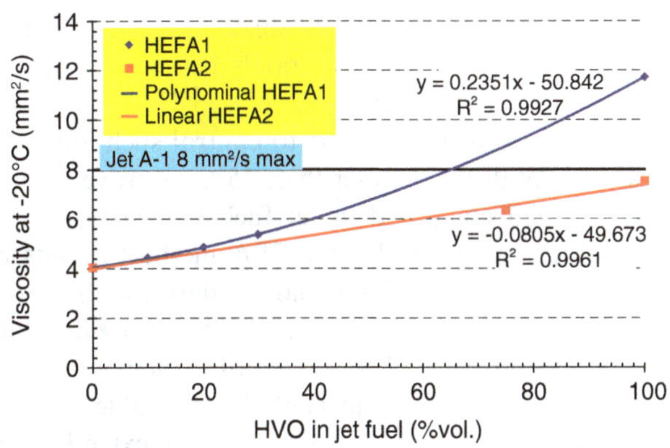

Figure 6

Impact of HEFA in Jet fuel A blending rate on final blend viscosity at −20°C.

Following the blend production and test results discussed above, blends were made with another Jet A-1, Jet fuel B, that has excellent low temperature results; as shown in Table 3, it has a significantly better freeze point than average and represents a close to "best case" blend partner for a poorer freezing point HEFA

TABLE 6
Characteristics of blends of Jet fuel A with HEFA2

Analysis	Unit	Jet A-1 specifications	Jet fuel	HEFA2	Jet + 75% HEFA2
Density 15°C	kg/m^3	775-840	803.3	765.9	775.0
Total aromatics (hplc)	vol%	26.5 max	18.9	0.0	4.8
Distillation Initial	°C		147.6	188.0	171.8
T10	°C	205 max	167.0	201.6	192.2
T20	°C		175.5	206.7	199.2
T50	°C		199.0	224.7	218.9
T90	°C		245.3	279.9	276.0
Final	°C	300 max	270.6	287.3	287.2
T90-T10	°C	22 min	78.3	78.3	83.8
Freezing point	°C	−47 max	−49.6	−57.5	−56
Smoke point	mm	25 max or 19 max (*)	21	> 50	41
Viscosity at −20°C	mm^2/s	8.0 max	4.04	7.517	6.335
Flash point	°C	38 min	40.5	68.0	58.0
Specific energy, net	MJ/kg	42.8 min	43.25	44.11	43.89

(*) and Naphthalenes < 3.0 wt% (D1840).

component. Its freezing point of −64.9°C is the highest reported for the fuel; analysts sometimes prefer not to determine freezing points accurately below an arbitrary temperature in the range −60 to −75°C as this takes more time and/or more effort to control such low temperatures. With this greater than 15°C improvement in freezing point compared to Jet fuel A, it was possible to consider higher levels of HEFA in the two sets of blends created. Table 7 summarises data for the seven blends created with HEFA1 and Table 8 provides data for the 5 blends with HEFA2; data for the neat HEFA and Jet fuel A and B are in Table 3, and Table 4.

As shown in Figure 7, Jet fuel B has extended the range over which HEFA1 can be blended and still achieve Jet A-1 requirements. The best fit for HEFA1 is now polynomial, with freezing points being higher (poorer) than a linear blend rule would expect. Based on the fit the highest level of HEFA1 to pass the Jet A-1 freezing point would be about 35% volume (or 30% based on actual data points). This is higher than seen with Jet fuel A. If only a Jet A freezing point needed to be met (−40°C maximum), the fit and actual data points would both suggest upper limits of 51% volume. For the HEFA2 blends the results are best described as "scattered"; there may or may not be a sweet spot around 60% volume with an exceptionally low freezing point (similar to behaviour seen with some GTL SPK [14]); many of the results are likely to be affected by the analyst's observation preference and the overriding message is that all the HEFA2 blends in Jet fuel B have very good freezing points, with no limits on HEFA2 content being caused by freeze point.

Viscosity data repeat the patterns seen with Jet fuel A: Figure 8 shows that the HEFA2 blends follow an approximately linear by volume behaviour while the HEFA2 blends are better fitted by a polynomial fit. HEFA1 blends above 63% volume fail the 8 mm^2/s limit, while all HEFA2 blends pass.

The limiting factor for HEFA2 blends would be set by the 8.8% volume total aromatics by hplc method, if the 50% maximum were not applied.

In summary, the maximum volume percentages of the HEFA in the two Jet A-1 are as in Table 9, Figure 9 and Figure 10. Given that in a country using Jet A rather than Jet A-1 there may not be many very good freezing point fuels like Jet fuel B, this does indicate that HEFA1 will probably have an upper limit of 30 to 35% volume; more typically HEFA1 be limited to values in the range 15 to 20% volume for Jet A-1 production.

TABLE 7

Properties of blends of HEFA1 with Jet fuel B

%vol. HEFA1 in Jet fuel B		20	30	40	50	60	70
Analysis	Unit						
Density 15°C	kg/m^3	811.3	806.7	802.1	797.4	792.7	788.1
Total aromatics (hplc)	vol%	17.5	15.1	12.9	10.8	8.8	6.7
Distillation Initial	°C	153.9	155.4	157.1	159.6	161.9	165.7
T10	°C	172.5	174.4	176.3	179.1	182.2	185.7
T20	°C	179.9	182.6	185.6	188.5	192.9	197.0
T50	°C	208.9	213.9	219.4	227.0	234.3	243.1
T90	°C	264.3	273.2	277.4	282.1	283.7	285.4
Final	°C	283.5	286.7	289.1	289.6	290.6	290.9
T90-T10	°C	91.8	98.8	101.1	103.0	101.5	99.7
Freezing point	°C	−58.8	−47.8	−42.8	−40.2	−37.1	−34.6
Smoke point	mm	21.5	22.0	24.5	25.5	30.0	34.0
Viscosity at −20°C	mm^2/s	5.063	5.504	6.059	6.100	7.412	8.229
Flash point	°C	44.0	46.0	47.0	49.5	51.0	53.5
Specific energy, net	MJ/kg	43.215	43.325	43.426	43.535	43.636	43.741

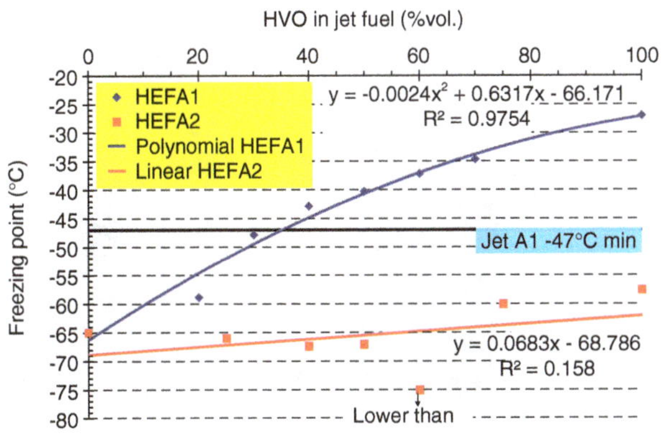

Figure 7

Freezing points for blends of HEFA1 and HEFA2 with Jet fuel B.

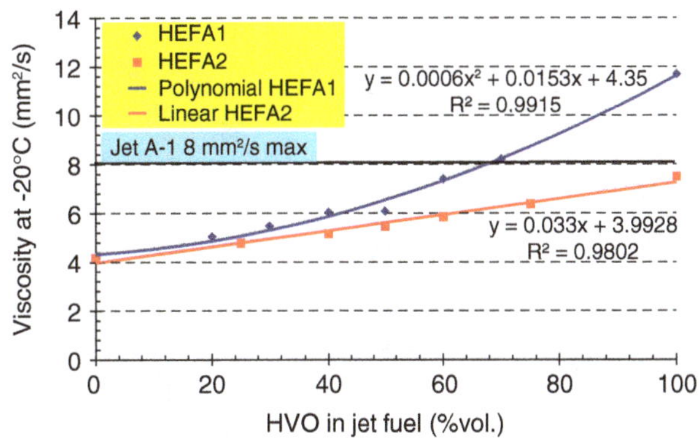

Figure 8

Viscosity for HEFA blends made with Jet fuel B.

2.3 Production Yield

Figure 11 shows the yield improvement for HEFA1 and HEFA2 production. It appears that lowering the severity of HIS operation, for production of HEFA1, leads to a gain of jet and Diesel fuel yield and particularly a gain of jet fuel yield to more than +10 points of mass yield relative to the oil.

For a given quantity of jet fuel produced, this potentially allows a reduction in the land surface area necessary for cultivation of the HEFA feedstock by at least 10%.

SUMMARY AND CONCLUSIONS

World wide air traffic has been steadily increasing for many years and is predicted to grow in the future.

TABLE 8
Properties of blends of HEFA2 with Jet fuel B

%vol. HEFA2 in Jet fuel B		25	40	50	60	75
Analysis	Unit					
Density 15°C	kg/m^3	807.0	799.0	793.6	788.0	779.6
Total aromatics (hplc)	vol%	16.2	13.1	11.0	9.1	5.5
Distillation Initial	°C	157.0	159.9	165.1	168.2	175.7
T10	°C	174.9	178.5	181.7	185.4	191.0
T20	°C	182.3	186.4	189.5	192.4	197.2
T50	°C	205.6	209.1	211.3	213.9	217.4
T90	°C	251.7	260.3	263.5	268.2	272.8
Final	°C	275.7	280.1	282.2	283.2	284.5
T90-T10	°C	76.8	81.8	81.8	82.8	81.8
Freezing point	°C	−65.9	−67.3	−67.0	< −75.0	−60.0
Smoke point	mm	20.5	22.0	26.0	33.0	34.0
Viscosity at −20°C	mm^2/s	4.811	5.190	5.495	5.876	6.412
Flash point	°C	45.5	47.5	49.5	52.0	56.0
Specific energy, net	MJ/kg	43.262	43.423	43.531	43.642	43.819

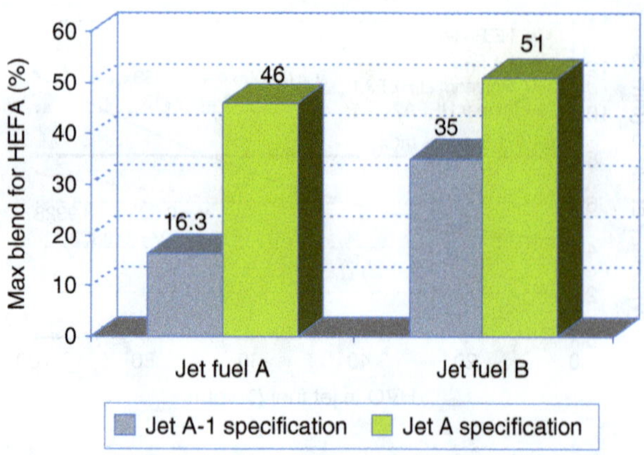

Figure 9

Maximum blend % volumes for HEFA1 / Jet fuels combinations neglecting current 50% specification limit.

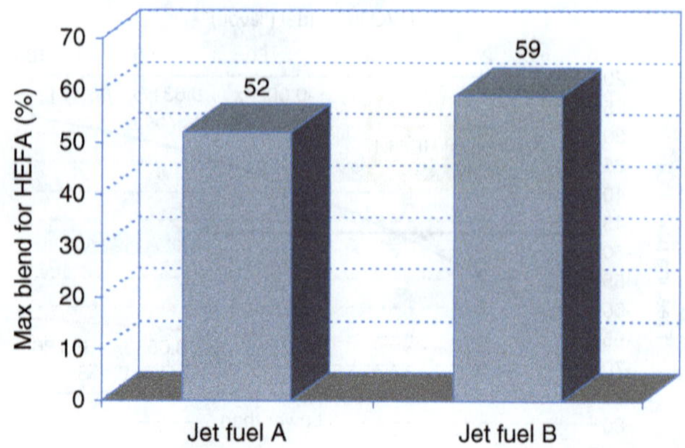

Figure 10

Maximum blend % volumes for HEFA2 / Jet fuels combinations neglecting current 50% specification limit.

Besides, the increased focus on climate change over the last decade has created pressure to reduce greenhouse gases emissions. One solution for the aeronautics sector to satisfy the objective of CO_2 reduction is the use of alternative fuel such as HEFA.

HEFA are produced from hydrogenation of mainly vegetable oils. This process generally comprises two steps: hydrotreatment and hydroisomerization. The process yield for HEFA remains limited because of the isomerisation step that allows production of a biofuel with very good cold flow properties. This work proposes to relax the freezing point requirement for the neat HEFA component (by decreasing the severity of the isomerisation step) in order to minimize the

TABLE 9

Maximum blend % volumes for HEFA/Jet A-1 combinations neglecting current 50% specification limit

HEFA	Jet A-1	Limit %vol.	Limiting property	Blend rule
HEFA1	Jet fuel A	16.3	Freezing point Jet A-1	
		46	Freezing point Jet A	
HEFA1	Jet fuel B	35 (fit) or 30 (data point)	Freezing point Jet A-1	
		51	Freezing point Jet A	
HEFA2	Jet fuel A	52	Total aromatics	Linear by volume
HEFA2	Jet fuel B	59	Total aromatics (hplc)	Linear by volume

production of less valuable light compounds. This strategy could lead to a significant additional biofuel yield with respect to the oil. Then this experimental study determines if a combination of processing/blending is possible to optimise HEFA incorporation for industry benefit.

A rapeseed oil was used for this study. The choice of this oil has been done for reasons of feedstock availability. It was decided to set two distinct freezing point targets for the jet fuel candidate HEFA: one HEFA to have a freezing point close to −20°C and the other one HEFA to have a freezing point below −47°C. Moreover two different jet fuels were used for the blends. A first Jet A-1 with conventional characteristics and a second Jet A-1, with some more extreme but still "on spec" properties (in particular, exceptionally good freezing point).

It has been shown that these HEFA can be blended in significant volumes in a conventional Jet A-1. The upper limits in terms of incorporation are:

- the freezing point for the HEFA1 (−27°C). HEFA1 can be incorporated at up to 30-35% volume before failing Jet A-1 freezing point. The incorporation ratio is dependant of the jet fuel quality regarding cold flow behavior;
- the aromatic content of the final blend for the introduction of HEFA2 (−57.5°C). However, HEFA2 can be incorporated at up to 59% before failing to achieve the minimum total aromatics content. With a higher aromatics jet fuel that has excellent freezing point characteristics, higher levels of HEFA can be incorporated, though the limiting properties remain the same.

These preliminary tests have shown that by combining processing conditions in combination with blending strategies there is a potential to enhance the incorporation of HEFA in conventional jet fuel. The blending of a high amount of a low freezing point HEFA could allow higher incorporation rates in the long term, while the incorporation of high freezing point HEFA in smaller %vol. could allow HEFA incorporation to be started in the shorter term, while optimizing the process yield. This strategy is more in accordance with biomass availability. Nevertheless, it has to be stressed that these are only preliminary results and need important validation work, such as fitness for purpose tests recommended in ASTM D4054, before going into the certification process and being industrially used.

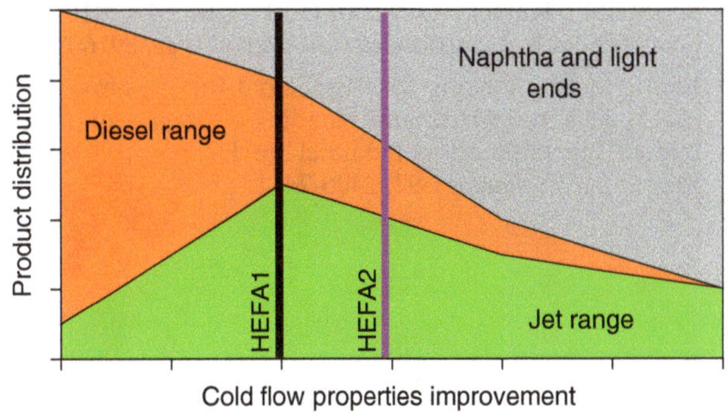

Figure 11

Yield improvement for HEFA production (source: IFP Energies nouvelles).

ACKNOWLEDGMENTS

This work was done as an additional contribution to the SWAFEA Project and the contributors are *Shell* Global Solutions and IFP Energies nouvelles.

REFERENCES

1 Advisory Council on Aeronautics Research in Europe (ACARE (2010) *Aeronautics and air transport: beyond vision 2020 (towards 2050)*.

2 http://www.airbus.com, Airbus Global market forecast 2010-2029.

3 Pachauri R.K., Reisinger A. (2007) *Contribution of working groups I, II and III to the fourth assessment report of the intergovernmental panel on climate change*, Intergovernmental Panel on Climate Change (IPCC), Geneva Switzerland.

4 American Society for Testing and Materials (2009) Standard Specification for Aviation Turbine Fuel Containing Synthesized Hydrocarbons, ASTM D7566 – 10.

5 Moses C.A. (2008) Comparative evaluation of semi-synthetic jet fuels, *CRC project No. AV-02-04a*, Sept.

6 American Society for Testing and Materials) (2011) *Standard Specification for Aviation Turbine Fuel Containing Synthesized Hydrocarbons*, ASTM D7566 – 11a.

7 http://www.swafea.eu.

8 Simacek P., Kubicka D., Sebor G., Pospisil M. (2010) Fuel properties of hydroprocessed rapeseed oil, *Fuel* **89**, 609-615.

9 Lapuerta M., Villajos M., Agudelo J.R., Boehman A.L. (2011) Key properties and blending strategies of hydrotreated vegetable oil as biodiesel for Diesel engines, *Fuel Processing Technology* **92**, 2406-2411.

10 Ballerini D. (2011) *Les Biocarburants – Répondre aux défis énergétiques et environnementaux des transports*, Éditions Technip, Paris, France.

11 Huber G.W., O'Connor P., Corma A. (2007) Processing biomass in conventional oil refineries: production of high quality Diesel by hydrotreating vegetable oils in heavy vacuum oil mixtures, *Applied Catalysis A: General* **329**, 120-129.

12 Simacek P., Kubicka D., Kubicková I., Homola F., Pospisil M., Chudoba J. (2011) Premium quality renewable Diesel by hydroprocessed of sunflower oil, *Fuel* **90**, 2473-2479.

13 Bloch M. (2010) State-of-the-Art BTL and HVO Production Processes and the Challenges for the Production of Bio-based Jet Fuel, *SWAFEA 2nd European Stakeholder Conference*, Munich Germany, July.

14 Bauldreay J.M., Heins R.J., Smith J. (2003) Synthetic jet fuels and their role in the future, *8th International Conference on Stability, Handling and Use of Liquid Fuels*, Steamboat Springs, CO, USA, 14-19 Sept.

7

CO_2 Storage Feasibility: A Workflow for Site Characterisation

Manuel Nepveu[1], Filip Neele[1]*, Florence Delprat-Jannaud[2], Maxine Akhurst[3], Olivier Vincké[2], Valentina Volpi[4], Ane Lothe[5], Suzanne Brunsting[6], Jonathan Pearce[3], Anne Battani[2], Axelle Baroni[2], Bruno Garcia[2], Cor Hofstee[1] and Jens Wollenweber[1]

[1] TNO, Princetonlaan 6, 3584 CB Utrecht - The Netherlands
[2] IFP Energies Nouvelles, Rueil-Malmaison - France
[3] BGS, Nottingham - United Kingdom
[4] OGS, Trieste - Italy
[5] SINTEF Petroleum Research, Trondheim - Norway
[6] ECN, Amsterdam - The Netherlands
e-mail: filip.neele@tno.nl

* Corresponding author

Abstract — *In this paper, we present an overview of the SiteChar workflow model for site characterisation and assessment for CO_2 storage. Site characterisation and assessment is required when permits are requested from the legal authorities in the process of starting a CO_2 storage process at a given site. The goal is to assess whether a proposed CO_2 storage site can indeed be used for permanent storage while meeting the safety requirements demanded by the European Commission (EC) Storage Directive (EU, 2009, Storage Directive 2009/31/EC). Many issues have to be scrutinised, and the workflow presented here is put forward to help efficiently organise this complex task.*
Three issues are highlighted: communication within the working team and with the authorities; interdependencies in the workflow and feedback loops; and the risk-based character of the workflow. A general overview (helicopter view) of the workflow is given; the issues involved in communication and the risk assessment process are described in more detail. The workflow as described has been tested within the SiteChar project on five potential storage sites throughout Europe. This resulted in a list of key aspects of site characterisation which can help prepare and focus new site characterisation studies.

Résumé — Faisabilité du stockage géologique du CO_2 : une méthodologie pour la caractérisation des sites de stockage — Dans cet article, nous présentons un aperçu de la méthodologie développée dans le projet SiteChar pour la caractérisation et l'évaluation des sites de stockage de CO_2. Caractériser et évaluer le site potentiel de stockage sont deux étapes nécessaires dès lors qu'un permis de stockage est requis par des autorités légales dans le processus de lancement d'un projet de stockage de CO_2 sur un site donné. L'objectif est d'évaluer si le site de stockage de CO_2 proposé peut en effet être utilisé à des fins de stockage permanent tout en répondant aux exigences de sécurité imposées par la Directive sur le stockage de CO_2 de la Commission Européenne (CE) (UE, 2009, *Storage Directive* 2009/31/EC). De nombreux aspects doivent

être investigués, et la méthodologie présentée dans ce papier apporte une aide pour organiser efficacement cette tâche complexe.
Trois points sont soulignés dans cet article : la communication au sein de l'équipe de travail et avec les autorités, les interdépendances entre les différentes étapes de caractérisation et les retours d'information et la spécificité de l'approche basée sur une analyse de risques. Une vue d'ensemble de la méthodologie est donnée ; les questions liées à la communication et au processus d'évaluation des risques sont décrites plus en détail. La méthodologie, telle que décrite dans le papier, a été testée dans le projet SiteChar sur cinq sites de stockage potentiels répartis en Europe. Il en a résulté des éléments clés de la caractérisation de site de stockage de CO_2 susceptibles d'aider à préparer et investiguer de nouvelles études de caractérisation de site.

INTRODUCTION

The large-scale introduction of Carbon Capture and Storage (CCS) for fossil-fuelled electricity generation and at large industrial plants is needed in order to curtail CO_2 emissions and help prevent future adverse consequences due to the effects of climate change (IEA, 2013). The storage capacity of deep geological (sedimentary) formations is sufficient to store CO_2 emission for several decades into the future (IEA, 2013). Storage capacity is available in depleted gas and oil fields and in deep saline formations. It is essential for the development of large-scale CCS that a sufficient reserve of proven and qualified storage capacity is available at any time, to provide certainty of storage for capture plants (Bachu *et al.*, 2007; Vangkilde-Pedersen *et al.*, 2009; Neele *et al.*, 2012).

The development of a storage site, which includes exploration, characterisation and infrastructure development for CCS, is a time-consuming and complex process. While the development and building of a capture plant is the most capital-intensive part of a CCS project, the long lead time associated with the development of a storage site is likely to constrain the timing of its development. It is therefore essential to start characterising the storage sites as early as possible in the development of CCS projects.

For any practical case the emphasis naturally lies on the operational issues. The more abstract workflow and its issues are not always described in full. For instance, in the Goldeneye project (ScottishPower CCS Consortium) documentation on purely technical issues is abundant, while workflow issues are barely addressed (UK Carbon Capture and Storage Demonstration Competition, 2011). Also, in the QUEST project (Canada) by *Shell* no information of this more abstract nature was found (www.//shell.ca/en/aboutshell/our-business-tpgk/upstream/oil-sands/quest.html).

In this article, then, we address the more generic points of such a workflow.

One of the central goals of the European Commission 7th Framework Research (EU FP7)-funded project SiteChar was to develop a workflow for site characterisation and risk assessment studies for the store of CO_2 under the EU Storage Directive (EU, 2009)[1]. It defines the work to be done to comply with the EU Storage Directive (EU, 2009), resulting in efficient site characterisation studies, to be presented to the authorities when applying for a storage permit. This workflow was applied to five potential storage sites (Akhurst *et al.*, 2015; Beaubien *et al.*, 2015; Gruson *et al.*, 2015; Brunsting *et al.*, 2015; Volpi *et al.*, 2015); the experience thus acquired was used to update and improve the workflow.

This paper provides a summary of the workflow, highlighting issues that are of paramount importance in a well-conducted characterisation and assessment study. The lessons learnt during the actual application to potential storage sites are discussed. This article presents an overview; for details the reader is referred to the original report (Neele *et al.*, 2013).

1 BACKGROUND

Several studies have been completed that address site characterisation for geological CO_2 storage (SAMCARDS, 2003; CO2CRC, 2008; DNV, 2009; NSBTF, 2009; NETL, 2010; Neele *et al.*, 2011; Wollenweber *et al.*, 2013). The studies describe, to a varying level and degree, the work to be done to include all aspects relevant to safe and secure geological storage of CO_2 in a specific formation. In these reports, the central role of risk assessment and management is generally emphasised.

As part of the efforts to streamline the development of CO_2 storage sites and the development of CCS in

[1] This workflow was presented in deliverable D1.4 of the EU FP7 project SiteChar (Neele *et al.*, 2013) (henceforth called the SiteChar report).

general, standards for the storage of CO_2 have been published, in Canada (CSA, 2012) and in Europe (DNV, 2012). Both standards cover the complete lifetime of a storage project, from screening until and including site closure.

However, a number of aspects of the site characterisation process are not or only partly covered:

- the sequence of the different steps and the timing of the process. Some of the steps in a site characterisation study are time-consuming and likely to determine the critical path;
- interdependencies and feedback loops within the process. The activities in a site characterisation study are strongly interlinked, creating an iterative process. This calls for an efficient flow of information between the different disciplines;
- the coverage of the different requirements of the EU Storage Directive (EU, 2009) in the process. The EU Storage Directive lists a number of aspects to be assessed for a storage site. While some explanation is given in the Guidance Documents (EU, 2011), no clear explanation is available of how to address the EU Storage Directive aspects with results obtained in a site characterisation study.

The above-mentioned SiteChar report (Neele *et al.*, 2013) explicitly addresses these points.

2 THREE FOCAL POINTS

Three issues that play a significant role within the workflow are described in the following sections.

2.1 Interplay between the Operator/Competent Authorities

Apart from the technical aspects of defining a workflow there is an important non-technical issue; the interplay between the operator of a prospective site and the "Competent Authorities" (CA) as mentioned in the EU Storage Directive (EU, 2009). The identity of the operator who will perform a site characterisation is clear enough; who the CA are depends on the national laws in force at the site under scrutiny. In the following, it is assumed that the CA are clearly identified in any specific situation. In a formal document such as the EU Storage Directive (EU, 2009) the CA feature at formal moments in the process that may lead to CO_2 storage, for instance when the site operator hands in documents required for a specific license.

Indeed, the characterisation and assessment is required by the EU Storage Directive (EU, 2009) in a formal permit application to the CA for CO_2 storage at a given site. The guidance documents on "Implementation of Directive 2009/31/EC on the geological storage of carbon dioxide" (EU, 2011) are a much welcomed addition as they advise the CA on how to perform their tasks at such moments, and what issues should have their full attention. However, it has become increasingly clear that the contacts between the operator and the CA cannot remain restricted to formal moments in time. Section 4 elaborates on this recommendation.

2.2 Interdependencies and Feedback Loops

The screening process results in a list of potential candidates for CO_2 storage, and in the end one or more specific sites are chosen, assumed to be suitable for storage.

Especially when the various geo-scientific disciplines are invoked in the quantitative aspects of site characterisation, a strictly linear workflow (*i.e.* without loops) is an illusion. For instance, new data or intermediate results may necessitate changes in the 3D static earth model, which then may have consequences for the reservoir engineer studying the fluid dynamic properties of the potential storage strata. These new results may raise further questions for the geologist, leading again to adaptations. The iterations depend on the specific properties of the site and cannot be listed at a generic level.

2.3 Risk-Based Workflow

Risk Assessment is the driver for the characterisation process. It is the process of identifying risks/scenarios that might be adverse to storage site performance and human safety and the environment. The risks as perceived by a multidisciplinary team of experts guide the work to be done. A qualitative overview is the starting point, based on general and site-specific knowledge concerning features, events and processes that may play a role. Qualitative Risk Assessment accompanies the characterisation process all along. If new and unexpected risks turn up in the course of the quantitative work it will feature as part of a feedback loop. The qualitative and quantitative aspects of risk assessment are intertwined. Section 5 deals with the Risk Assessment (RA) process.

3 THE WORKFLOW IN HELICOPTER VIEW

Figure 1 shows a graphical representation of the workflow. The arrows represent the flow of the work activities and of information. The figure contains a number of iterations. These are feedback loops, schematically shown in the figure through arrows that point back towards an

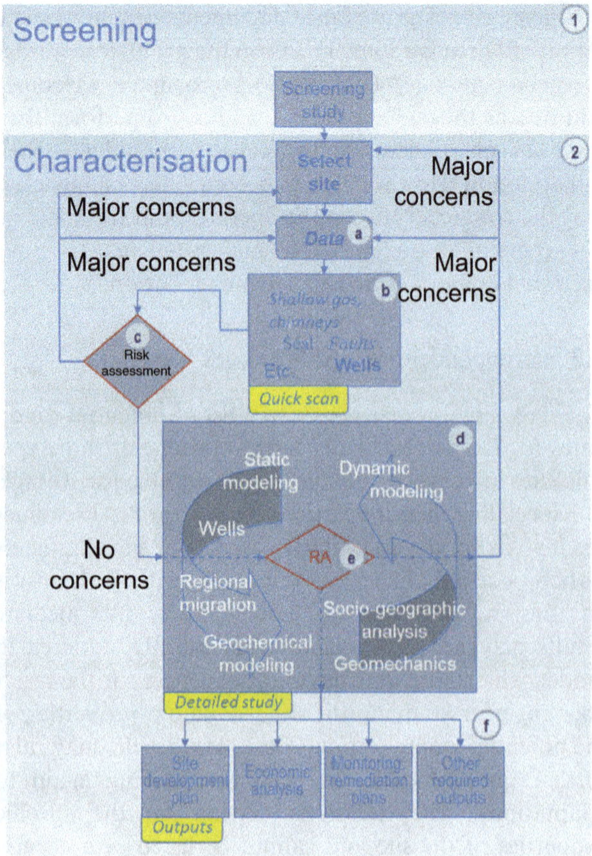

Figure 1

Workflow of site screening and characterisation. The numbers and letters in circles refer to the list in Section 3.

'earlier' stage in the general flow. Diamonds represent RA decision points.

A site characterisation study generally starts with a screening and selection study of potential sites, in which the options for storage in a given area or region are investigated.

A site characterisation study is similar to work done in oil and gas exploration, although in the case of CCS the focus and area of study are different. While in oil and gas exploration the emphasis is put on the reservoir, a CO_2 storage feasibility study must qualify the storage complex, which includes not only the reservoir, but also the cap rock and the overburden and side burden strata. In the case of CO_2 storage, the ability of a geological structure to trap and retain CO_2 permanently must be demonstrated. Given the geological uncertainties, the aim of a site characterisation study is to estimate the risks that accompany CO_2 storage in a given storage complex and whether remediation programmes can be conducted.

The areas of expertise that must be covered by the team (and that are shown in the large box in *Fig. 1*) include:

- structural geology/sedimentology/petrophysics,
- reservoir engineering,
- geomechanical modeling,
- geochemical analysis and geochemical modeling,
- well engineering,
- risk assessment,
- social analysis,
- environmental impact assessment.

Apart from these areas, additional areas of expertise may be required to obtain further results to prove a site's suitability for storage:

- economic analysis,
- engineering and design of injection facilities.

3.1 Screening Study

This is a high-level investigation of all options for CO_2 storage in a specific area or region. This screening may be undertaken by operators or by the CA in preparation for leasing potential areas for storage. Typical screening criteria are derived from CO_2 storage itself (such as depth of the formation), from the capture installation (volume of CO_2 to be stored, rate, timing), and economic considerations (distance from the capture plant, cost of storage, other uses of the pore space). RA starts already in the screening phase, as any risks perceived at this stage must be taken into account; these include the existence of old and/or abandoned wells and interference with other activities in the subsurface and at the surface. Other aspects should also be included at this stage, such as environmental and societal restrictions. Experts form an opinion on available data and use knowledge of a general nature. Overall geoscientific knowledge of the region is an important part of the input and of the decision-making. At present some general rules of thumb are available that allow some rough preliminary estimates; see, for example (Ramirez *et al.*, 2010).

The expected output of the screening phase is a list of promising potential storage sites. The next step, characterisation and assessment, is intended to either elevate such sites to the status of "suitable", or dismiss them.

3.2 Site Characterisation Study (Including Assessment)

The elements of the site characterisation are described in some detail, following the numbering in Figure 1.

a) The first step in the characterisation study is to collect all available data on the site, in addition to the data collected for the screening phase. For a depleted hydrocarbon field, there is usually not much shortage of existing data. Well data, production data and reservoir models may be available. For saline formations, the situation may be quite different. In some cases, the saline formation is associated with hydrocarbon production and wells may penetrate the formation, with well logs and other data available. In the case of a virgin formation, with few or even no wells penetrating the formation, this first step might involve active data collection; shooting a seismic survey; and collection of data from publications or observations of the formation, where exposed, or of similar formations. The role of the CA is to ensure that the data collected are sufficient to give potential evidence of the storage prospect. The CA should view the data with respect to their completeness for the characterisation and assessment as intended;

b) A quick analysis then follows. The aim of this step is to identify any problems related to the site before the study is continued. All available data are scrutinised, so as to find anything that could impede safe and secure storage, or that could affect the site's ability to meet the storage requirements;

c) A qualitative RA has to be done as a first step in the characterisation phase. The quick analysis is followed by a workshop with the specialists from the team, who define the risks associated with the site. These risks are related to the safety and security of storage, as well as the conformity with storage requirements. The aim of this step in the workflow is to identify whether there are aspects that render storage at the site unviable, and whether additional data is to be collected. A semi-quantitative step based on expert input may point to the risks with the highest probability of occurrence (Nepveu *et al.*, 2009);

d) When this first qualitative RA is passed, the site is studied and modeled in the different areas of expertise. This is represented by the large box labelled 'Detailed study' in Figure 1. The figure lists a number of highlights from the respective areas. This is the most time-consuming and also the most complex part of the study, requiring intensive interaction within the team;

e) Once all aspects of safe and secure storage have been studied and once internal consistency in results and data is reached, the risk analysis can be made quantitative, *i.e.* HSE analysis can be performed (HSE stands for Health, Safety, Environment; HSE analysis concerns itself with quantifying risks in these areas). Risks are compared with an *a priori* determined risk threshold. Adequate mitigation actions are then envisaged so as to reduce risks. However, if risks are too high and mitigation measures cannot be taken or are too expensive, the site shall be discarded[2].

f) When the risks are deemed acceptable, the last elements of a site characterisation study can start. These elements include setting up a monitoring plan and baseline studies, drafting a site development plan, and analysing the costs of storage. The monitoring plan is a requirement for a storage site, defined in the Storage Directive (EU, 2009) and so is the development of a "corrective measures" plan, based on hazards that might occur. The site development plan is part of the activities of the future operator, but not formally required by the Storage Directive (EU, 2009).

The characterisation and assessment study is **risk-based** as well as **site-specific**. The (site-specific) qualitative risk assessment will act as a guide that pervades the study in all respects; it is "leading". Those scenarios that may lead to significant irregularities and are quite possible in the given, site-specific situation have to be investigated in detail.

The work of the expert team is to produce a so-called risk matrix. This is a diagram in which risks are positioned along two axes: probability *versus* severity. This graphic method indicates which risks have to be scrutinised in full measure; Figure 2. Produced in the first version in step 'c)' above, this risk matrix will guide the numerical work in the subsequent steps.

The required level of detail of scrutiny should be clear, as well as which theories and approximations of the different parts of the investigation are appropriate to reduce the uncertainties to acceptable levels.

When quantitative detailed analyses are undertaken, it is quite possible that new risks are discovered. In fact, any numerical investigation is not only directed at "getting numbers", but also at getting a fuller picture of what happens, and which processes unfold. If and when such new risks arise, the risk assessment and characterisation process has to be reiterated. For obvious reasons, the CA should be informed of the status and of changes in the risk matrix. In any case, the CA and the operator should agree on the scope and focus of the site characterisation work, so as to smooth the process, and avoid undue delays at the formal moments in the process of working towards a permit application.

[2] Note: in the SiteChar workflow HSE analysis is not described. It is a stand-alone activity that follows the RA and characterisation work, using CO_2 fluxes and timescales of surface emissions as input. Its outcome will not influence the other parts of the workflow by feedback loops.

Probability \ Severity	Very low	Low	Medium	High	Very high
Very high	0	0	0	0	0
High	0	2	4	2	0
Medium	0	4	18	8	0
Low	0	3	20	13	0
Very low	0	0	4	1	0

Figure 2

Risk matrix: probability *versus* severity. As an example, the number of risks is reproduced; in a full risk matrix the risks are described in detail.

4 COMMUNICATION ISSUES

4.1 Interplay between the Operator/Competent Authorities

In the EU Storage Directive (EU, 2009) the CA feature at formal moments in the process that may lead to CO_2 storage. However, it has become increasingly clear that the contacts between the operator and the CA cannot remain restricted to formal moments in time. The annexes to the EU Storage Directive (EU, 2009) suggest a sizeable programme of research to be conducted, However, in any specific situation some aspects of the storage complex need to be studied in more detail than others. It has to be decided in an interplay between the operator and CA which research is deemed necessary and sufficient in order to comply with the requirement of the Storage Directive (EU, 2009) that a geological formation and its surroundings shall be characterised and assessed as suitable for storage. No significant risk of leakage, and no significant environmental or health risk should exist for such a formation to be eligible as a storage site. For this reason, the interplay between the operator and CA should have a more continuous character, the formal moments in the process remaining as they are. Continuous interplay will lead to a better understanding by the CA of the site-specific characteristics; it will lead to a focus on the activities to carry out for a permit application – what the operator should deliver at the formal moments in the process. The risk matrix will act as the central item to be discussed between the CA and the operator.

Characterising and assessing a prospective site is rather different from characterising and assessing a chemical plant. Whereas the construction and structure of such a plant can be known to the finest detail, the subsurface clearly cannot. No amount of data can render a 3D earth model undisputed to the finest detail. There is still the well-known dichotomy of high-resolution/strictly localised data (well-logging data) and low-resolution/global data (seismic survey data). Much, then, has to be filled in from the geologist's general background knowledge. The consequence is that a degree of subjectivity enters the process which can never be eradicated in full measure. Therefore, it is advocated to consider the characterisation process as a mutual concern of the operator and CA, even though these parties have different formal roles. The CA must build up trust that the operator is "doing the right things", given what is known of a site. The CA must have experts at their disposal who can take up such a communicative role.

4.2 Interplay of Disciplines

Interdependencies and feedback loops in the workflow play a significant role. This entails that communication between the different disciplines (their experts) must be attributed great significance. The interaction starts at the qualitative RA, which is necessarily a multidisciplinary investigation. When the quantitative work starts, there is the threat of different scientific disciplines becoming too isolated. Certainly, there are necessarily contacts between the disciplines on questions of desired input and output, but after that has been cleared, there is a phase where every discipline acts on its own. Here lies the danger of experts working in "splendid isolation".

The best way to mitigate this danger is to have a small committee installed that organises regular meetings with all the workers. These meetings will keep everybody informed on the development of the site characterisation, provide the information on whether the study is properly focused, and consider implications from results of one discipline on another field of survey. It should be decided beforehand how often such meetings are held.

These regular meetings act as a bond and facilitate the contact between different disciplines. This streamlines the workflow. Action plans must be formulated, as the discussions should lead to specific activities for the specialists involved. At the meetings the risk matrix (see *Sect. 3*) plays a central role. Are the known risks being brought down to acceptable levels? Have new risks been found as a result of the quantitative work? The whole team involved in the workflow thus stays informed of results and considers their implications as soon as they are obtained.

5 THE RISK ASSESSMENT PROCESS

5.1 Risk-Based Character of Site Characterisation

This section describes technical aspects of the workflow on a generic basis. The workflow is risk-based, with three generic risks: leakage through the seal, the wells, and along the faults. The workflow must first model the subsurface and the man-made structures (wells) in "standard mode" in which no leakages of any kind occur. Only when this preliminary phase has led to some understanding, and model(s) have been successfully obtained can the effects of the uncertainties established in the qualitative RA be studied. This leads to the definition of some particular risks.

5.2 Data Completeness, Sensitivity Analysis

Input data are never enough to fully constrain a model of the 3D subsurface. Uncertainties are generally large, as data derived from wells represent point measurements in a (very) large 3D volume, or represent measurements of subsurface properties taken from the surface (in the case of seismic data). Hence, we cannot rely on just a single digital representation of the storage complex. The key tool to address this issue is sensitivity analysis. However, by conducting a sensitivity analysis we touch on a generic problem: how many runs have to be performed, and how many 3D models are required? A provisional answer can be given with the help of the theory of dimensional analysis, displaying the order of magnitude of the work, *i.e.* the number of models to be run (Hanche-Olsen, 2004). Here this amount of work will become prohibitive, if each uncertainty has to be taken into account in full measure. Choices must necessarily be made. Hence, proper discussions between the operator and CA about the depth of the sensitivity analysis are recommended.

5.3 Continuous RA during Site Characterisation

Site characterisation will be undertaken in roughly this order:

1. data acquisition,
2. quick analysis,
3. qualitative RA,
4. geological assessment (static model building),
5. geomechanical assessment and potential for induced seismicity,
6. dynamic behaviour,
7. geochemical assessment,
8. migration path analysis,
9. well integrity analysis,
10. consequence analysis (health safety and environment, HSE).

Step 4 leads to steps 5 to 9, and steps 7, 8 and 9 can largely be performed in parallel. Steps 5 and 6 tend to be strongly linked. The given ordering is not carved in stone, but more important: returning to an earlier issue might be necessary on account of later emergence of unanticipated risks. Hence: we have to deal with a continuous risk assessment process.

5.4 Quantitative RA

In a quantitative RA, each of the different scenarios has to be considered and "carried" all the way through the steps identified above. The computations and results in one area of expertise should deliver relevant input for the next step. For each step the following questions are to be addressed:

- Which description ("theory", "mathematical model") applies in the different fields of expertise? What effects can be neglected? What effects should be included in any case? Which degree of accuracy is consistent with the available knowledge?
- What is the uncertainty associated with each parameter? It may become clear that many essential parameters are not well known. It is important, then, to perform a sensitivity analysis to find out which parameters most significantly affect the simulation results, and therefore merit further characterisation aiming at reducing their uncertainty. Intimately connected is the question of how many runs should be performed to cover the parameter space adequately;
- Which tools ("software packages") must be used? Tools must be robust and precise. The actual models may be only global in nature, nevertheless the calculations should give trustworthy results. Actually, global models, displaying trends rather than many, often uncertain details, are easier to handle in flow calculations. In addition, these models, when adequate, have the added bonus of producing precise results with much less computational effort.

The different steps have to interface with each other. If unexpected phenomena turn up within one of the disciplines it might be decided that a feedback loop should be established between two or more disciplines, complicating the simple linearity suggested above. In order to make this possible, it is vital to heed the points made in Section 4 on communication among the various disciplines.

5.5 Various Remarks

First, a few technical matters are addressed:

- in the end each model (each representation of the 3D subsurface) is analysed in the various steps. Thus, ultimately the information obtained on subsurface CO_2 migration and pressure development yields CO_2 flux densities and relevant timescales at the surface, if they occur at all; at that point HSE questions have to be addressed. Ideally many computations would have to be performed so that for all parameters that are indispensable for this task, one can establish a (joint) probability density function. This density function contains all information on uncertainties of the relevant parameters and their correlations. For instance, atmospheric CO_2 dispersion simulations can be done on the basis of the information in that density function about flux densities and the surface area and timescales involved;
- the concept of "probability" is important for the work presented here; it introduces itself naturally into the RA process that should deal with uncertainties. The SiteChar workflow implicitly adheres to the Bayesian interpretation of probability which accounts for our current state of knowledge, rather than "frequencies" of occurrence. The Bayesian interpretation is most natural within the workflow as proposed.

Finally, there is an important legal issue that should be pointed out: Article 4.4 of the Storage Directive (EU, 2009) states that "A geological formation shall only be selected as a storage site, if under the proposed conditions of use there is no significant risk of leakage, and if no significant environmental or health risks exist". Clearly, the intention is permanent storage. How the assessors should demonstrate this status is not specified in the directive. In fact, it is not specified what "significant risk" should mean in practical terms. The level of abstraction of the Storage Directive (EU, 2009) forces the national authorities to interpret and understand the results of the site characterisation study in terms of 'significant risk'. For a more complete list of points that need a "translation" in practical terms the reader is referred to Lako *et al.* (2011).

5.6 Critical Path of Site Characterisation Study

The critical path of a RA project is defined by the acquisition of the necessary data plus the modeling sequence of static geological assessment and dynamic simulation of the CO_2 injection process.

The experience from the SiteChar partners is that steps 1 (data acquisition), 4 (static model building) and steps 5 and 6 (geomechanical assessment and dynamic behaviour, respectively) often define the duration of the site characterisation process.

6 KEY ISSUES IN A SITE CHARACTERISATION PROCESS

The Sitechar workflow describes the tasks done within the different disciplines, the required input and output, the interconnections, and the interactions among the various disciplines. The lessons learnt by actually performing site characterisation on the different SiteChar sites are summarised here. In part they reiterate issues already discussed, in part they are technical.

6.1 Key Points Related to Static Model Building

Static model building is particularly important, since its output is the basis of all further steps in site characterisation. The better the static model, the more relevant the subsequent steps in the various disciplines.

Data. High-quality data are needed in sufficient abundance and with a spatial distribution that allows characterisation of the various geological components of the storage site. Data from hydrocarbon wells are of particular importance as well as the coverage and vintage of the seismic data. Data on the petrophysical characteristics of lithological formations obtained from the wells are very important. Spatially unequal coverage is a serious drawback.

How many static models? Multiple static models might be called for if the structural configuration of the site is not entirely clear (*e.g.*, are there faults clearly seen from the data, or is it a matter of interpretation?). This clearly adds to the workload. An alternative is to successively update the static model as deemed necessary from the inputs/requirements of the other disciplines.

Model extent. The model should encompass the volumes that might contain CO_2 in steady state. A practical way might be to construct a detailed model of the reservoir (containment volume) with its overburden/underburden, whereas the laterally equivalent strata are modeled on a coarser grid. Potential CO_2 migration paths should be modeled with different tools, and the results compared and discussed.

Model resolution. The resolution of the model depends on the data resolution. However, it should be kept in mind that if lateral resolution is not an issue, simulation of CO_2 migration calls for a grid with coarse-scale horizontal resolution. Grid refinement at and around wells/faults should be possible. It would be a bonus if the static model could be updated easily in this respect. Vertical gridding could be coarse in the overburden/underburden, depending on the geological variability.

Software used. Commercially produced standard industry software is strongly recommended in the different fields of expertise (such as static model building and dynamic modeling). That software is widely used, has been developed and tested extensively and the export facilities to different tools and formats are well developed. Nevertheless, exchange of data between different software tools remains a crucial issue. This should be tested and proven at the very beginning of the RA activities.

Collaboration between different experts. This issue has been dealt with earlier on a general level. A number of practical, related issues are listed here:

1. different software as used by different research groups can require model manipulation for obtaining compatible formats. Distortions of the original can result;
2. the static model has to be built by working together with the dynamic modelers from the start of the project. Fluid simulations can be quite sensitive to vertical grid and model cell attribution; this point should be taken up in this cooperation;
3. exchange of information on different software proposed to be used has to take place as soon as possible. Important in this respect are the issues on input/output formats;
4. importantly: the CO_2 storage concept should be well known by the team, and the overall research should be planned carefully, and this requires the relevant site-specific questions to be asked. This emphasises the need for a good basic understanding of the proposed storage site, and this again shows the importance of a thorough qualitative risk assessment early on.

Data acquisition:

1. licensing and acquisition of data can take a long time, up to several months, and should be started as soon as possible. Significant effort may be needed to review and select the data;
2. significant cost may be involved in acquiring new data (such as seismic survey data, drilling a well and acquiring well logs);
3. data might have unequal areal coverage;
4. there are always conversion issues (formats), especially with legacy data. Converting data to formats used by modeling tools may require significant effort;
5. in depleted oil/gas fields there are abundant data, but sifting them as to relevance requires quite some effort. In saline aquifers, there are not so much data already available generally. They have to be obtained and in that process some assumptions must be made (*e.g.* on spatial sampling resolution).

Altogether the static model building appears a most critical phase: data gathering and contacts with different specialists from other disciplines make it a time-consuming step in the site characterisation process.

6.2 Key Points Related to Dynamic Modeling

The following points merit special attention in dynamic modeling:

1. Coordination between different experts. There is a need for close interaction with the static modelers, in order to streamline and iterate parameters for the dynamic simulation scenarios. Close interaction with geomechanical modelers is also needed, to agree on injection scenarios, and the use of compatible modeling software;
2. It is advisable to consider dynamic and geomechanical modeling as coupled processes that must be modeled simultaneously;
3. The injection scenarios should be discussed and agreed with the operator. There is a direct link between cost of injection and the choice of injection scenarios, through the number of wells and injection sites involved;
4. It is important to define appropriate boundary conditions, especially when comparing regional and local model simulations;
5. Data availability is always a major concern:
   - the collection of data should start early, as it is a time-consuming process;
   - there is a considerable added value in the participation of the hydrocarbon field owner, to have access to detailed reservoir properties. Especially, relevant information on reservoir heterogeneity is much welcomed, as it is highly relevant input for the static modeling, pressure history and fluid properties for the dynamic modeling;
   - fault behaviour is key information. Here sensitivity analysis is strongly advised as fault properties are not generally well known.

6.3 Key Points Related to Geomechanical Modeling

1. Appropriate pressure and temperature conditions at the reservoir boundaries (or perhaps farther away) need to be known; they serve as boundary conditions in the computations;

2. Availability of data is an issue. In particular:
   - information on fault properties,
   - initial stress state,
   - overburden properties,
   - proneness to seismicity, from past records;
3. Assumptions must be shared explicitly among the team of experts;
4. There is a need for close operation with the dynamic modelers:
   - geomechanical modeling establishes crucial seal rock fracture pressures, which is a constraint on the injection scenario(s). Such modeling may also inform on the likelihood of induced seismicity during injection, and perhaps after closure;
   - coupled dynamic and geomechanical modeling is needed;
   - consistency of models must be ensured;
   - software/format compatibility must be assured before commencing dynamic and geomechanical modeling.

6.4 Key Points Related to Geochemical Modeling

1. A good understanding of the mineralogy of the reservoir is essential for a reliable geochemical study. Core analysis from the immediate vicinity is important. Formation water composition and property data may have to be assumed, if not available;
2. There is a need for the evaluation of geochemical modeling software that accommodates reactions in oil and gas and water-bearing strata (if relevant).

6.5 Key Points Related to Well Integrity Analysis

1. Availability of appropriate data:
   - analysis of the status of existing wells can only be done when real data are available (e.g., status of the cement, well casing);
   - *in situ* observations of wells are required. Such operations are expensive. Cost estimates and timing of such operations should be addressed as early in the process as possible;
   - information on well cement and other well material is often not available. This is problematic and hampers reliable well integrity modeling;
2. There may be a need to study a worst-case scenario where real data is not available and estimate the associated risks, to side-step the problem of a lack of data.

6.6 Key Points Related to Migration Path Analysis

1. The timing and the volume of CO_2 migration should be quantified for a proper risk assessment and to define preventative measures;
2. Information on fault properties is important;
3. An iteration may be needed with dynamic modeling that is related to spill points identified by the migration path analysis.

CONCLUSION

This paper describes the SiteChar project workflow for a site characterisation study for CO_2 storage, as required for a storage permit application under the EU Storage Directive (EU, 2009).

The key points in a characterisation and assessment study are the following:

1. The characterisation study intends to fulfil the obligations laid down in the EU Storage Directive (EU, 2009). Two parties are directly involved: the operator of the prospective site and the so-called Competent Authorities. Next to the formal moments of contact between them, as indicated by the Storage Directive (EU, 2009), it is essential that the parties have regular contact. These contacts will inform the operator on what is expected from them in the study, on the basis of the national implementation of the EU Storage Directive (EU, 2009). They must also lead to a fuller understanding of the prospective site by the CA, who are to define the actions to be performed by the operator. The interaction should speed up the process that will lead to exploration and storage permits when appropriate;
2. The process is risk-based. A preliminary (qualitative) risk assessment is performed in the screening phase; if the potential site meets the requirements, a more thorough investigation of the risks and uncertainties (moving towards a quantitative risk assessment) is undertaken during the site characterisation study. The expert team defines risks and associated adverse scenarios and the work should always be based on the risk assessment and risk ranking. The risk matrix will form the central issue in the contacts between the CA and the operator. The detailed site characterisation, numerical in nature, should be expected to uncover relevant new risks that were not anticipated earlier. These risks must lead to reiteration. It is advisable that the parties involved agree on a protocol to be followed in such cases;
3. The characterisation study is multidisciplinary. It should encompass a quick scan, qualitative risk assessment, static modeling, dynamic modeling, geochemical analyses and modeling, geomechanical modeling, well integrity analysis, migration path analysis, socio-geographic analysis, and quantitative risk analysis. The focus of the activities in each

discipline is strongly site-specific and should be based on the risk assessment;

For completeness the following point is added.

4. Further activities that follow from the characterisation and assessment are drawing up a monitoring plan, a corrective measures plan and a site development plan together with cost estimates. It is to be noted that the monitoring plan is also risk-based and site-specific.

The prime keywords in site characterisation are "risk-based" and "site-specific". This makes it difficult to generically specify all the actions to be undertaken by the investigators as if they are carved in stone: they are not. For the reasons above regular communication between the operator and CA is a practical necessity. In order to speed up the process of site characterisation and assessment such contacts are important as well.

ACKNOWLEDGMENTS

The research leading to these results received funding from the European Union Seventh Framework Programme (FP7/2007-2013) under grant agreement No. 256705 (SiteChar project) as well as from *Enel*, *Gassnova*, *PGNiG*, *Statoil*, *Vattenfall*, *Véolia Environnement* and the Scottish Government.

REFERENCES

Akhurst M., Pearce J., Hannis S., Shi J.Q., Delprat-Jannaud F., Bossie-Codreanu D. (2015) Risk-Led Characterisation of a Site for the Geological Storage of CO_2, *Oil & Gas Science and Technology* **70**, 4, 567-586.

Bachu S., Bonijoly D., Bradshaw J., Burruss R., Holloway S., Christensen N.P., Mathiassen O.M. (2007) CO_2 storage capacity estimation: Methodology and gaps, *Int. J. Greenhouse Gas Contr.* **1**, 430-443.

Beaubien S., Ruggiero L., Annunziatellis A., Bigi S., Ciotoli G., Deiana P., Graziani S., Lombardi S., Tartarello M. (2015) The Importance of Baseline Surveys of Near-Surface Gas, Geochemistry for CCS Monitoring, as Shown from Onshore Case Studies in Northern and Southern Europe, *Oil & Gas Science and Technology* **70**, 4, 615-633.

Brunsting S., Mastop J., Kaiser M., Zimmer R., Shackley S., Mabon L., Howell R. (2015) CCS acceptability: Social Site Characterization and Advancing Awareness at Prospective Storage Sites in Poland and Scotland, *Oil & Gas Science and Technology* **70**, 4, 767-784.

CO2CRC (2008) Storage capacity estimation, site selection and characterisation for CO_2 storage projects, CO2CRC report RPT08-1001.

CSA Group (2012) Geological storage of carbon dioxide, Standard Z741-12.

DNV (2009) CO_2 Qualstore – Guideline for selection and qualification of sites and projects for geological storage of CO_2 DNV report 2009-1425.

DNV (2012) Geological storage of carbon dioxide, DNV-RP-J203.

EU (2009) Storage Directive 2009/31/EC, http://eur-lex.europa.eu/LexUriServ/LexUriServ.do?uri=OJ:L:2009:140:0114:0135:EN:PDF.

EU (2011) Implementation of Directive 2009/31/EC on the geological storage of carbon dioxide–Guidance document 2 – Characterisation of the storage complex, CO_2 stream composition, monitoring and corrective measures (ec.europa.eu/clima/policies/lowcarbon/docs/gd2_en.pdf).

Gruson J.-F., Serbutoviez S., Delprat-Jannaud F., Akhurst M., Nielsen C., Dalhoff F., Bergmo P., Bos C., Volpi V., Lacobellis S. (2015) Techno-Economic Assessment of Four CO_2, Storage Sites, *Oil & Gas Science and Technology* **70**, 4, 753-766.

Hanche-Olsen H. (2004) Buckingham's pi-theorem. http://www.math.ntnu.no/~hanche/notes/buckingham/buckingham-a4.pdf.

IEA (2013) Technology roadmap – carbon capture and storage, Paris, available at http://www.iea.org/publications/freepublications/publication/CCS_Roadmap.pdf.

Lako P., van der Welle A.J., Harmelink M., van der Kuip M.D.C., Haan-Kamminga A., Blank F., de Wolff J., Nepveu M. (2011) Issues concerning the implementation of the CCS Directive in the Netherlands, *Energy Procedia* **4**, 5479-5486.

Neele F., Nepveu M., Hofstee C., Meindertsma W. (2011) CO_2 storage capacity assessment methodology, TNO report TNO-060-UT-2011-00810, report available at http://www.globalccsinstitute.com/publications/co2-storage-capacity-assessment-methodology.

Neele F.P., ten Veen J., Wilschut F., Hofstee C. (2012) Independent assessment of high-capacity offshore CO_2 storage options, TNO report TNO-060-UT-2012-00414/B, report available at http://www.globalccsinstitute.com/publications/independent-assessment-high-capacity-offshore-co2-storage-options.

Neele F., Delprat-Jannaud F., Akhurst M., Vincke O., Volpi V., Nepveu M., Hofstee C., Wollenweber J., Lothe A., Brunsting S., Pearce J., Battani A., Baroni A., Garcia B. (2013) SiteChar, Characterisation of European CO_2 Storage, Deliverable No D1.4 SiteChar Characterisation Workflow.

Nepveu M., Yavuz F., David P. (2009) FEP Analysis and Markov Chains, *Energy Procedia* **1**, 2519-2523.

NETL (2010) Site screening, selection and initial characterisation for storage of CO_2 in deep geological formations, DOE report DOE/NETL-401/090808.

NSBTF (North Sea Basin Task Force) (2009) Monitoring Verification Accrediting and Reporting (MVAR) Report for CO_2 storage deep under the seabed of the North Sea.

Ramirez A., Hagedoorn S., Kramers L., Wildenborg T., Hendriks C. (2010) Screening CO_2 options in the Netherlands, *Int. Journal of Greenhouse Gas Control* **42**, 367-380.

SAMCARDS (2003) Safety Assessment Methodology for CO_2 Sequestration, DOE DE-FC26-01NT41145.

UK Carbon Capture and Storage Demonstration Competition (2011) ukccs-kt-s7.24-shell-001-operations-philosophy.pdf, ScottishPower CCS Consortium.

Vangkilde-Pedersen T., Lyng Anthonsen K., Smith N., Kirk K., Neele F., van der Meer B., Le Gallo Y., Bossie-Codreanu D., Wojcicki A., Le Nindre Y.-M., Hendriks C., Dalhoff F., Christensen N.P. (2009) Assessing European capacity for geological storage of carbon dioxide – the EU GeoCapacity project, *Energy Procedia* **1**, 2663-2670.

Volpi V., Forlin E., Baroni A., Estublier A., Donda F., Civile D., Caffau M., Kuczynsky S., Vincké O., Delprat-Jannaud F. (2015) Evaluation and Characterization of Potential CO_2 Storage Site in the South Adriatic Offshore, *Oil & Gas Science and Technology* **70**, 4, 695-712.

Wollenweber J., Busby D., Wessel-Berg D., Nepveu M., Bossie Codreanu D., Grimstad A.-A., Sijacic D., Maurand N., Lothe A., Wahl F., Polak S., Boot H., Grøver A., Wildenborg T. (2013) Integrated Carbon Risk Assessment (ICARAS), *Energy Procedia* **37**, 4825-4832.

8

Experimental Study on Hydrate Induction Time of Gas-Saturated Water-in-Oil Emulsion using a High-Pressure Flow Loop

X.F. Lv[1], B.H. Shi[1], Y. Wang[1], Y.X. Tang[2], L.Y. Wang[1] and J. Gong[1]*

[1] Beijing key laboratory of Urban Oil and Gas Distribution Technology, China University of Petroleum (Beijing), Beijing 102249 - China
[2] SINOPEC Luoyang Petrochemical Engineering Corporation, Luoyang 471003 - China
e-mail: ydgj@cup.edu.cn

* Corresponding author

Abstract — *Hydrate is one of the critical precipitates which have to be controlled for subsea flow assurance. The induction time of hydrate is therefore a significant parameter. However, there have been few studies on the induction time of the natural gas hydrate formation in a flow loop system. Consequently, a series of experiments were firstly performed, including water, natural gas and Diesel oil, on the hydrate induction time under various conditions such as the supercooling and supersaturation degree, water cut, anti-agglomerant dosage, etc. The experiments were conducted in a high-pressure hydrate flow loop newly constructed in the China University of Petroleum (Beijing), and dedicated to flow assurance studies. Then, based on previous research, this study puts forward a method for induction time, which is characterized by clear definition, convenient measurement and good generality. Furthermore, we investigated the influences of the experimental parameters and analyzed the experimental phenomena for the hydrate induction time in a flowing system.*

Résumé — Étude expérimentale sur le temps d'induction d'hydrate d'une émulsion eau-en-huile saturée en gaz en utilisant une boucle à circulation sous haute pression — Les hydrates font partie des principaux précipités qui doivent être contrôlés pour assurer le débit d'écoulement. Le temps d'induction des hydrates est donc un paramètre significatif. Toutefois, peu d'études ont été réalisées sur le temps d'induction de formation d'hydrates de gaz naturel dans un système à boucle de circulation. Par conséquent, une série d'expériences a tout d'abord été effectuée, comprenant de l'eau, du gaz naturel et du gasoil, pour évaluer le temps d'induction des hydrates dans différentes conditions, telles que le degré de surréfrigération et de sursaturation, le watercut, le dosage d'anti-agglomérants, etc. Les expériences ont été réalisées dans une boucle de circulation d'hydrates sous haute pression nouvellement réalisée à l'Université Chinoise du Pétrole (Beijing) et dédiée aux études de maintiens d'écoulements. Puis, sur la base de recherches précédentes, la présente étude met en avant une méthode pour évaluer le temps d'induction, qui est caractérisée par une définition claire, un système de mesure pratique et une bonne généralisation. En outre, nous avons étudié les influences de paramètres expérimentaux et analysé le phénomène expérimental pour le temps d'induction des hydrates dans un système à écoulement.

LIST OF SYMBOLS

t_i	Induction time
V	System volume
J	Hydrate nucleation rate
α	Proportional constant
r	Power of the growth rate of the hydrate nuclei
t_e	Time needed to reach an equilibrium state
t_t	Required time for turbidity
t_g	Required time for the critical hydrate nuclei to grow into visible crystals
t_n	Required time to form the critical nuclei
S	Supersaturation degree
A	Kinetic parameters
c	Shape parameter
σ_{ef}	Effective specific surface energy
$\Delta\mu$	Driving force for the nucleation
k	Boltzmann constant
G	Growth rate when the nuclei grow into visible crystals
β	Crystalline volume fraction
α_n	Shape factor
f_{gv}	Gas fugacity
f_{eq}	Equilibrium fugacity
ΔT_{sub}	Supercooling degree
T_{eq}	Three-phase equilibrium temperature
T_{exp}	Temperature of the flowing fluid in the experimental system

INTRODUCTION

Since the first discovery of hydrate in the natural gas transmission lines in the 1930s (Sloan, 2005), hydrate formation and blockage in long deep-water pipelines have long been a problem for offshore petroleum production. As oil and gas production moves towards the deeper water, and the deep sea oil and gas gathering and transportation tend to adopt mixed-transport technology (Shi, 2012), the low temperature and high pressure conditions necessary for hydrate formation between natural gas and water are frequently satisfied in petroleum pipelines. Upon formation, hydrate accumulation and agglomeration ultimately form a plug, blocking the flow through the pipeline. It is costly and dangerous to remove these plugs in the pipeline, which will lead to a significant loss in the production (Sloan and Koh, 2008). Therefore, the formation mechanism of gas hydrate in a flowing system has become a focal point in the hydrate research field, in order to effectively inhibit or prevent the hydrate formation.

The formation process of the gas hydrate is similar to the crystallization process (Sloan and Koh, 2008), including the hydrate nucleation stage and the hydrate growth stage, in which meta-stable crystallization is mainly caused by supercooling and/or supersaturation. The stable hydrate crystal nuclei that exceed the critical dimensions can be formed in the hydrate nucleation stage; then, these stable nuclei grow to the solid state in the hydrate growth stage. There exists quite a period of time, before the burst and the rapid growth of the hydrate nuclei, during which the macroscopic characteristics of the system will not change significantly. This period is defined as the hydrate induction phenomenon, which can be described by the hydrate induction time.

In other words, the induction time reflects the latency time before the hydrate begins to grow massively and rapidly. A long induction time would allow transport of fluids through the pipelines without the hydrate formation in the system; while a shorter period of the induction time might lead to a pipeline plug. Accordingly, the induction time in gas hydrate crystallization is an important characteristic of the kinetics of the process. The induction time is a measure of the ability of a supersaturated system to remain in the state of meta-stability; it could also indicate safe operation for the submarine multiphase pipelines (here safe means there is no hydrate blockage). According to the research reported by Kashchiev and Firoozabadi (2003), Vysniauskas and Bishnoi (1983), and Englezos *et al.* (1987), the occurrence and significance of the induction time have been evidenced and recognized.

However, due to the high meta-stable state existing in the course of the hydrate reaction, and the infinitesimal phenomenon that involves thousands of molecules, hydrate nucleation can hardly be observed accurately through usual experimental methods and/or installations (Mullin, 1993). As a result, there has been some experimental research on the hydrate induction time in autoclave cell systems, but with their experimental data differing greatly from each other (Muller-Bongartz *et al.*, 1992; Parent and Bishnoi, 1996; Bansal, 1994; Nerheim *et al.*, 1994; Cingotti *et al.*, 1999; Kelland *et al.*, 2000). For example, Cingotti *et al.* (1999) observed that the distribution of induction time data was scattered in their experiments in clean conditions, and they designed a specific procedure (1st formation/dissociation/ 2nd formation) to increase the reproducibility of the induction time. It is undeniable that the nucleation is stochastic from a theoretical point of view, but with different experimental procedures and conditions, a narrower induction time distribution can be obtained.

Although this research has made some achievements in this field, there is still a demand for well-performing predictive theoretical models, or a unanimous definition/measuring method for the hydrate induction time. Consequently, there are various definitions of the induction time in view of different experimental methods (Kashchiev, 2000; Kashchiev *et al.*, 1991; Natarajan *et al.*, 1994; Natarajan, 1993; Kashchiev and Firoozabadi, 2002a; Söhnel and Mullin, 1988; Christiansen and Sloan, 1995).

One of the definitions of the induction time started from the equilibrium state to the time point at which the first crystal nucleus, with the critical size and stable property, came into being (Kashchiev, 2000; Kashchiev *et al.*, 1991; Natarajan *et al.*, 1994). Based on this, Kashchiev (2000) proposed a mononuclear theory of gas hydrate nucleation and established a corresponding mathematical model for the mononuclear induction time:

$$t_i = 1/(JV) \tag{1}$$

where t_i is the induction time in the mononuclear theory, and V is the system volume. J is the hydrate nucleation rate, *i.e.* the amount/number of the critical nuclei produced per unit time and volume.

However, Natarajan *et al.* (1994) argued that the induction time should be defined as the time period from the equilibrium state to the time point where a large number of critical nuclei (not the first one) formed. Also, they provided the determination method for a typical induction time based on gas consumption (*Fig. 1*), and a corresponding mathematical model:

$$t_i = \alpha/J^r \tag{2}$$

with the induction time defined as:

$$t_i = t_t - t_e - t_g \tag{3}$$

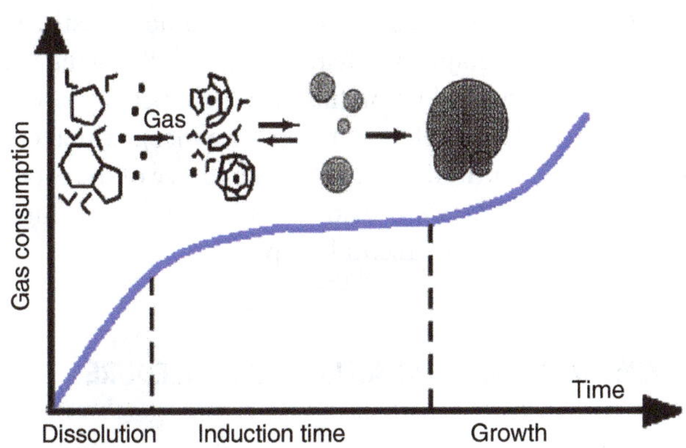

Figure 1

Typical gas consumption plots (Lederhos *et al.*, 1996).

where α is a proportional constant, and r is the power of the growth rate (J) of the hydrate nuclei. t_e is the time needed to reach an equilibrium state for the system, and t_t is the time needed for turbidity. t_g is the required time for the critical hydrate nuclei to grow into visible crystals. At the same time, the experimental deviations are easily calculated from the determination of t_t and t_g in Equation (3).

Furthermore, Natarajan (1993) described the nucleation rate J in Equation (2) using an empirical equation (*Eq. 4*), in which J is a function of the supersaturation degree, according to the crystallization kinetics study (Mullin, 1993).

$$J = k(S-1)^n \tag{4}$$

where k and n are constants, and S is the supersaturation degree.

Natarajan (1993) measured the induction time of gas hydrate crystallization. The result indicated that it could obtain repeatable data from either the structured water or unstructured water under relatively high pressure (>3.5 MPa). Kashchiev and Firoozabadi (2002a) derived an arithmetic expression (*Eq. 5*) for the hydrate nucleation rate through the analysis of nucleation kinetics of single-component gas hydrate in a rich aqueous solution:

$$J = Ae^{\frac{\Delta\mu}{kT}} \exp\left(-\frac{4c^3 v_h^2 \sigma_{ef}^3}{27kT\Delta^2\mu}\right) \tag{5}$$

where A is the kinetic parameters, and c is the shape parameter. v_h is the structure unit volume of hydrate. σ_{ef} is the effective specific surface energy. $\Delta\mu$ is the driving force for the nucleation. k is the Boltzmann constant.

In another definition, the induction time should be counted from the equilibrium state to the time point where plenty of visible crystals appear for the first time in a system (Mullin, 1993; Söhnel and Mullin, 1988). Kashchiev (2000) named this definition the polynuclear theory to distinguish it from mononuclear theory.

Söhnel and Mullin (1988) gave the expression of the induction time as:

$$t_i = t_n + t_g \tag{6}$$

where t_i is the induction time in the polynuclear theory, and t_n is the required time to form the critical nuclei.

Kashchiev *et al.* (1991) combined the mononuclear and polynuclear theories to generate an integrated model to calculate the induction time for the single-component gas hydrate:

$$t_i = 1/(JV) + \left[\beta/\left(\alpha_n J G^{n-1}\right)\right]^n \tag{7}$$

where G is the growth rate when the nuclei grow into visible crystals. β is the crystalline volume fraction and α_n is the shape factor.

However, Christiansen and Sloan (1995) derived another expression according to which the induction time is more sensitive to the growth rate than to the nucleation rate of the hydrate crystallites. They also proposed an empirical formula for the induction time (t_i). The authors cautioned that this formula should not be used outside the range of the experimental data correlated by them.

Maeda *et al.* (2012) experimentally studied the induction time under different supercooling degrees, and obtained a longer induction time at the slower cooling rate.

In view of the fact that the aforementioned research was mainly conducted in autoclave cells under static or stirring conditions, Sun (2001) and Sun *et al.* (2004) fitted an empirical correlation (*Eq. 8*) of the induction time for methane hydrate in a flowing system. The experiments adopted the shielding ratio and pressure drop measurement method. The results showed that the hydrate induction time in a flowing system is not only related to pressure, but also to the flow velocity of the fluid:

$$t_i = K\left(\frac{f_g^v}{f_{eq}}\sqrt{Q/Q_0} - 1\right)^{-m} \quad (8)$$

where Q/Q_0 is the mixing factor. f_g^v is the gas fugacity (calculated by MRK EOS (Chen *et al.*, 2000)). f_{eq} is the equilibrium fugacity at the experimental temperature (calculated by the Chen-Guo hydrate model (Chen and Guo, 1996, 1998)).

Gainville and Sinquin (2011) researched the effect of the water cut, flow rate and flow regime on the induction time in the Lyre loop located at *IFP Energies nouvelles* Lyon. The experimental results indicated that the induction time was reduced with an increase in the water cut and flow rate, and the flow pattern's influence on the induction time was of great significance.

Gaillard (1996) researched the nucleation and growth of methane hydrate in a lab loop based both on crystallization theory and on measurements of gas consumption. The results indicated that the hydrate induction time in a flowing system relied on the influence of pressure, temperature and liquid velocity.

Turner (2005) defined the induction time for hydrate formation in the flow loop as the difference between the time that the loop temperature achieved the hydrate equilibrium temperature at the experimental pressure and the time of hydrate appearance (visual appearance of hydrate in the sight glasses, and a sudden shift in the particle size analyzer's total particle number registered were coincident and were interpreted at the point of hydrate appearance in the loop), and investigated the influence of the pump motor speed, water cut and fractions of gas on the induction time in an *ExxonMobil* flow loop. The experimental results showed that the induction time decreased as the impeller speed increased, but there was an exception to the general trend at the higher shear rates in all low gas fraction experiments. Meanwhile, no strong trend was apparent for the effect of the water cut on the flow loop induction time.

Therefore, the research on the induction time of multicomponent gas hydrate in an axial flow system is still at the experimental stage, which requires a lot of reliable experimental data so that the regularity of the hydrate induction time can be performed.

In summary, the existing definitions of the induction time can be classified from the microscopic view or the macroscopic view, respectively. Let the micro-view denote the definitions based on the critical nuclei; and the macro-view those according to the visible crystals. It can be found that these two measurement methods present different ranges for the induction time. In detail, the induction time ends at the critical nuclei's occurrence in the micro-view, while the broader induction time in the macro-view additionally covers the period from the critical nuclei to the visible crystals. However, it is hard to determine the critical nuclei uniformly under micro-view conditions owing to the diversity of the experimental equipment and/or measuring methods. Consequently, in view of the feasibility, there is limited applicability as well as uncertainty existing in the induction time in the micro-view; and the induction time in the macro-view still has major observation errors even with relatively easier determination.

Therefore, an explicit easily-measured definition of the hydrate induction time is proposed in this paper, which is applicable in an axial flow system. It is expected to supplement the studies in a flowing system with respect to the prevalent/current autoclave cells under static or stirring conditions in the majority. Meanwhile, the influencing factors of this defined induction time, *i.e.* the supersaturation/supercooling degree, flow rate (disturbance), water cut, additive concentration, the geometry and size of the experimental loop, etc., were also investigated in the high-pressure hydrate experimental loop.

1 EXPERIMENTAL APPARATUS AND PROCEDURE

1.1 High-Pressure Hydrate Experimental Loop

Experimental tests were carried out in the high-pressure hydrate experimental loop for Flow Assurance studies (*Fig. 2*). Natural gas and the liquid phase are injected separately by a plunger compressor (2 200 Nm^3/h) and a custom-made magnetic pump (flow rate up to 12 m^3/h) into the loop. It should be noted here that the physical interaction of the magnetic pump with the hydrate particles still cannot be quantitatively characterized due to its complication in the present experimental conditions; and if the particles are smaller than the clearance of the pump, the particles should not be affected by the pump.

Two sight glasses sit in the test sections. The gas injection point is the test section inlet. At the outlet of the test section, gas and liquid are collected in an insulated separator and are redirected toward the test section compressor (from the upper part) and pump (from the bottom), respectively. Several tanks allow maintenance of the loop and separator pressure as hydrate forms.

The 30-m stainless steel test section consisted of two rectilinear horizontal lengths joined together to form a pipe with 2.54 cm (1 inch) internal diameter, and a 5.08-cm (2-inch)-diameter jacket circulating a water-glycol blend surrounded the test section. The process temperature control ranged from −20° to 100°C.

1.2 The Hydrate Experimental Loop Instrumentation

This experimental flow loop is equipped with several sensors. Thermocouples are regularly set along the pipe, inside the separator, inside the water/glycol system and on the different gas utilities. A Coriolis flowmeter measures the density of the liquid mixture and the flow rate. The mean density of the multiphase fluid can also be measured using two FM1000 gamma ray densitometers. Differential pressure sensors are located along the loop to follow the evolution of the linear pressure drop along the loop. Rapid data acquisition is used to detect rapid phenomena.

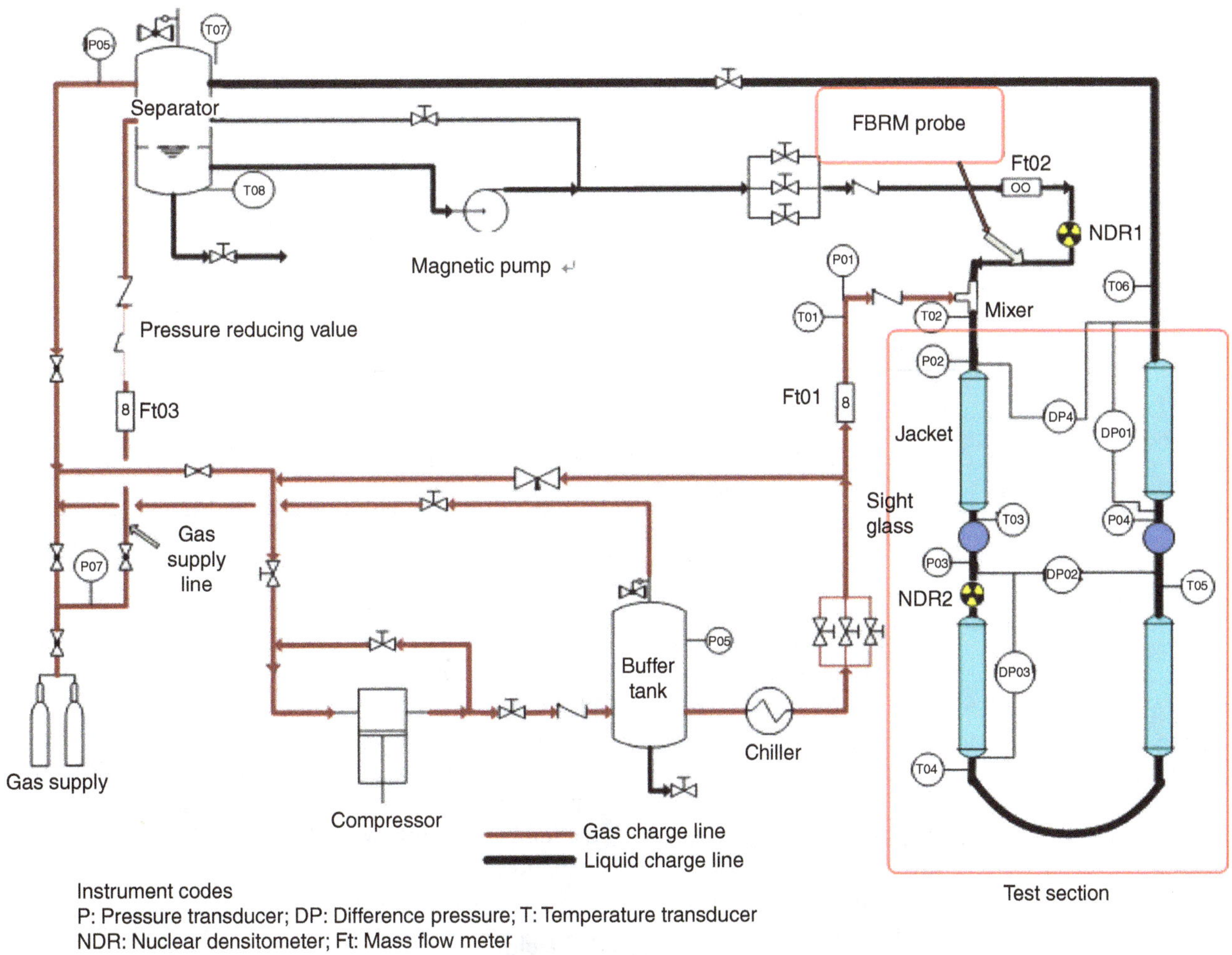

Figure 2

Schematic of the high-pressure hydrate experimental loop.

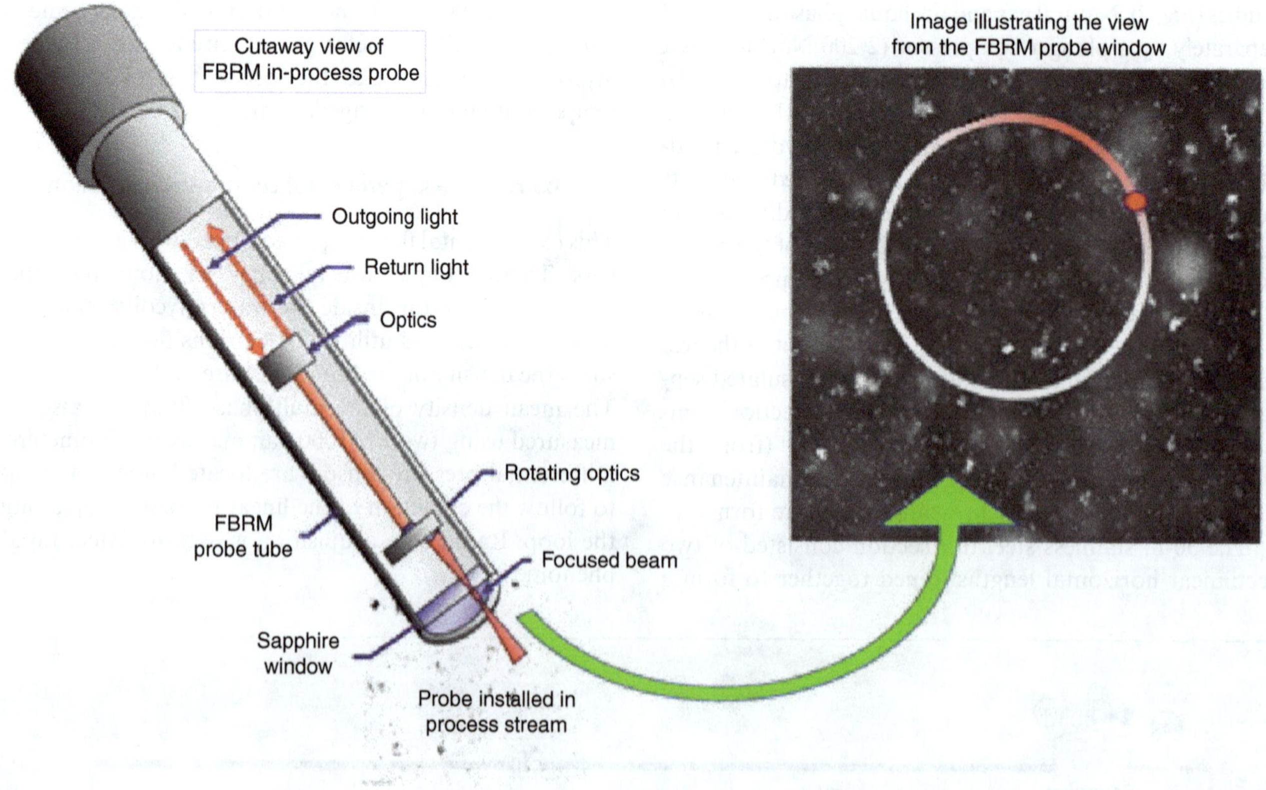

Figure 3

Principle of the FBRM particle size analyzer.

TABLE 1

The composition of gas samples (mol%)

Composition	Mol%	Composition	Mol%
N_2	1.53	C_3	3.06
CO	2.05	iC_4	0.33
CO_2	0.89	iC_5	0.04
C_1	89.02	nC_6+	0.01
C_2	3.07	–	–

TABLE 2

The composition of −20 # Diesel oil

Composition	Mol%	Composition	Mol%
C_{11}	0.89	C_{16}	6.83
C_{12}	3.36	C_{17}	7.99
C_{13}	5.38	C_{18}	7.46
C_{14}	6.2	C_{19}	6.38
C_{15}	6.78	$C_{20}+$	48.73

A Focused Beam Reflectance Measurements (FBRM) probe (*Mettler Toledo Company*) which was installed on the loop, with a measurement range from 0.5 ~ 1 000 μm, measured the evolution of objects (droplets, bubbles and solid particles) carried inside the flow. This apparatus comprises a low-intensity rotating LASER beam (*Fig. 3*) which is reflected when intercepting a particle. From the reflection time a chord length is deduced. A Chord Length Distribution (CLD) and a mean chord length are given every 10 s. This CLD gives the chord length evolution *versus* time of the objects carried by the flow. As recommended for a representative sampling of the particle size distribution, this particle size analyzer was installed on the straight vertical pipe which is ahead of the inlet of the experimental loop. The probe window cuts the flow streamlines at an angle of 45° and reaches the center of the pipe. The FBRM probe was used to estimate the initial water droplet (*Dp*) size inside the fluid and to follow the hydrate particle agglomeration with time.

1.3 Fluid

Testing used deionized water, civil natural gas and −20# Diesel (*Tab. 1, 2*). The Anti-Agglomerants (AA) are a

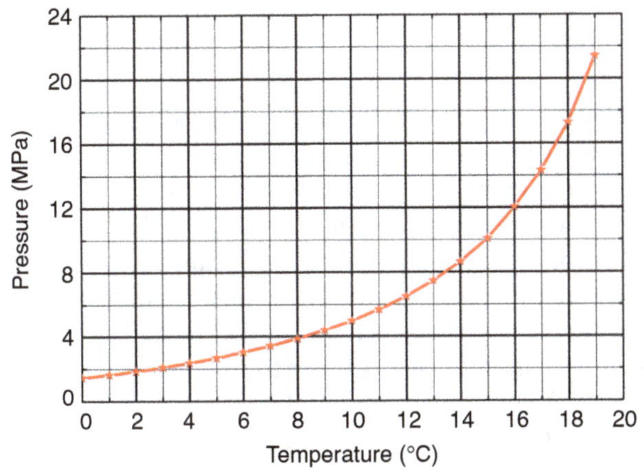

Figure 4

Hydrate formation curve of test natural gas.

formulation of some polymers and surfactants, which can be used in the water and oil mixture to form a W/O emulsion where the water is dispersed as droplets in the oil phase. In this sense, the added AA are chemicals that are designed to allow hydrate formation but prevent adhesion of hydrate particles to each other. The AA (Chen *et al.*, 2011) used in this work are from the China University of Petroleum's (Beijing) hydrate research laboratory of the College of Chemical Engineering.

An electronic balance weighed the quantity of AA (decrement method), with a measuring error of ± 0.01 g, which is calculated according to various water cut situations. A high-pressure piston pump is used to inject the AA into the flow system. The preliminary Chen-Guo model (Chen and Guo, 1996, 1998) determined the curve of hydrate formation (*Fig. 4*) for the defined natural gas composition.

1.4 Test Protocol

1. The entire experimental loop is vacuumed until the vacuum degree reaches 0.9 bar;
2. The loop is loaded with Diesel and water (100 vol% liquid loading) considering the set water cut for each test. Here, water cut is defined as the volume fraction/ratio of water to the loading liquid (Diesel + water), with the fixed Diesel volume being 70 L during all these experiments. The gas-supply unit begins to inject gas into the separator at room temperature (20°C) until achieving the required experimental pressure (4, 5 and 6 MPa in this paper);
3. The water and oil mixture is circulated at a constant flow rate to form a stable emulsion with the set AA dosage for each test. The stability of the water/oil emulsion referred to a relatively stable process (dynamic stability) according to the measured data from FBRM under shearing action. This is to say that the emulsion was regarded as stable when the average chord length of droplets fluctuated by ± 0.2 μm within 2 hours;
4. Under the initial pressure and flow velocity, the temperature gradually decreases to the set value. During the hydrate formation process, a data acquisition system continuously collects the temperature, pressure, pressure drop, flow rate, density and chord length;
5. A round of experiments finishes at the end of the formation process when all measured data are stable, such as the system pressure and temperature. The dissociation of hydrate is carried out by increasing the temperature up to 30°C. The system is kept in these conditions for 24 hours with the aid of the particle size analyzer preparing for the next round.

2 RESULTS AND DISCUSSION

Currently, the majority of studies of the hydrate induction period typically performed with autoclave cells are either under static or stirring conditions (Bansal, 1994; Nerheim *et al.*, 1994; Cingotti *et al.*, 1999; Kelland *et al.*, 2000); relatively speaking, there is fewer research carried out in axial flow systems (Sun, 2001; Sun *et al.*, 2004; Gainville and Sinquin, 2011; Gaillard, 1996; Gaillard *et al.*, 1996; Turner, 2005; Sarshar *et al.*, 2010). Consequently, an explicit general method, which is easy to adopt in field measurement, used to delimitate the induction time is proposed according to the combination of previous research and the hydrate slurry flow experiments in this work.

Based on the macro-concept, this method for the induction time defines a start point (an unanimous timing point) and an end point where the hydrate formation can be easily observed; the start point thereof should be typical as well as universal for one system or one process. Therefore, the time point when the system temperature dropped to the equilibrium temperature of the hydrate formation denotes the start point (t_s) for the system in this paper, while the end point (t_e) is the time point where the system temperature was about to rise due to the massive hydrate formation, as illustrated in Figure 5. It can therefore be perceived that this definition of the hydrate induction time involves both the process of hydrate nucleation and the growth stage of hydrate. All the influencing factors of both hydrate nucleation (Natarajan *et al.*, 1994; Lederhos *et al.*, 1996; Natarajan, 1993; Kashchiev and Firoozabadi, 2002a) and hydrate growth (Kashchiev *et al.*, 1991) would thus affect this defined hydrate induction time.

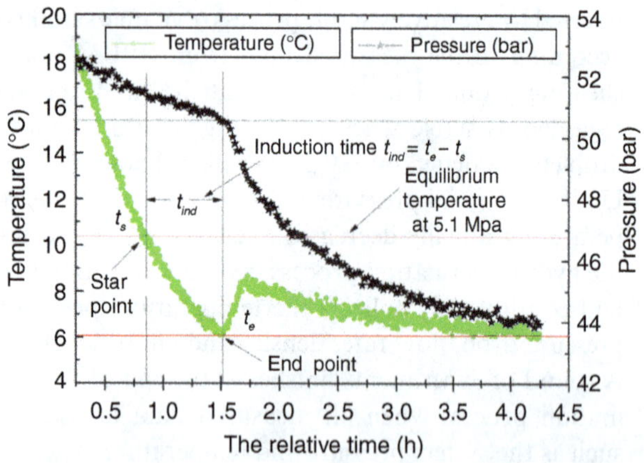

Figure 5

The definition method of the hydrate induction time.

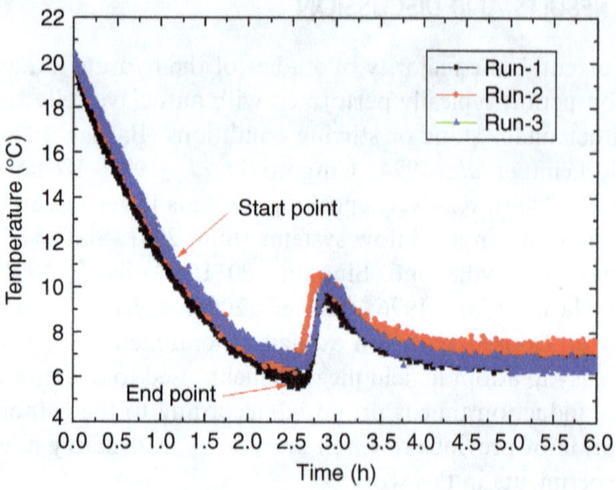

Figure 6

Natural gas induction time during three reproduced formation experiments in the flow loop (4.7 MPa, 0.8 m/s).

The flow loop experiment was performed three times to determine the reproducibility of the natural gas hydrate induction time defined in this paper, as shown in Figure 6. Induction time was remarkably reproducible at 1.6 ± 0.1 h.

2.1 The Influence of the Supercooling Degree on the Induction Time

The supercooling degree (ΔT_{sub}) is the difference between the three-phase equilibrium temperature (T_{eq}) in the experimental conditions and the temperature of the flowing fluid (T_{exp}) in the experimental system, *i.e.* $\Delta T_{sub} = T_{eq} - T_{exp}$.

Hydrate induction time was investigated under the conditions of gradually cooling and depressurization in the paper. These isochoric experiments are different from the reported isobaric/isothermal ones in the literature (Skovborg *et al.*, 1993), which was meant to simulate the actual conditions of the deep-water pipeline transportation. Therefore, the supercooling degree in this paper is defined as the difference between the thermodynamic equilibrium temperature corresponding to the average system pressure before hydrate formation and the temperature bottom point which was about to rise owing to the hydrate formation, *i.e.* the temperatures corresponding to the t_s and t_e points in Figure 5, respectively. There has been research considering the hydrate supercooling degree under isobaric/isothermal conditions.

Skovborg *et al.* (1993) found under various experimental conditions that the hydrate induction time has an exponent relation with the supercooling degree at a relatively larger subcooling degree. More specifically, the larger the supercooling degree, the shorter the induction time will be. The main reason for this phenomenon is that the greater subcooling degree would lead to a more regular structure of water molecules (Sloan and Koh, 2008), then the faster growth rate of hydrate nuclei, thus the shorter hydrate induction time.

The hydrate flowing experiments with the Diesel + water + natural gas system in this work (*Fig. 7, 8*) showed a similar experimental law to that of Skovborg *et al.* (1993) with respect to the influence of the supercooling degree on the induction time. It can be seen from Figure 7 that the hydrate induction time decreases with the increasing supercooling degree at both 1 200 kg/h and 1 500 kg/h flow rates, under the 4 MPa system pressure. Similarly, the variation trend of the hydrate induction time could be observed at 1 200 kg/h, 1 500 kg/h and 1 820 kg/h flow rates, under 5 MPa system pressure (*Fig. 8*).

Meanwhile, it can also be seen from Figures 7 and 8 that the induction time increases with the increasing flow rate in the same supercooling conditions. This also revealed the influence of the flow rate on the cooling effect in an axial flowing system. In detail, the relatively higher flow rate extended the hydrate induction time in several ways, such as generating heat from the friction, decreasing the heat exchange efficiency, etc. The overall results showed that the hydrate induction time in a flowing system is not only related to the supercooling degree, but also to the flow velocity of the fluid. It should be noted that the variation trend of induction time with the flow rate in this paper is different from the result of Sun (2001) and Sun *et al.* (2004).

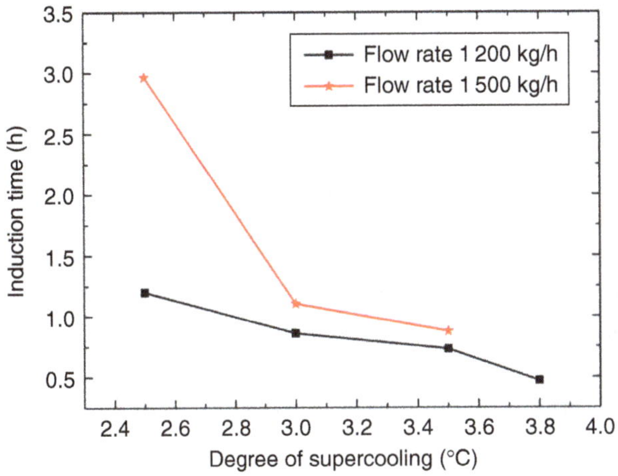

Figure 7

The influence of the supercooling degree on the hydrate induction time (4 MPa).

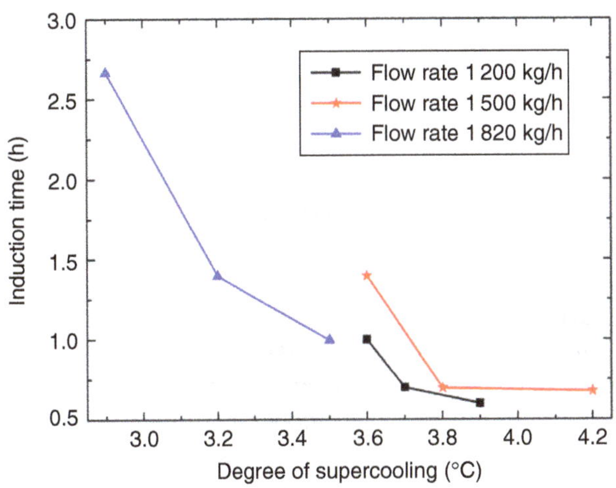

Figure 8

The influence of the supercooling degree on the hydrate induction time (5 MPa).

2.2 The Influence of the Supersaturation Degree on the Induction Time

The term "supersaturation" is used for the driving force that induces hydrate formation. In the current literature, supersaturation has been expressed in many forms, such as the ratio of system pressure to equilibrium pressure, $S_P = P/P_{eq}$ (Herri *et al.*, 1999) and the system to equilibrium fugacity ratio, $S_f = f/f_{eq}$ (Natarajan, 1993). Other supersaturations could be conceived to include the ratio of system to equilibrium hydrate former concentrations ($S_C = C/C_{eq}$) or chemical potentials (Kashchiev and Firoozabadi, 2002a) ($S_\mu = \mu/\mu_{eq}$). Each of the supersaturation types above are defined to equal unity in equilibrium conditions.

Similar to the definition of the supercooling degree in Section 2.1, supersaturation has been defined as the ratio of the experimental equilibrium pressure (at the time point t_s) in the cooling process to the equilibrium pressure of the hydrate formation point, *i.e.* the ratio between the equilibrium pressure at the time point *ts* and the corresponding equilibrium pressure of t_e (experimental temperature) in Figure 5.

Kashchiev and Firoozabadi (2002b, 2003) considered that the hydrate induction time decreases significantly with the increasing supersaturation degree. Turner (2005) confirmed this conclusion that there is an inverse relationship between the hydrate induction time and the supersaturation degree. The interpretation of this phenomenon in the autoclave experiments is that the larger supersaturation degree enhances the driving force of hydrate nucleation, and thus speeds up the hydrate nucleation rate and shortens the hydrate induction time. This paper studied the influence of the supersaturation degree on the hydrate induction time in an axial flow system (*Fig. 9, 10*).

As shown in Figure 9, the induction time decreases generally as the supersaturation degree increases at the flow rates of 1 200 kg/h and 1 500 kg/h. The exception that the induction time increases with the supersaturation degree at the 1 820 kg/h flow rate differs from the known experimental laws derived from the autoclave experiments. In Figure 10, the tendency of the induction time curve at the 1 200 kg/h flow rate agrees with that in an autoclave still, while the curves present a v shape (first decrease and then increase) at both the 1 500 kg/h and the 1 820 kg/h flow rates.

The research achievements of the following researchers could provide a reference for interpretation of the experimental phenomena. For instance, Makogon (1981) reported that the rate of formation of the methane hydrate nuclei increased with the subcooling degree and it presented a maximum value at a $\Delta T_{sub} = 2$ K. Gaillard *et al.* (1999) also found that the methane hydrate growth rate presented a maximum value that depended on the following parameters: the nature of the liquid phase, water/hydrocarbon condensate mixtures, flow velocity in the loop and the strongly correlated temperature-pressure subcooling variables.

Therefore, combined with the experimental law stemming from the autoclave, flow loop, and the actual operating conditions of the experiments in this work, the phenomenon that the hydrate induction time increases with the increasing supersaturation degree at relatively higher flow rates could be explained as follows:

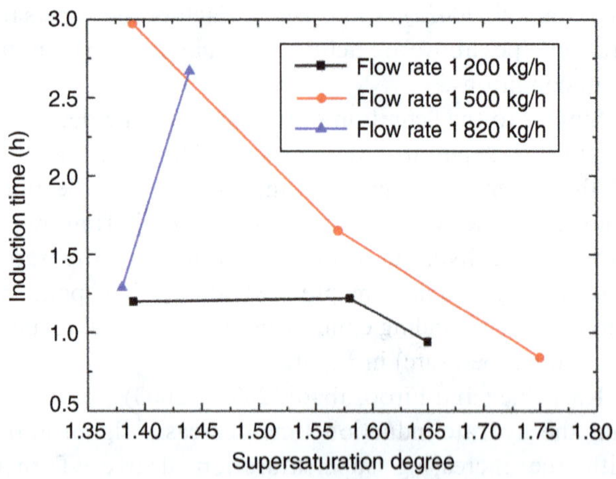

Figure 9

The influence of the supersaturation degree on the hydrate induction time (4°C temperature control).

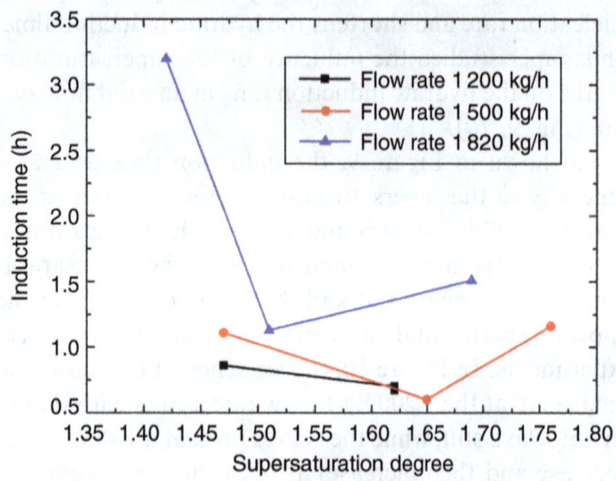

Figure 10

The influence of the supersaturation degree on the hydrate induction time (2°C temperature control).

1. On one hand, the larger supersaturation degree enhances the driving force of the mass transfer, and facilitates the hydrate nucleation and growth. It raises the temperature at which the hydrate begins to form, however, which accordingly reduces the supercooling degree, weakens the structure of the water molecules (Sloan and Koh, 2008), and prolongs the induction time of hydrates in the system;
2. On the other hand, the enhanced shearing action at relatively higher flow rates forms a more distributed emulsion; however, the higher flow rate increases the flow friction (generated heat), and weakens the cooling effect from the temperature-control system, suppressing the hydrate nucleation and growth in this axial flow system. This extends the hydrate induction time as well.

Moreover, the external temperature control setting of the experimental loop influences the induction time to some extent. The combination of the above two interpretations could lead to the phenomenon that the hydrate induction time extends with an increasing supersaturation degree at relatively higher flow rates.

2.3 The Influence of the Flow Rate on the Induction Time

Perturbation is one of the important factors influencing the nucleation rate of the hydrates; thus it could affect the hydrate induction time to some extent. Englezos *et al.* (1987) found that the induction time changed significantly by changing only the stirring rate, keeping the other conditions fixed in a semi-batch stirred reactor. The induction time was in inverse proportion to the stirring rate (300 ~ 450 RPM), *i.e.* the maximum stirring rate generated the shortest induction time in the experimental conditions. Jensen *et al.* (2008) studied the formation of propane hydrate with/without additives under different stirring rates, observing that high-speed stirring can distinctly shorten the hydrate induction time and facilitate the rate of formation. In addition, a similar law was extracted from the experiments of the methane induction time by Skovborg *et al.* (1993) (stirring rate from 200 to 380 RPM). Sun (2001) performed the experiments for the hydrate induction time in a (R12 + H_2O) system, and found that the flow rate of the liquid is crucial to the induction time under a certain pressure; that is, the higher flow rates can lead to a shorter hydrate induction time. Moreover, Turner (2005) concluded that the induction time of hydrate basically tends to diminish with the increasing Revolutions Per Minute (RPM) speed in both the autoclave still and the flow loop.

The probable reason for the above experimental phenomena might be that the higher stirring rate stands for enhanced disturbance intensity, and thus gives rise to a larger gas-water interfacial area. The larger interfacial area permits more gas to be in contact with the water phase so as to accelerate the hydrate nucleation rate. Thus, the induction time becomes shorter with respect to the higher concentration of gas in the liquid within a certain time. However, it needs to be further verified whether the above conclusion can be used as a universal principle under the existing stirring rate.

Therefore, a series of experiments focusing on the hydrate induction time were conducted in order to validate the regularity of the aforementioned phenomena in an axial flow system, including water, natural gas and

Diesel oil. The influence of the flow rate on the hydrate induction time is shown in Figures 11 and 12.

As shown in Figure 11, the experimental rule is disparate from the prevailing one that the hydrate induction time would decrease remarkably with the increasing flow rate. Instead, the hydrate induction time reduces first then increases with the increasing flow rate. In detail, the induction time decreases in the flow rate range from 950 kg/h to 1 200 kg/h, and then increases progressively when the flow rate exceeds 1 200 kg/h. This can also be deduced from Figure 12.

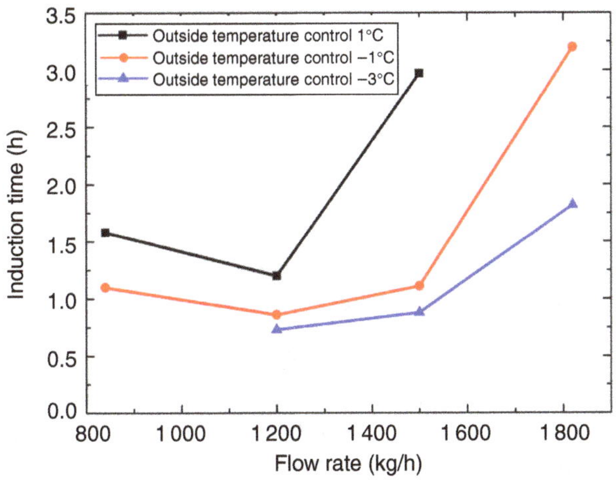

Figure 11

The influence of the flow rate on the hydrate induction time (4 MPa).

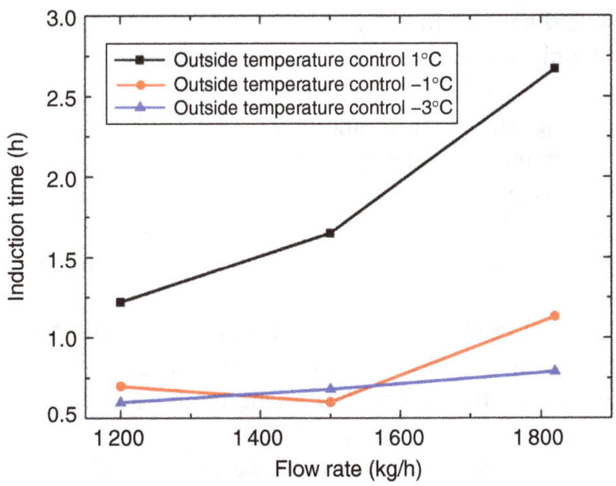

Figure 12

The influence of the flow rate on the hydrate induction time (5 MPa).

Firstly, the difference in the experimental apparatus and test protocols accounts for the disparate experimental phenomena to a certain degree. Secondly, the above phenomena can be mainly determined from the following two aspects:

1. The greater shearing action produced by the higher flow rate causes smaller water droplets, a larger contact area, and enhanced mass transfer by decreasing interfacial diffusion boundary layer thickness. So, the number of the nucleation sites increases, lessening the induction time of the hydrates. On one hand, the greater shearing action leads to a more uniform emulsion;
2. On the other hand, the greater flow rate brings about a higher friction pressure drop which raises the temperature of the fluids; it also reduces the cooling effect from the temperature-control system. Thus, it could extend the hydrate induction time.

The combination of these aspects results in the experimental phenomena presented in this work. It can be seen accordingly that the mass transfer between the guest molecules will significantly affect the hydrate induction time in the process of hydrate nucleation or formation. Therefore, the influence of the mass transfer on the hydrate induction time should be taken into consideration, with regard to the hydrate nucleation and growth process, which is dominated by the mass transfer.

2.4 The Influence of the Water Cut on the Induction Time

Turner (2005) studied the influence of the water cut on the hydrate induction time in the autoclave as well as the flow loop. The results indicated that the hydrate induction time decreased with the increasing water cut in the autoclave, while it changed little at various water cuts, *i.e.* the influence of the water cut was not obvious in an axial flow system.

The influence of the water cut on the hydrate induction time in this work is shown in Figure 13. Unlike the results obtained by Turner (2005), it can be seen that the induction time decreases first then increases with the increasing water cut during the experiments. The reasons for this difference are proposed as follows:

1. On one hand, the higher water cut produces a larger gas-water interfacial area (*Fig. 14*), leading to more nucleation sites and an accelerated nucleation rate;
2. On the other hand, the increased water cut could also result in the decreased gas solubility per unit volume of the oil-water emulsion. In the nucleation and growth

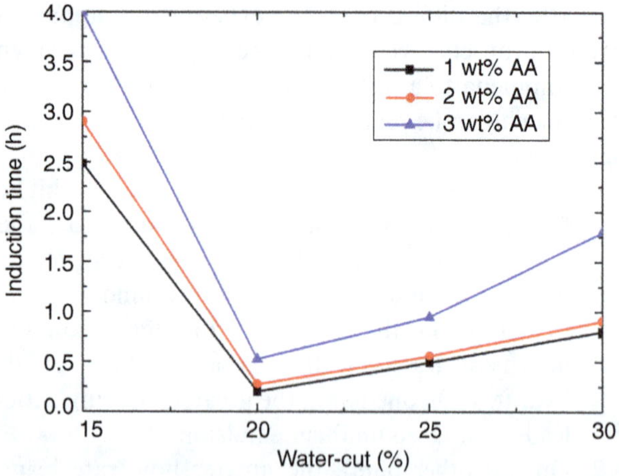

Figure 13

The influence of the water cut on the hydrate induction time.

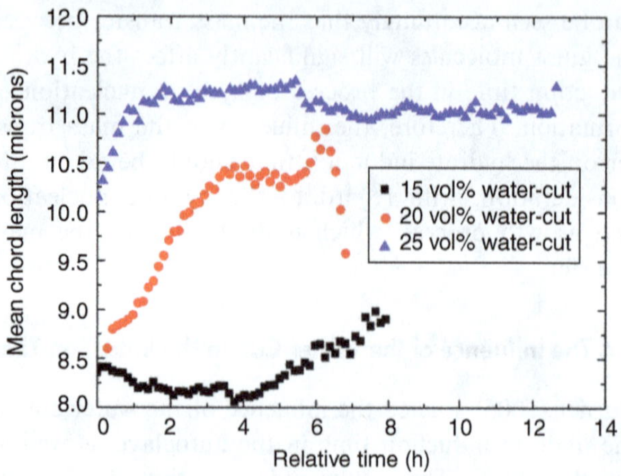

Figure 14

The influence of the water cut on chord length distribution (1.25 m/s flow velocity, 4 MPa, 1 wt% AA).

process, the confined gas mass transfer decreases the nucleation and growth rate of the hydrates.

In general, the experimental phenomenon, that the hydrate induction time first decreases then increases with the increasing water cut, results from the combined impact of the above two explanations. It also proves that, in the hydrate formation process, the mass-transfer efficiency of the gas directly affects the formation of hydrates.

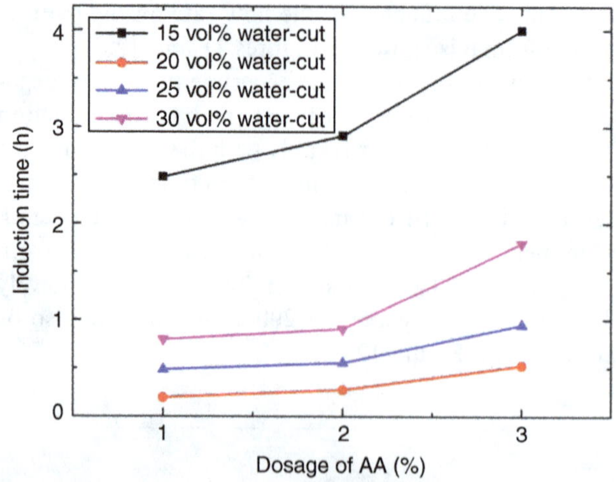

Figure 15

The influence of the AA concentration on the hydrate induction time.

2.5 The Influence of AA Concentration on the Induction Time

The AA (Peng, 2012) are a formulation of some polymers and surfactants, which can be used in the water and oil mixture to form a W/O emulsion where the water is dispersed as droplets in the oil phase. In this sense, the added AA would allow hydrate formation but prevent adhesion of hydrate particles to each other. The result is supposedly hydrate slurry that flows without plugging.

Turner (2005) conducted experiments in a flow loop to investigate the influence of the AA concentration on the induction time. He concluded that the AA could not only prevent the aggregation between the hydrate particles, but also exhibited kinetic inhibitor behavior, delaying the hydrate induction time. Therefore, the hydrate induction time will be extended with increasing AA dosage. The experiments in this work confirmed this conclusion that enhanced AA concentration prolongs the hydrate induction time, as shown in Figure 15.

Figure 16 illustrated the inhibition effect of the AA on the hydrate nucleation and formation. As shown in Figure 15, the mean chord length of water droplets decreased with increasing AA dosage, which caused the larger contact area between water and oil phases. So, in the heterogeneous nucleation, the nucleation rate should be proportional to the interfacial area; however, this experimental result presented a different trend (*Fig. 15*). Therefore, the addition of the AA might not

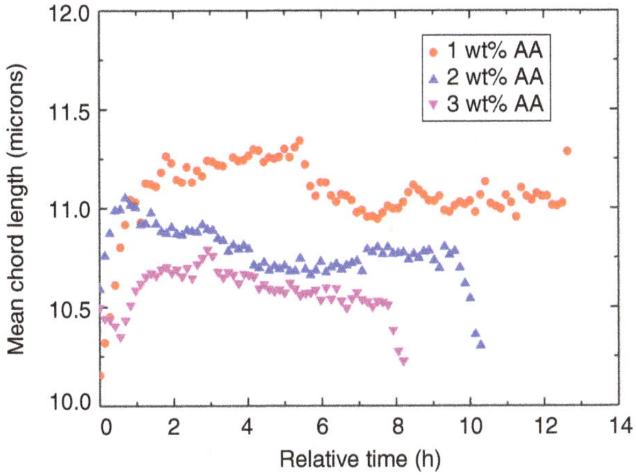

Figure 16

The influence of AA doses on chord length distribution (25 vol% water cut, 1.25 m/s flow velocity, 4 MPa).

only restrain the agglomeration of the hydrate particles, but also play a role as a hydrate kinetic inhibitor.

CONCLUSIONS

1. The hydrate induction time decreases remarkably with the increasing supercooling degree in an axial flow system (an experimental loop in this paper), which is similar to the published experimental results conducted in the autoclave stills;
2. The variation tendency of the hydrate induction time along with the supersaturation degree in an axial flow system differs slightly from that in an autoclave still, *i.e.* the induction time would either present a v-shape curve or increase progressively with the increasing supersaturation degree at a relatively high flow rate. These phenomena could be construed as that the enhanced supersaturation degree reduces the structure of the water molecules and that the flow rate inhibits the hydrate nucleation and growth;
3. The hydrate induction time decreases first, and then increases (or gradually increases) with the increasing flow rate in the high-pressure flow loop;
4. There is a little discrepancy between the influence of the water cut on the hydrate induction time in this work and that in the literature. In other words, the hydrate induction time decreases first and then increases with the increasing water cut in these experiments;
5. The influence rule of the AA concentration resembles that in the literature; that is, an increased AA dose prolongs the hydrate induction time within a certain range.

ACKNOWLEDGMENTS

This paper was sponsored by the National Natural Science Foundation of China (51274218, 51306208 and 51134006), National Science & Technology Specific Project (2011ZX05026-004-03), and Science Foundation of China University of Petroleum-Beijing (No. 2462013 YXBS010 and No. 2462014YJRC006).

REFERENCES

Bansal V. (1994) Kinetic Study of Clathrate Hydrates, *MS Thesis*, Colorado School of Mines, Golden, CO.

Christiansen R.L., Sloan E.D. (1995) *Proceedings of the 74th Gas Processors Association Convention*, San Antonio, TX, 13-15 March.

Chen G.J., Guo T.M. (1996) Thermodynamic Modeling of Hydrate Formation Based on New Concepts, *Fluid Phase Equilibria* **112**, 43-65.

Chen G.J., Guo T.M. (1998) A New Approach to Gas Hydrate Modeling, *Chemical Engineering Journal* **71**, 145-151.

Chen G.J., Li W.Z., Li Q.P., Sun C.Y., Mu L., Chen J., Peng B.Z., Yang Y.T., Meng H. (2011) China Patent Number 201110096579.2.

Chen G.J., Sun C.Y., Guo T.M. (2000) A Theoretical Revision of the Derivation of Liquid Property Expressions from an Equation of State and Its Application, *Chemical Engineering Science* **55**, 4913-4923.

Cingotti B., Sinquin A., Pic J.S., Herri J.M., Cournil M. (1999) *Proceedings SPE International Symposium on Oilfield Chemistry*, Houston, TX, 16-19 Feb.

Englezos P., Kalogerakis N., Dholabhai P.D., Bishnoi P.R. (1987) Kinetics Formation of Methane and Ethane Gas Hydrates, *Chemical Engineering Science* **42**, 11, 2647-2658.

Gaillard C.. (1996) Kinetic of Methane Hydrate Formation in a Lab Loop, *PhD Thesis*, Institut National Polytechnique de Toulouse, France.

Gaillard C., Monfort J.-P., Peytavy J. (1996) Formation and Growth Kinetics of Natural Gas Hydrate, *The 2th International Conference on Gas Hydrates*, Toulouse, France, 2-6 June.

Gaillard C., Monfort J.-P., Peytavy J.L. (1999) Investigation of Methane Hydrate Formation in a Recirculating Flow Loop: Modeling of the Kinetics and Tests of Efficiency of Chemical Additives on Hydrate Inhibition, *Oil & Gas Science and Technology – Rev. IFP* **54**, 3, 365-374.

Gainville M., Sinquin A. (2011) Hydrate Slurry Characterization for Laminar and Turbulent Flows in Pipelines, *The 7th International Conference on Gas Hydrates*, Edinburgh, Scotland, United Kingdom, 17-21 July.

Herri J.M., Pic J.S., Gruy F., Cournil M. (1999) Methane hydrate crystallization mechanism from *in-situ* particle sizing, *AIChE Journal* **45**, 590-602.

Jensen L., Thomsen K., von Solms N. (2008) Propane hydrate nucleation: Experimental Investigation and Correlation, *Chemical Engineering Science* **63**, 3069-3080.

Kashchiev D. (2000) *Nucleation: Basic Theory with Application*, Butterworth-Heineman, Oxford.

Kashchiev D., Firoozabadi A. (2002a) Driving Force for Crystallization of Gas Hydrates, *Journal of Crystal Growth* **241**, 220-230.

Kashchiev D., Firoozabadi A. (2002b) Nucleation of Gas Hydrates, *Journal of Crystal Growth* **243**, 476-489.

Kashchiev D., Firoozabadi A. (2003) Induction Time in Crystallization of Gas Hydrates, *Journal of Crystal Growth* **250**, 499-515.

Kashchiev D., Verdoes D., Von Rosmalen G.M. (1991) Induction Time and Metastability Limit in New Phase Formation, *Journal of Crystal Growth* **110**, 373-380.

Kelland M., Svartaas T., Ovsthus J., Namba T. (2000) A new Class of Kinetic inhibitor, *Annals of the New York Academy of Sciences*, New York, January.

Lederhos J.P., Long J.P., Sum A., Christiansen R.L., Sloan E.D. Jr (1996) Effective kinetic inhibitors for natural gas hydrates, *Chemical Engineering Science* **51**, 8, 1221-1229.

Maeda N., Wells D., Hartley P.G., Kozielski K.A. (2012) Statistical Analysis of Supercooling in Fuel Gas Hydrate Systems, *Energy & Fuels* **26**, 1820-1827.

Makogon Y.F. (1981) *Hydrates of Natural Gas*, Penn Well Book, Tulsa, United States.

Muller-Bongartz B., Wildeman T.R., Sloan E.D. (1992) *Proceedings Second International Offshore and Polar Engineering Conference*, San Francisco, USA, 14-19 June.

Mullin J.W. (1993) *Crystallization*, 3rd Edition, Butterworth-Heinmann, Oxford, UK.

Natarajan V. (1993) Thermodynamic and Nucleation Kinetics of Gas Hydrates, *PhD Dissertation*, University of Calgary, Calgary.

Natarajan V., Bishnoi P.R., Kalogerakis N. (1994) Induction Phenomena in Gas Hydrate Nucleation, *Chemical Engineering Science* **49**, 2075-2087.

Nerheim A.R., Svartaas T.M., Samuelsen E.J. (1994) *Proceedings Fourth International Offshore and Polar Engineering Conference*, Osaka, Japan, 10-15 April.

Parent J.S., Bishnoi P. (1996) Investigations into the Nucleation Behavior of Methane Gas Hydrates, *Chemical Engineering Communication* **144**, 51-64.

Peng B.Z. (2012) Study on Hydrate Slurry Transportation in Multi-phase Pipeline, *PhD Dissertation*, China University of Petroleum, Beijing.

Sarshar M., Smaeilzadeh F., Fathikalajahi J. (2010) Induction Time of Hydrate Formation in a Flow loop, *Theoretical Foundations of Chemical Engineering* **44**, 201-205.

Shi B.H. (2012) Study on the Characteristics of Hydrate Growth and Flow in Gas-Condensate Multiphase Pipelines, *PhD Dissertation*, China University of Petroleum, Beijing.

Skovborg P., Ng H.J., Rasmussen P., Mohn U. (1993) Measurement of Induction Times for the Formation of Methane and Ethane Gas Hydrates, *Chemical Engineering Science* **48**, 3, 445-453.

Sloan E.D. (2005) A Changing Hydrate Paradigm-from Apprehension to Avoidance to Risk Management, *Fluid Phase Equilibra* **228-229**, 67-74.

Sloan E.D., Koh C.A. (2008) *Clathrate Hydrates of Natural Gas*, 3rd Edition, CRC Press, Taylor & Francis Group, Boca Raton, FL.

Sun C.Y. (2001) The Related Basic Research to the Development of Gas Mixtures Separation Technology by Forming Hydrate, *PhD Dissertation*, China University of Petroleum, Beijing.

Sun C.Y., Chen G.J., Yue G.L. (2004) The Induction Period of Hydrate Formation in Flow System, *Chinese Journal Chemical Engineering* **12**, 527-531.

Söhnel O., Mullin J.W. (1988) Interpretation of Crystallization Induction Periods, *Journal Colloid Interface Science* **123**, 43-50.

Turner D.J. (2005) Clathrate Hydrate Formation in Water-in-oil Dispersions, *PhD Dissertation*, Colorado School of Mines, Colorado.

Vysniauskas A., Bishnoi P.R. (1983) A Kinetic Study of Methane Hydrate Formation, *Chemical Engineering Science* **38**, 1061-1072.

Cite this article as: X.F. Lv, B.H. Shi, Y. Wang, Y.X. Tang, L.Y. Wang and J. Gong (2014). Experimental Study on Hydrate Induction Time of Gas-Saturated Water-in-Oil Emulsion using a High-Pressure Flow Loop, *Oil Gas Sci. Technol* **70**, 6, 1111-1124.

9

Effect of Pilot Injection Timings on the Combustion Temperature Distribution in a Single-Cylinder CI Engine Fueled with DME and ULSD

Joonho Jeon[1], Yong Hee Park[2], Sang Il Kwon[2] and Sungwook Park[3]*

[1] Department of Mechanical Convergence Engineering, Graduate School of Hanyang University, Seoul 133791 - Republic of Korea
[2] National Institute of Environmental Research, Incheon 404170 - Republic of Korea
[3] School of Mechanical Engineering, Hanyang University, 17 Haengdang-dong, Seongdong-gu, Seoul 133791 - Republic of Korea
e-mail: parks@hanyang.ac.kr

* Corresponding author

Abstract — *Many studies of DiMethyl Ether (DME) as an alternative fuel in Compression-Ignition (CI) engines have been performed. Although diverse DME engine research has been conducted, the investigation of combustion behavior and temperature distribution in the combustion engine has not progressed due to the fact that there is no sooting flame in DME combustion. In order to investigate the combustion characteristics in this study, the KIVA-3 V code was implemented to research various pilot injection strategies on a single-cylinder CI engines with DME and Ultra-Low-Sulfur Diesel (ULSD) fuels. The combustion distribution results obtained from the numerical investigation were validated when compared with the measurement of flame temperature behaviors in the experimental approach.*

This study showed that long intervals between two injection timings enhanced pilot combustion by increasing the ambient pressure and temperature before the start of the main combustion. Different atomization properties between DME and ULSD fuels contributed to the formation of a fuel-air mixture at the nozzle tip and piston lip regions, separately, which strongly affected the temperature distribution of the two fuels. In addition, the pilot injection timing played a vital role in regard to ignition delay and peak combustion temperatures. Exhaust emissions, such as NO_x and soot, are related to the local equivalence ratio and temperature in the combustion chamber, also illustrated by the contrary result on a Φ (equivalence ratio) – T (temperature) map.

Résumé — Effet des réglages à l'injection sur la répartition de la température de combustion dans un moteur monocylindre à allumage par compression et alimenté avec du DME et du Diesel à ultra basse teneur en soufre — De nombreuses études concernant le diméthyléther (DME) en tant que carburant alternatif dans des moteurs à allumage par compression (CI) ont été effectuées. Bien que différentes recherches sur les moteurs DME aient été réalisées, l'étude du comportement à la combustion et des répartitions de température dans le moteur à combustion n'a pas progressé du fait qu'il n'existe aucune flamme d'encrassement lors de la combustion du DME. Afin d'étudier dans le présent travail les caractéristiques de combustion, le code KIVA-3 V a été appliqué pour rechercher différentes stratégies d'injection pour des moteurs CI monocylindres avec comme carburants du DME et du Diesel à teneur en soufre

ultra basse (ULSD). Les résultats obtenus dans l'étude numérique de la distribution de la combustion, ont été validés par rapport aux mesures de la température de flamme de l'approche expérimentale.
Cette étude a montré que de longs intervalles entre deux réglages de l'injection amélioraient la combustion en augmentant la pression ambiante et la température avant le début de la combustion principale. Différentes propriétés d'atomisation entre les carburants DME et ULSD ont contribué à la formation séparée du mélange combustible-air à la pointe du gicleur et dans des zones du piston, ce qui affecte fortement la répartition des températures des deux carburants. En outre, le réglage à l'injection a joué un rôle essentiel quant aux retards à l'allumage et aux pics de températures de combustion. Les émissions à l'échappement, tels que les NO_x et les suies, sont liées au rapport d'équivalence local et à la température dans la chambre de combustion, comme cela est illustré par un résultat contradictoire sur une cartographie Φ (rapport d'équivalence) – *T* (température).

NOMENCLATURE

ALE	Arbitrary Lagrangian-Eulerian
ATDC	After Top Dead Center
BTDC	Before Top Dead Center
CA	Crank Angle
CI	Compression-Ignition
deg.	Degree (°)
DICI	Direct Injection Compression-Ignition
DME	DiMethyl Ether
EGR	Exhaust Gas Recirculation
HRR	Heat Release Rate
HSDI	High-Speed Direct Injection
IMEP	Indicated Mean Effective Pressure
KH	Kelvin-Helmholtz
$P_{inj.}$	Injection pressure
Φ	Equivalence ratio
RNG	ReNormalization Group
RT	Rayleigh-Taylor
T	Temperature
$\tau_{inj.}$	Injection timing
TDC	Top Dead Center
ULSD	Ultra-Low-Sulfur Diesel

INTRODUCTION

As the quality of human life increases, many people are concerned with their health and the environment. Naturally, they try to improve their life conditions and keep their surroundings clean. As a result, very strict regulations have emerged. These regulations have led to highly technical evolution in various industrial fields. For instance, the Euro VI regulation has high standard values for noxious emissions emitted from Diesel engines. Although Diesel engines are able to generate high power, harmful emissions such as HydroCarbons (HC), Carbon monoxide (CO) and soot are also made. Therefore, many researchers and engine manufacturers have developed various technologies; combustion strategies, after-treatment devices, EGR systems, and alternative fuels to allow engine manufacturers to meet stringent standards. Among the various methods used for the reduction of noxious emissions, alternative fuels have been highlighted because they can possibly reduce both emission problems and a dependency on fossil fuels. DME fuel is one of the best alternative fuels for the Compression-Ignition (CI) engine due to its high self-ignition characteristics, oxygen content, and no soot combustion [1].

Numerous experiments for the adaptation of DME fuel to the CI engine have been conducted. Kim *et al.* [2] reported that DME combustion showed a higher Indicated Mean Effective Pressure (IMEP) and a better complete combustion process than those of Diesel fuels based on a constant energy. In addition, DME fuel showed a more stable combustion event than Diesel combustion. Due to the low lubricity and high evaporation rate of DME fuel, Sato *et al.* [3] designed a new type of injection system called a jerk-type in-line DME injection system, optimized for the light-duty truck engine. The DME engine with the jerk-type in-line system achieved better fuel consumption and lower emission levels than those of Diesel combustion under similar, sustained output conditions. DME fuel has a low energy density of about 55% of Diesel fuel, which requires a greater injection quantity and longer injection duration compared with Diesel fuel. Despite the increase in mass and injection duration, the DME engine exhausts a smaller amount of

NO_x (by 40%), CO_2 (by 15%) and extremely low THC levels. An *et al.* [4] reported that the high injection pressure of DME fuel using a common-rail injection system shows low energy consumption and active atomization characteristics. Although the increased injection pressure improved the combustion performance, it was indicated that high NO_x emissions were also generated because of the active combustion event. Moreover, with no soot combustion from DME fuel, it is easily possible to increase engine torque in low speed conditions.

DME combustion in the CI engine produces high engine performances and low emissions such as CO, HC and soot. However, it cannot be ignored that a significant concentration of nitrogen oxides is emitted from DME engines. In order to decrease NO_x emissions, various combustion strategies have been applied in the CI engine fueled with DME. Yoon *et al.* [5] studied the effects of pilot injection on DME combustion and emission in a single-cylinder Direct-Injection Compression-Ignition (DICI) engine. They discovered that DME pilot injection reduces the NO_x emission concentrations, along with other emissions (CO, HC). An investigation of the reduction of NO_x emissions using the DME engine with an EGR system was conducted by Song *et al.* [6]. Due to the fact that NO_x emissions are actively formed under high oxygen concentration and combustion temperature conditions, the EGR system suppressed both of these conditions, and thus the formation of NO_x emissions. Their research demonstrated the optimized EGR rate under the ESC test cycle to meet the Euro IV standards. In their study, the DME engine with an EGR system achieved lower NO_x emissions than the Euro IV limit, and suggested that an additional oxidation catalyst can also reduce CO and HC simultaneously. Among these methods to lower exhaust emissions, a pilot injection strategy is easily available in conventional Diesel engine fueled with DME without any modification.

However, there have been few detailed studies on the effects of the pilot injection on DME combustion. Therefore, this study focuses on DME combustion and emission characteristics under various injection regimes using the numerical approach. A previous study [7] carried out analytical research on the DME combustion and emission under single-injection conditions, and showed high prediction results as compared with experiments. The same simulation code and methods were used in the current research on the effects of the pilot injection timing. The numerical experiment analyzed the processes of the fuel-air mixture formation and combustion temperature distribution for combustion characteristics. For an analysis of emissions for DME and ULSD combustion, an Φ (equivalence ratio) – T (temperature) map was introduced.

1 NUMERICAL AND EXPERIMENTAL METHODOLOGIES

In the present study, a modified version KIVA-3 V release 2 which was a combination of the KIVA-3 V code [8, 9] and the CHEMKIN-II code [10], for spray and combustion simulations, respectively, was employed to enhance the prediction in the CI engine fueled with ULSD and DME fuels. In addition, an experiment with the AVL 5402 research single-cylinder engine was carried out for simulation validation. A multi-dimensional grid was generated based on the AVL 5402 engine.

1.1 Numerical Models for Engine Simulation

Spray which occurred inside the internal combustion engine was simulated by the KIVA-3 V code which takes the Arbitrary Lagrangian-Eulerian (ALE) method into account to solve the complex multi-phase flow. For the atomization calculation, the hybrid break-up KH-RT (Kelvin-Helmholtz and Rayleigh-Taylor) model [11] was applied. Also, the in-cylinder turbulent flow was considered with the modified ReNormalization Group (RNG) k-ε model [12]. For the interactions between spray droplets and the wall, a spray-wall interaction model was used, which enhanced the prediction of spray characteristics in this study. Although the KIVA-3 V code included fuel information of a gaseous DME fuel as a function of temperature, it was required to supplement it with the liquefied DME properties depending on factors such as temperature, viscosity, latent heat, and vapor pressure for the KIVA-3 V code. The liquefied DME properties were obtained from Teng *et al.*'s [13-15] and Teng and McCandless's [16] investigations and added to the fuel library. For the ULSD spray simulation, tetradecane ($C_{14}H_{30}$) was chosen as a surrogate fuel because tetradecane not only has analogous physical properties with those of the ULSD fuel, but also perfectly reproduces the behavior of ULSD spray in the combustion engine. The CHEMKIN-II code, which was a chemistry solver, coupled with the KIVA-3 V code was used for chemical reactions of fuels in their self-ignition and combustion processes. In the present study, a detailed DME mechanism [16], including 79 species and 351 reversible elementary reactions, was introduced to simulate ignition delay and fuel oxidation processes under various engine operating conditions. Because the mechanism contained low and high temperature reactions, the overall DME oxidation reactions could be depicted in the CI engine. The simulation for the ULSD combustion was conducted using the n-heptane (C_7H_{16}) mechanisms as the ULSD surrogate fuel, which was composed of 30 species and 65 irreversible

TABLE 1
Validation engine specifications

Item	Description
Engine type	Naturally-aspirated CI single cylinder engine
Injection system	*Bosch* common-rail direct injection system
Bore × stroke	85 mm × 90 mm
Displacement volume	510.7 cm^3
Compression ratio	17.1:1
Number of injection holes	5
Spray angle	142°

reaction steps [17]. The emission models were also included for NO_x and soot emissions in the modified KIVA-3 V code. The reduced GRI NO mechanism [18] in combination with the detailed mechanism predicted the formation of NO_x emissions during the combustion processes. Additionally, a two-step soot model [19], which considered the competitive relation between formation and oxidation reactions, was used to compute the rate of soot emissions under various experimental conditions.

1.2 Experimental Apparatus

In order to validate the ULSD and DME fuel combustion and emission models, experimental research was performed using an AVL 5402 single-cylinder research engine equipped with a common-rail injection system. The detailed specifications and dimensions of the test engine are summarized in Table 1. The AVL single-cylinder compact test bed consists of a dynamometer (AMK DW 13-170-4), with a rated power up to 38 kW, and a rated torque up to 120 Nm, which is a three-phase asynchronous motor that can supply and absorb torque. A prototype ETAS engine control unit, which is an open-loop fuel injection strategy designed by AVL, was able to adjust the fuel injection parameters. The injection parameters such as the injection timing, pressure, and quantity could be altered respectively using INCA software that communicates with the ETAS system. In the fuel supply system, two high pressure pumps (HSF-300, *Haskel*) linked in parallel were used to stably supply fuel from a fuel tank to the common-rail, as seen in Figure 1. A HORIBA MEXA-554JKNOx exhaust gas analyzer was used to measure the CO (±0.06% vol.), HC (±12 ppmvol), and NO_x (±20 ppmvol) emissions under steady-state conditions. The ESC sensor, using a chemical method, estimates NO_x emissions and the HC and CO emissions

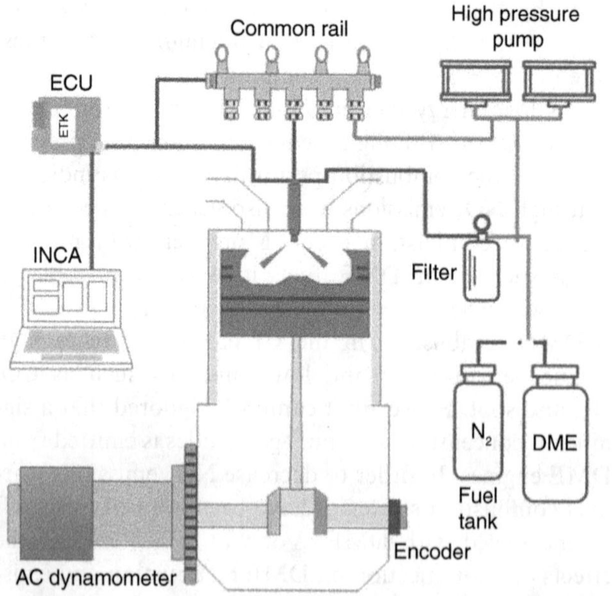

Figure 1
Schematic of experimental apparatus for the validation test.

are detected by the non-dispersive infrared ray method. In addition, the soot meter (Micro Soot Sensor, *AVL*) determined the soot concentration directly using the photo-acoustic measurement method (accuracy of ~5 μg/m^3). To visualize the combustion behavior and temperature distribution, a visualization system was introduced in this research. The visualization system consists of two parts: endoscope visualization equipment (Visioscope, *AVL*) and an image post-processing program (ThermoVision, *AVL*). The Visioscope is composed of an endoscope, CCD camera (1 280 × 1 024 pixels), and light unit. The endoscope was installed on the cylinder head with the CCD camera, and performed electric recording of the processes that took place in the combustion engine. The CCD camera and light unit were synchronized with a crank angle encoder (365C, *AVL*) to capture combustion images at an exact timing (1 CA resolution). The ThermoVision program allowed captured images to calculate flame temperature using electromagnetic radiation [20].

1.3 Experimental Procedures

The numerical experiment conditions were fixed to validation test conditions which were obtained from a single-cylinder CI engine. The injection timing was expanded from one condition to three main timings and seven pilot timings for DME and ULSD fuels.

Detailed test conditions were tabulated in Table 2. In the validation test, the two test fuels matched the same power condition. Fuel injection quantities used the same value as the experimental results.

2 RESULTS AND DISCUSSION

2.1 Model Validation

For an accurate prediction of multi-injection cases in the KIVA/CHEMKIN code, a comparison between experimental and numerical results was conducted. The two different injection conditions; single-injection and double-injection cases were chosen for validation with regard to the combustion pressure, heat release rate, and temperature distribution. The engine operating conditions for the validation were 1 500 rpm engine speed, 50 MPa injection pressure, and 18 Nm engine load. Under theses engine conditions, test fuels were injected at BTDC 60 CA and BTDC 10 CA, respectively. The computational grid modeled only one-fifth of the AVL 5402 research single-cylinder engine based on symmetry assumptions to reduce computation time.

TABLE 2

Detail simulation test conditions

Description		Conditions
Fuel		DiMethyl Ether (DME), Ultra-Low-Sulfur Diesel (ULSD)
Injection pressure (MPa)		50
Injection strategy		Pilot injection, main injection
Pilot injection quantity (mg)		2
Main injection quantity	DME (mg)	24.5
	ULSD (mg)	13.5
Pilot injection timing (BTDC, CA)		30, 40, 50, 60, 70, 80, 90
Main injection timing (BTDC, CA)		5, 10, 15
Engine speed (rpm)		1 500

The combustion pressure and the heat release rate for all validation cases are illustrated in Figure 2. In the ULSD case, the calculated pressure was predicted accurately compared with the measured pressure. Although the heat release rate shows a slight difference between the simulation and the experiment, this is because KIVA did not reflect the heat transfer effect at the cylinder wall. This effect is also observed in the DME combustion case. For the heat release rate of the DME simulation, the overall heat release rate was overestimated compared with the measured one. However, the peak pressure trends showed good agreement with the experimental results. The ignition delay of the DME case in the simulation was a little slower than that of the measured result. This is because a delayed first combustion was caused by the pilot injection, which led to the disagreement of the main combustion starting points with the experiment. On the other hand, the ULSD pilot combustion started earlier, raising the combustion pressure condition before the main combustion. Thus, the main combustion event

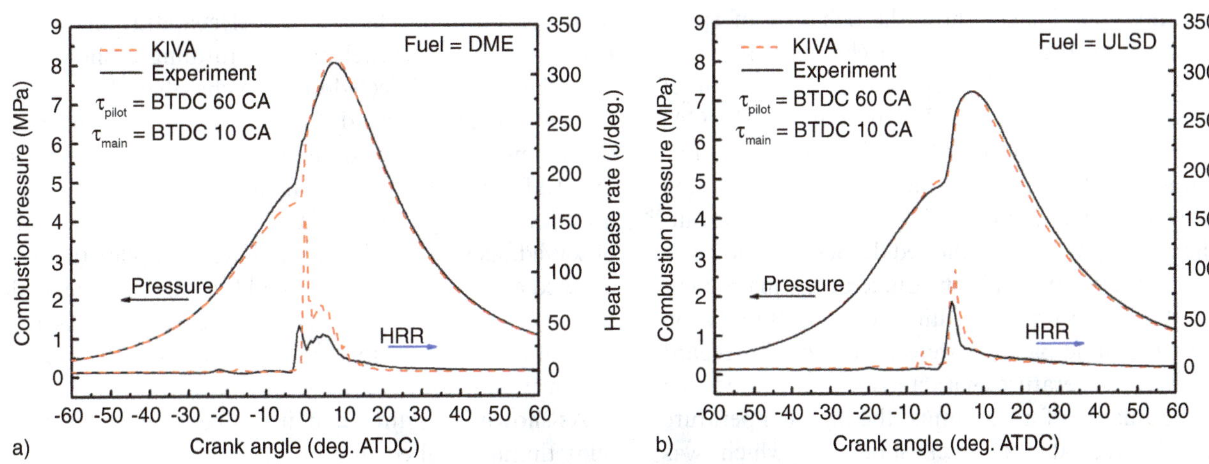

Figure 2

Combustion characteristics of a) DME and b) ULSD fuels in the comparison between the simulation and experiment.

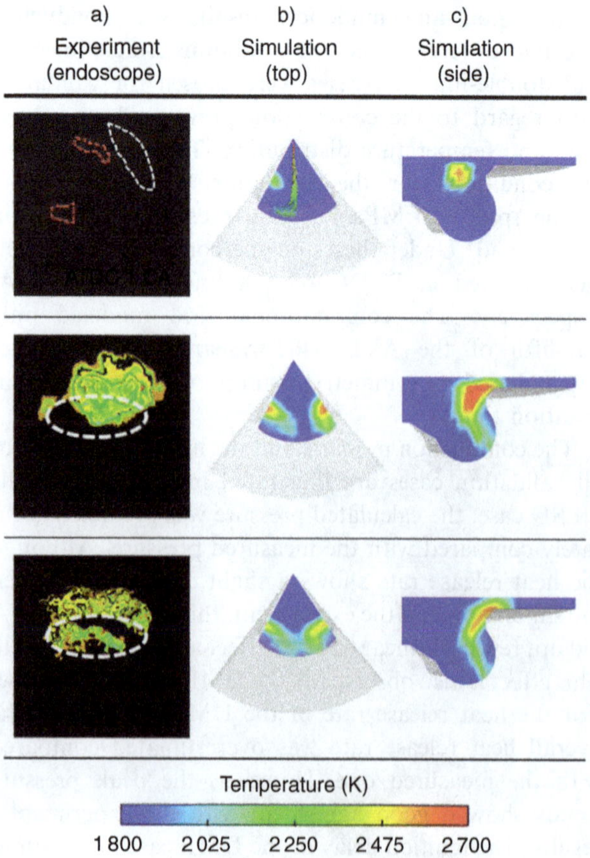

Figure 3

Comparison of spray and temperature distribution between the experiment and simulation for ULSD combustion.

happened at the same time as that of the experimental case.

The endoscopic images show the behavior of spray and combustion in the combustion chamber. Figure 3 represents the spray and temperature distribution of the ULSD pilot injection (BTDC 60 CA and 10 CA) case, compared with the KIVA results under the same conditions. The first image in the left column captured the main spray jet and sparks. The white dashed line outlines the main spray jet and the red dashed line indicates sparks which were the ignition source of the main combustion. In the second and third columns (simulation contour), the same patterns were observed with regard to spray and temperature characteristics. As can be seen in the first images of all columns, the high-temperature region appeared at the outer spray jet, which was induced by a rapidly forming mixture between fuel droplets and charged air in the middle of the jet stream. The images in the first column were obtained using the ThermoVision program, and illustrate temperatures from 1 800 K to 2 700 K. Flame temperature distribution images are shown in the second and third pictures. There is a piston bowl center in the endoscopic pictures that is expressed by the white dashed line. USLD combustion was mainly developed near the piston lip and bowl in the experimental case. For the numerical case, the high combustion temperature region occupied the same position.

Due to the fact that DME fuel has no carbon-to-carbon bond in the chemical structure, it is not possible to detect the DME flame and temperature distribution using the visualization system [7]. However, a numerical approach can allow flame and temperature analysis with accurate prediction capability. The above verified numerical model was also used in the DME combustion study and it distinctly showed the effects of pilot injection strategies on DME combustion characteristics including temperature distributions.

2.2 Combustion Characteristics of Pilot Injection

Figure 4 shows the combustion pressure and the heat release rate for DME and ULSD fuels under various pilot injection timings. As can be seen in Figure 4, the first injected fuel increased the in-cylinder pressure and temperature before the main combustion started. Although pilot combustion events were observed at the same time regardless of the first injection timing, they affected combustion pressure differently at the time when the main combustion events began. For the case of BTDC 15 CA (main injection) in Figure 4a, DME combustion shows similar combustion characteristics regardless of pilot timings. This is because the main oxidation was started before the formation of the high pressure and temperature in the combustion chamber by the pilot combustion. The active fuel evaporation rate of DME led to an air/fuel mixture forming a short time after the fuel was injected into the chamber. As a result, the ignition delay seemed shorter and the premixed combustion phase occurred dominantly compared with those of ULSD fuel. On the other hand, the effect of the pilot injection timing on ULSD main combustion was observed, as seen in Figure 4a. When the pilot fuel was injected earlier, the longer air and fuel mixing time made more homogeneous mixtures, which caused the discrepancy of the auto-ignition starting points and the peak of combustion pressure.

As shown in Figure 4b, c, it can be observed that the pilot timing resulted in contrary combustion behavior between DME and ULSD fuels. When the pilot injection timing was retarded in the DME combustion case, higher in-cylinder pressure and a shorter ignition delay

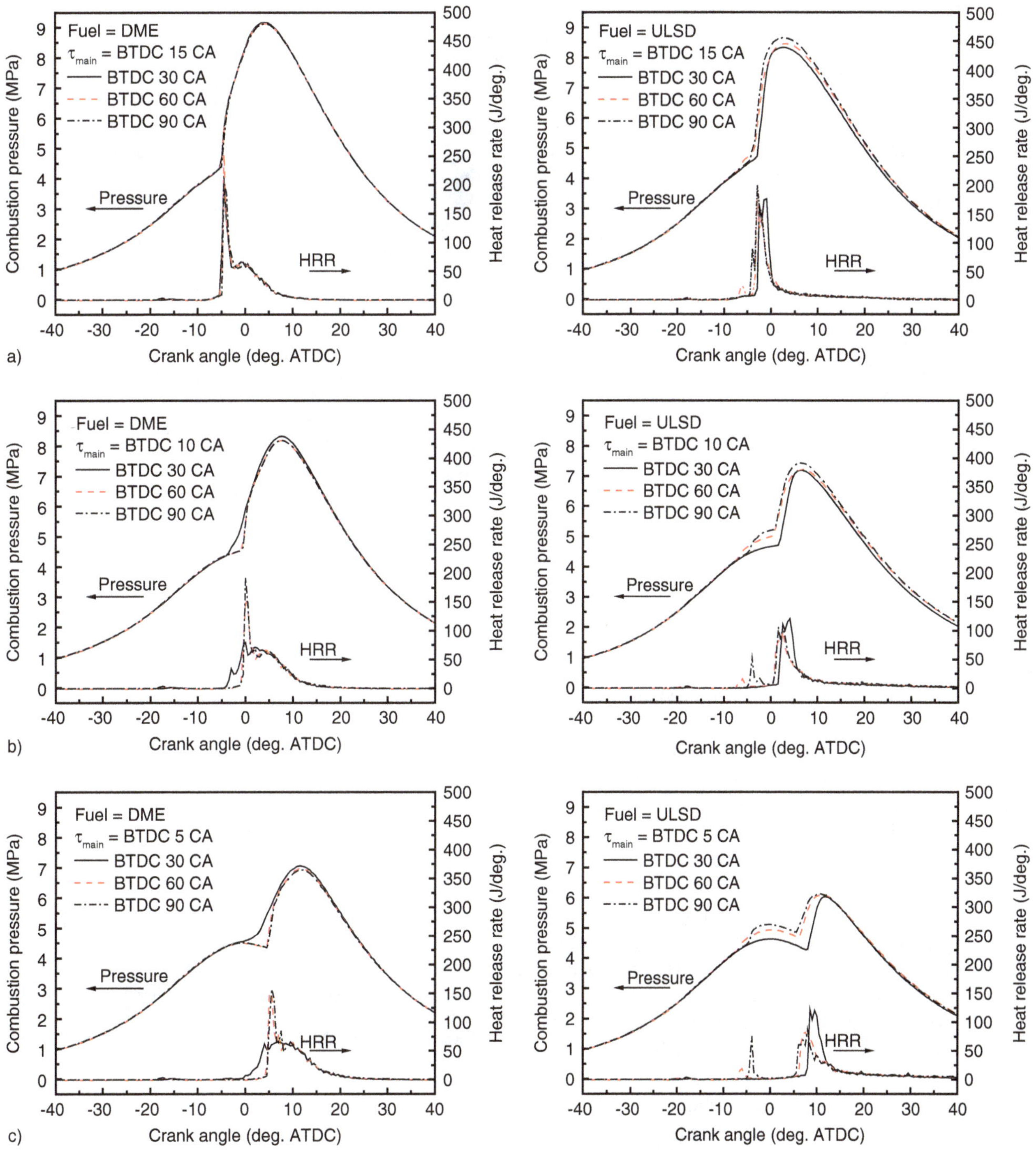

Figure 4

Combustion pressure and heat release rate of various pilot injection timings for DME and ULSD fuels (main injection timing at: a) BTDC 15 CA, b) BTDC 10 CA, and c) BTDC 5 CA).

were observed in the main combustion processing. On the other hand, the peak pressure and ignition delay of ULSD combustion were lower and longer, respectively. Due to the fact that the fast oxidation reaction of the DME fuel suppressed the influence of earlier pilot injections, advanced pilot timings (BTDC 90 CA and 60 CA) caused a similar result in the single-injection case. Whereas, the oxidation of the fuel injected at

BTDC 30 CA continued until the second combustion started, which induced an increased combustion pressure and lower premixed heat release rate.

Although the mass for the pilot injection was fixed at 2 mg in all cases, the ULSD pilot combustion shows strong effects, as seen in the pressure-heat release rate-crank angle diagram. In comparison with the DME properties, the ULSD shows poor atomization and long ignition delay characteristics. For DME fuel properties, low boiling temperature and high vapor pressure make it a gaseous state under lower pressure conditions [13]. Thus, liquefied DME fuel which is pressurized during the injection into the combustion chamber evaporates fast after fuel is emitted from the injector. When DME fuel turns into the gaseous state, the formation of the fuel/air mixture is accelerated and then the ignition starts fast. In addition, the high cetane number (>55) and oxygen contents in the DME lead to a better compression-ignition quality and short ignition delay [21]. On the other hand, the evaporation rate of ULSD fuel is lower and fuel breakup processing takes more time than those of DME fuel. Although the gas oil boils in low temperature conditions (~248 K), the conventional Diesel fuel vaporizes under high ambient temperature (over 450 K), which suppresses the atomization response relatively.

These features of ULSD fuel induced a two-stage combustion phenomenon. One part of the fuel injected in the first step was consumed at around BTDC 18 CA, and the other, which was left in the squish region, was burned rapidly at the end of the compression process. This pilot combustion made the in-cylinder pressure higher than the motoring pressure, thereby shortening the ignition delay of the main injection. In addition, the combustion pressure and the heat release rate had a high peak value during the first-stage combustion because the earliest pilot timing case took a long time to form homogeneity in the fuel-air mixture.

As can be seen in Figure 4, the dwelling time between the pilot and the main injection strongly influenced the combustion characteristics. Figure 4a shows that the advanced main timing dominantly produced a premixed combustion phase. This is because the unburned air-fuel mixture from the first stage was oxidized with the injected fuel from the second injection. When the dwelling time increased, the peak value of the heat release rate in the main combustion event decreased for the ULSD due to the fact that most of the pilot fuel was consumed before the main combustion. For the DME fuel, the well-mixed fuel/air mixtures by the aforementioned fuel properties induced a high premixed combustion phase.

The development of the spray procedure and the equivalence ratio in the pilot injection is illustrated in Figure 5. It is confirmed that the better atomization characteristics of the DME fuel than the ULSD fuel result in smaller droplet particles and a higher equivalence ratio region downstream of the spray jet. These easily broken particles quickly formed a fuel-rich area, mixing with the ambient air. As can be seen in Figure 5b, however, particles which had a large size penetrated farther into the combustion chamber and arrived at the squish area. Moreover, the fuel-rich mixture by the pilot injection appeared at the end of the squish zone in the later time due to the poor atomization properties of the ULSD fuel. During the compression process,

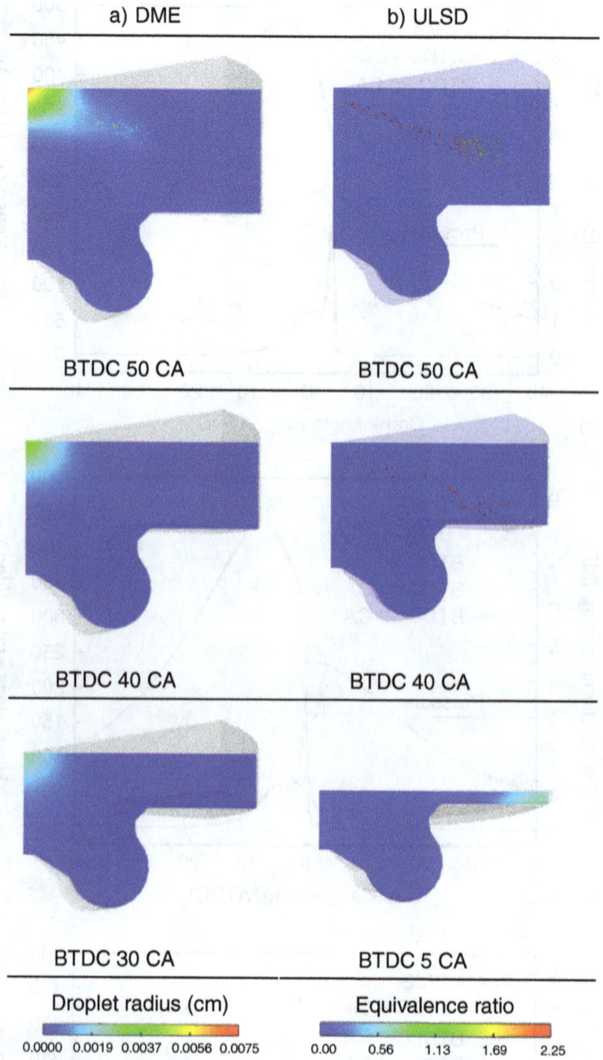

Figure 5

Pilot spray development process and equivalence ratio distribution with respect to the crank angle (τ_{pilot} = BTDC 60 CA, τ_{main} = BTDC 10 CA).

the pressure in the narrow squish space was increased to ignite the fuel-air mixture, which allowed an interesting thing to happen in the ULSD cases: two fast pilot fuels were burned at the same timing (around BTDC 7 ~ 5 CA) in different main timings, as shown in Figure 4. Such fuels formed their mixture at similar squish areas even under various main injection conditions, which generated the same ignition timing. However, the BTDC 90 CA condition showed a high heat release rate due to longer mixing time.

2.3 Combustion Temperature Distribution

The comparison of the temperature distribution for DME and ULSD fuels during the combustion process is illustrated in Figure 6. As there are no carbon-to-carbon bonds in the chemical structure of DME, the combustion temperature distribution is unable to be detected by the experimental equipment. However, it can be predicted by the KIVA code. The main DME combustion started at the side of the spray jet and at an earlier time than ULSD. Furthermore, the combustion process developed rapidly following the trace of the spray penetration. This active combustion characteristic was caused by a good atomization process and oxygen contents in the DME fuel. A high temperature zone spread widely in a short time when the DME fuel burned. On the contrary, the ULSD combustion reaction was observed at the piston lips where the spray impinged. Despite a smaller quantity of ULSD than DME fuel, a slow break-up process generated a longer penetration length and relatively low combustion temperature. For the ULSD combustion temperature, the small amount of entrained air was used for oxidation in the restricted region (at the end of the injection fuel jet) where the premixed fuel-air mixture was contained, resulting in decreased combustion temperatures, as can be seen in Figure 7a. In the average temperature diagram, it shows that the ULSD temperature increased faster and had a lower peak value compared with that of the DME fuel. Although the combustion starting point of DME was advanced, the ULSD temperature increased faster and higher because of the strong pilot combustion.

Figures 7b and 8 represent the effects of the pilot timing on the average temperature and the combustion temperature distribution with top and side views. The injection conditions state that the main injection timing was at BTDC 5 CA, and pilot timings from BTDC 90 CA to BTDC 30 CA in 30 CA steps were employed. There was little difference in temperature values between the BTDC 90 CA and BTDC 60 CA conditions. For comparison of the two pilot injection

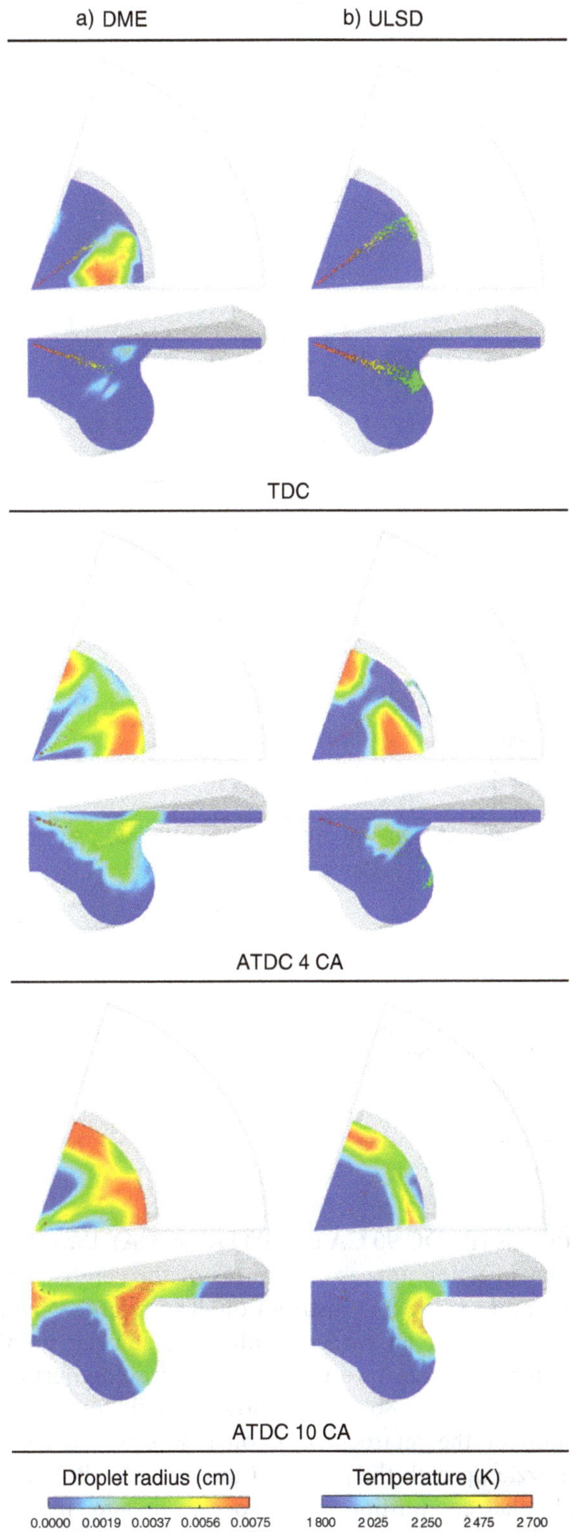

Figure 6

The development of combustion and temperature distribution for DME and ULSD fuels at the τ_{pilot} = BTDC 60 CA and τ_{main} = BTDC 10 CA.

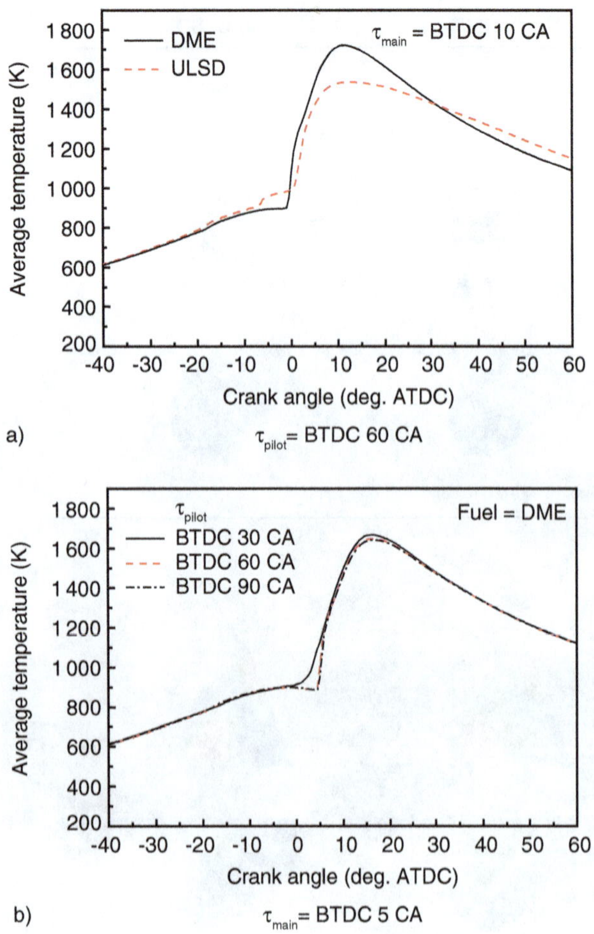

Figure 7

Effect of: a) fuel types and b) pilot injection timings on the average combustion temperature.

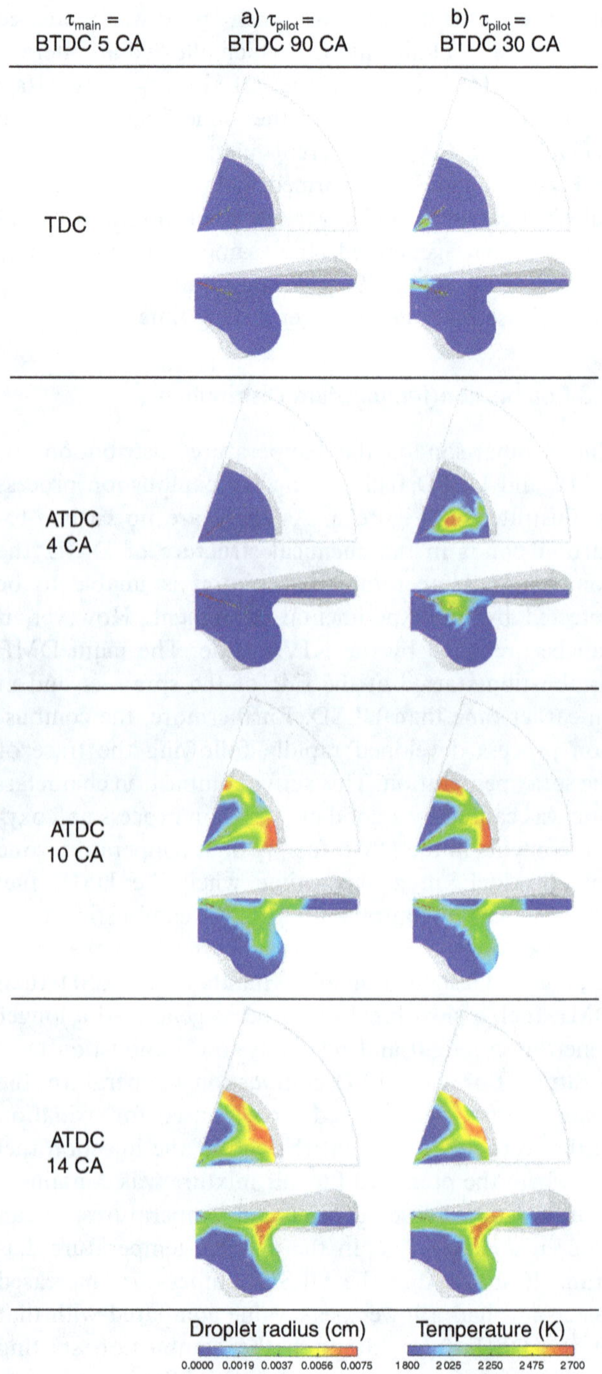

Figure 8

Effect of the pilot injection timing on combustion temperature distribution of DME fuel at τ_{main} = BTDC 5 CA.

conditions (BTDC 90 CA and BTDC 30 CA), the effect of pilot injection appeared in terms of the combustion temperature value and distribution. The retarded pilot timing case had an earlier combustion starting point and higher temperature during the entire combustion process. In Figure 8b, it also shows that the oxidation reaction of the retarded injection case emerged from the nozzle tip with the premixed mixture. On the other hand, the combustion event occurred late in the advanced injection case. After ATDC 10 CA in Figure 8, the temperature contours show very similar patterns between the two different timing conditions, with slightly different range of temperature values. This difference can be explained by the change in pilot injection timings causing different injection targeting points. As a result, the injection at BTDC 90 CA was spread out at the squish region, mixing with ambient air uniformly before the main fuel was injected. The first injected fuel in the BTDC 30 CA case was distributed to the squish

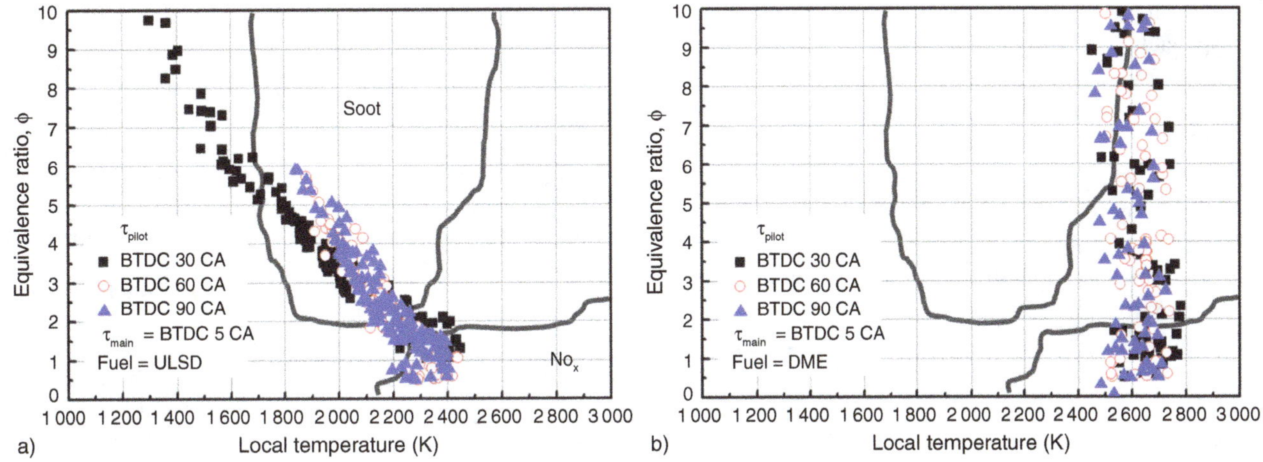

Figure 9

NO_x and soot emission of various pilot injection conditions related to the equivalence ratio and local temperature for: a) ULSD and b) DME fuels.

area and the piston bowl region, which formed a fuel-rich region in the piston bowl. The high equivalence ratio zone generated the main ignition with a high temperature, as illustrated in Figure 8b.

2.4 Exhaust Emission Characteristics

Figure 9 shows that the pilot injection timings affected exhaust emissions such as NO_x and soot on the Φ (equivalence ratio) – T (temperature) map. Each point represents the Φ and T values of calculated cells obtained from the KIVA-3 V at the crank angle at 90% of the maximum heat release rate. The $\Phi - T$ map has been widely used to theoretically analyze the effect of the equivalence ratio and local temperature on combustion products [22, 23]. Akihama *et al.* [24] expanded the concept of the $\Phi - T$ map for the investigation of smokeless Diesel combustion performing 0-D calculations, which was introduced as a reference guideline in this study. The solid line in Figure 9 indicates the soot formation region (shape of a peninsula) and NO_x formation region (right bottom corner), respectively. Furthermore, the concentration of each emission is increased when moving toward the center of the solid line.

As illustrated in Figure 9b, combustion products of the DME fuel occupied the high NO_x region and soot-free region. As described above, due to the fact that there is no carbon-to-carbon bond in the chemical structure, DME combustion rarely creates soot emissions. Moreover, it was revealed that the oxygen content in the fuel and its lean-burn condition generate a high combustion temperature (2 500 K $< T <$ 2 800 K) and suppress soot formation. Due to high fuel-rich mixture cells, the ULSD combustion had relatively lower local temperature values and occupied the high soot formation area. Additionally, these low temperatures and the high soot area resulted in a small amount of NO_x emissions and a high concentration of soot emissions in the ULSD combustion.

The effect of the pilot injection timing on emissions is also represented in Figure 9 when the main injection timing was BTDC 5 CA and pilot injection timings were from BTDC 90 CA to BTDC 30 CA The retarded pilot timing condition shows a slightly higher temperature for both fuels. Especially, the BTDC 30 CA case in the ULSD combustion has a wide range of equivalence ratios and local temperatures. This is because the first injected fuel does not have sufficient time to be mixed with the charged air. On the other hand, the results of the other conditions are gathered in a specific range with $0 < \Phi < 6$ and 1 800 K $< T <$ 2 400 K. Pre-combustion reduced the main combustion temperature as well as premixing the first injection, which did not form a fuel-rich mixture.

CONCLUSIONS

This numerical investigation was performed to verify the effects of various pilot injection timings on the combustion behavior of DME and ULSD fuels in terms of combustion pressure, heat release rate, and temperature distribution, which were undetectable in the experimental approach. To heighten the accuracy of the simulation, model validation compared with the experimental results was conducted and showed good agreement. The KIVA-3 V code implemented numerical experiments of various injection conditions fueled with DME and ULSD and drew the results as follows:

1. The span between the pilot injection timing and the main timing strongly affected combustion characteristics. With a short span the advanced main timing (BTDC 15 CA) decreased the influence of the pilot injection to start the main combustion before pilot oxidation reacted. In contrast, a retarded second injection condition initiated under high ambient pressure and temperature decreased the peak combustion pressure and heat released by the first combustion;
2. Due to different fuel properties for DME and ULSD fuels, an opposite trend of combustion pressure and heat release rate occurred. Pilot-injected fuel formed a fuel-air mixture at different positions in the combustion chamber according to the test fuels. As a result, the first combustion occurred in different areas;
3. DME combustion occurred in a wide range of areas within the combustion chamber, with a higher combustion temperature compared with ULSD combustion. Although the pilot injection led to the change in the main ignition timing and temperature spectrum values, the overall temperature distribution pattern was determined by the main combustion;
4. The production of NO_x and soot emissions is represented on the expanded $\Phi - T$ map. DME combustion occupies soot-free combustion and high-temperature regions, which also suppressed the formation of soot emissions. On the other hand, ULSD combustion included high soot formation conditions and a low-temperature region in which a low concentration of NO_x is produced. Retarded pilot timing cases had a slightly higher local temperature than other conditions.

ACKNOWLEDGMENTS

This work (2012R1A2A2A01046859) was supported by Mid-career Researcher Program through NRF (National Research Foundation) grant funded by the MSIP (Ministry of Science, ICT and Future Planning).

REFERENCES

1 Teng H., McCandless J.C., Schneyer J.B. (2004) Thermodynamic properties of dimethyl ether – an alternative fuel for compression-ignition engines, *SAE Tech. Paper* 2004-01-0093.

2 Kim M.Y., Yoon S.H., Ryu B.W., Lee C.S. (2008) Combustion and emission characteristics of DME as an alternative fuel for compression ignition engines with a high pressure injection system, *Fuel* **87**, 2779-2786.

3 Sato Y., Nozaki S., Noda T. (2004) The Performance of Diesel Engine for Light Duty Truck Using a Jerk Type In-Line DME Injection System, *SAE Tech. Paper* 2004-01-1862.

4 An B., Sato Y., Lee S.W., Takayanagi T. (2004) Effects of Injection Pressure on Combustion of a Heavy Duty Diesel Engine with Common Rail DME Injection Equipment, *SAE Tech. Paper* 2004-01-1864.

5 Yoon H., Yeom K., Bae C. (2007) The Effects of Pilot Injection on Combustion in Dimethyl-ether (DME) Direct Injection Compression Ignition Engine, *SAE Tech. Paper* 2007-24-0118.

6 Song R., Li K., Feng Y., Liu S. (2009) Performance and Emission Characteristics of DME Engine with High Ratio of EGR, *Energ. Fuel.* **23**, 5460-5466.

7 Jeon J., Kwon S.I., Park Y.H., Oh Y., Park S. (2014) Visualizations of combustion and fuel/air mixture formation processes in a single cylinder engine fueled with DME, *Appl. Energ.* **113**, 294-301.

8 Amsden A.A. (1993) KIVA-3: A KIVA program with block-structured mesh for complex geometries, *Los Alamos National Laboratory Report* LA-12503-MS.

9 Amsden A.A. (1997) KIVA-3 V: A block-structured KIVA program for engines with vertical or canted valves, *Los Alamos National Laboratory Report* LA-13313-MS.

10 Kee R.J., Rumpley F.M., Miller J.A. (1989) CHEMKIN-II: A Fortran chemical kinetics package for the analysis of gas phase chemical kinetics, *Sandia National Laboratory Report* SAND89-8009.

11 Reitz R.D. (1987) Modeling atomization processes in high-pressure vaporizing sprays, *Atomization Spray.* **3**, 309-337.

12 Han Z., Reitz R.D. (1995) Turbulence modeling of internal combustion engines using RNG k-ε models, *Combust. Sci. Technol.* **106**, 267-295.

13 Teng H., McCandless J.C., Schneyer J.B. (2001) Thermochemical characteristics of dimethyl ether – an alternative fuel for compression-ignition engines, *SAE Tech. Paper* 2001-01-0154.

14 Teng H., McCandless J.C., Schneyer J.B. (2002) Viscosity and lubricity of (liquid) dimethyl ether – an alternative fuel for compression-ignition engines, *SAE Tech. Paper* 2002-01-0862.

15 Teng H., McCandless J.C., Schneyer J.B. (2003) Compression ignition delay (physical + chemical) of dimethyl ether – an alternative fuel for compression-ignition engines, *SAE Tech. Paper* 2003-01-0759.

16 Teng H., McCandless J.C. (2005) Comparative study of characteristics of Diesel-fuel and dimethyl-ether sprays in the engines, *SAE Tech. Paper* 2005-01-1723.

17 Patel A., Kong S.C., Reitz R.D. (2004) Development and validation of a reduced reaction mechanism for HCCI engine simulation, *SAE Tech. Paper* 2004-01-0558.

18 Kong S.C., Sun Y., Reitz R.D. (2007) Modeling Diesel spray flame liftoff, sooting tendency, and NO_X emissions using detailed chemistry with phenomenological soot model, *J. Eng. Gas Turb. Power* **129**, 245-251.

19 Patterson M.A., Kong S.C., Hampson G.J., Reitz R.D. (1994) Modeling the effects of fuel injection characteristics on Diesel engine soot and NO_X emissions, *SAE Tech. Paper* 940523.

20 Schmidradler D., Werlberger P. (1999) Engine Videoscope Thermovision: Vision Based Temperature Measurement for Diesel Engine, *ISATA 99ME039*.

21 Gill D., Ofner H. (1999) DiMethyl Ether – a clean Fuel for Transportation, *SAE Tech. Paper* 990059.

22 Kmimoto T., Bae M. (1988) High Combustion Temperature for the Reduction of Particulate in Diesel Engines, *SAE Tech. Paper* 880423.

23 Neely G.D., Sasaki S., Huang Y., Leet J.A., Stewart D.W. (2005) New Diesel Emission Control Strategy to Meet US Tier 2 Emissions Regulations, *SAE Tech. Paper* 2005-01-1091.

24 Akihama K., Takatori Y., Inagaki K., Sasaki S., Dean A.M. (2001) Mechanism of the Smokeless Rich Diesel Combustion by Reducing Temperature, *SAE Tech. Paper* 2001-01-0655.

10

Sensitivity of Technical Choices on the GHG Emissions and Expended Energy of Hydrotreated Renewable Jet Fuel from Microalgae

Laure Patouillard* and Pierre Collet

IFP Energies nouvelles, 1-4 avenue de Bois-Préau, 92852 Rueil-Malmaison - France
e-mail: e-mail:laure.patouillard@ifpen.fr - pierre.collet@ifpen.fr

* Corresponding author

Abstract — *Taking into account the environmental impacts of biofuel production is essential to develop new and innovative low-emission processes. The assessment of life cycle GreenHouse Gas (GHG) emissions of biofuel is mandatory for the countries of the European Union. New biomass resources that hardly compete with food crops are been developed increasingly. Microalgae are an interesting alternative to terrestrial biomass thanks to their high photosynthetic efficiency and their ability to accumulate lipids. This article provides an analysis of potential environmental impacts of the production of algal biofuel for aviation using the Life Cycle Assessment (LCA). Evaluated impacts are GHG emissions and the primary energy consumption, from extraction of raw materials to final waste treatment. This study compared two management choices for oilcakes generated after oil extraction from microalgae. In the first system, these cakes are treated by energetic allocation and in the second by anaerobic digestion. In both cases, the steps of cultivation and harvesting have the highest impact on the results. Sensitivity analyzes are performed on technical choices of operating systems (choice of the type of nutrients, mode of harvesting, drying and oil extraction) as well as a Monte-Carlo analysis on key parameter values for GHG emissions (concentration of microalgae in ponds, productivity and oil content). The results highlight the impact of the use of chemical fertilizers and the importance of the concentration of algae on GHG emissions and energy consumption.*

Résumé — **Analyse de sensibilité des paramètres techniques sur les émissions de gaz à effet de serre et les consommations d'énergie de la production de biocarburant pour l'aviation à partir de microalgues** — La prise en compte des effets sur l'environnement de la production de biocarburants est essentielle afin de développer de nouveaux procédés innovants et peu polluants. L'évaluation du bilan de gaz à effet de serre des biocarburants est à ce titre un prérequis obligatoire pour les pays de l'Union Européenne. L'exploitation de nouvelles ressources de biomasse induisant peu de compétition avec les cultures alimentaires est de plus en plus recherchée. Grâce à leur très haut rendement photosynthétique et leur capacité à accumuler les lipides, les microalgues représentent une solution alternative intéressante. Cet article fournit une analyse des impacts environnementaux potentiels de la production de biocarburant pour l'aviation à partir de microalgues à l'aide de l'Analyse de Cycle de Vie (ACV). Les impacts évalués sont les émissions de Gaz à Effet de Serre (GES) et les consommations d'énergie primaire, depuis l'extraction des matières premières jusqu'au traitement ultime des déchets. Cette étude compare deux choix de gestion des tourteaux générés après extraction de l'huile des microalgues. Dans le premier système, ces tourteaux sont traités par allocation énergétique et dans le second par digestion anaérobie. Dans les deux cas, les étapes de culture et de récolte sont les plus impactantes. Des analyses de sensibilité, portant à la fois sur des choix techniques d'exploitation des systèmes (choix du type de nutriments, du mode de récolte, de séchage et d'extraction d'huile) ainsi que sur une analyse de Monte-Carlo des valeurs de paramètres clés pour les émissions de GES (concentration en microalgues dans les bassins, productivité et teneur en huile) ont été réalisées. Elles ont permis de souligner le poids de l'utilisation de fertilisants chimiques et l'importance de la concentration en algues dans les émissions de GES et les consommations en énergie.

DEFINITIONS/ABBREVIATIONS

DAF	Dissolved Air Flotation
GHG	GreenHouse Gas
GWP	Global Warming Potential
HRJ	Hydrotreated Renewable Jet
HVO	Hydrotreated Vegetable Oil
ISO	International Standard Organization
LCA	Life Cycle Analysis
LCI	Life Cycle Inventory
LCIA	Life Cycle Impact Assessment
LHV	Lower Heating Value
NREP	Non Renewable Expended Primary (energy)
PBR	PhotoBioReactor
RED	Renewable Energy Directive
SWAFEA	Sustainable Way for Alternative Fuel and Energy in Aviation
TSS	Total Solid Suspended
WTW	Well to Wake

INTRODUCTION

In a context of climate change and fossil fuel depletion, there is a rising interest in the development of alternative and renewable sources of energy. Renewable energies from wind, sun or geothermal power are promising ways to produce electricity or heat, but these energy forms are hard to store. Consequently, in the transport sector, the development of biomass liquid biofuels in order to power cars and planes is crucial. The use of agroresources to produce biofuels generally induces a lower impact on climate change but can increase other environmental impacts, especially those linked to the consumption of fertilizers (*e.g.* eutrophication, acidification) and to the use of pesticides (*e.g.* toxicity).

Microalgae as feedstock to produce biofuel, often considered as third generation biofuels, can represent an interesting way to produce storable bioenergy [1, 2]. Their high photosynthetic yield, a better control of the ground emissions and the ability to use CO_2 directly from industrial emissions as a source of carbon are promising way to reduce environmental impacts of biofuels. However, this new biofuel production system should be assessed in order to analyze its environmental performances and to identify which processes should be improved.

This paper proposes an environmental assessment of the production of Hydrotreated Vegetable Oil (HVO) from microalgae with a Life Cycle Assessment (LCA) approach and compares it to a petroleum alternative. A sensitivity analysis on technical choices and on parameter values is carried out in order to discuss the variation of the results.

1 LCA METHODOLOGY

1.1 General Presentation of LCA

LCA is a modeling tool to consider and quantify the total environmental effects of a process or service, based on a "cradle to grave" inventory of all the emissions and all the consumed resources. All the different steps, from the production of raw materials to the disposal of waste and products at the "end of life", are therefore included in the scope of a LCA. According to the International Standard Organization (ISO) recommendations [3, 4], LCA is divided into four steps: goal and scope definition, Life Cycle Inventory (LCI), Life Cycle Impact Assessment (LCIA) and life cycle interpretation. LCA is an environmental tool, so social and economic aspects are not included in the scope of the publication. Nevertheless, decision making must be taken in light of the two other pillars of sustainable development.

Goal and scope definition is the step in which the main objectives of the work are defined. The perimeter of the considered system and the choice of the environmental impacts to be assessed are fixed. The functional unit, which is the reference flow adapted to the function of the assessed product, is determined. All inputs and outputs in the LCI, and consequently all the environmental impacts are related to this functional unit. Many systems lead to the production of several products with different functions. Hence, allocation rules are defined to distribute the environmental burdens between the main product and the coproducts. This is one of the most critical issues in LCA, and the work of Luo *et al.* have highlighted the sensitivity of the results of biofuels production to these allocation rules [5].

The LCI is the step where all the necessary data are collected and treated. This is generally the most time consuming step of the LCA [6], but with relatively few methodological problems.

During the LCIA step, all the environmental flows (consumed resources and emissions to the environment) are converted into a selected number of impacts. Depending on the chosen impact method, midpoint impacts (*e.g.* ozone depletion, acidification) or endpoint impacts (human heath, ecosystem quality) can be selected.

The last step of the ISO recommendations is the interpretation of the results, in which the obtained results are assessed and the limits of the study are set.

1.2 Goal and Scope Definition

This study is based on a previous work carried out for the Sustainable Way for Alternative Fuels and Energy for Aviation (SWAFEA) project. The European project SWAFEA aimed at the assessment of the feasibility and the impacts of introducing alternative fuels in the aviation sector from technical, economical and environmental points of view. A LCA study was conducted in the SWAFEA project to assess the environmental impacts of alternative fuels that were technically selected as promising for the aviation sector. The SWAFEA project selected several types of fuels as promising fuels for the aviation sector. Hydrotreated Renewable Jet fuel (HRJ) from biomass derived oils is one of them as it is a "drop-in" fuel meaning that HRJ is fully compatible with current aircraft engines and fuel supply infrastructures without any additional economic investment. In addition, HRJ production relies on a mature production process and can be blended with conventional jet fuel A1. Numerous flight demonstrations have proven its potential for being an alternative to jet fuel. On the other hand, algae have been selected in the SWAFEA project as a promising biomass feedstock for alternative jet fuel production as they promise higher yields than terrestrial crops and have modest requirements on land quality and avoid a direct competition with food.

TABLE 1

Global Warming Potentials at 100 years time horizon [3]

Flow	IPCC, 2007
CO_2	1 (1 g CO_2 = 1 gCO_2eq)
CH_4	25 (1 g CH_4 = 25 gCO_2eq)
N_2O	298 (1 g N_2O = 298 gCO_2eq)

Hence this study concentrates on the assessment of one pathway of this promising fuel: HVO from microalgae extracted from SWAFEA LCA study [7].

The goal of this LCA study, as defined in the SWAFEA project, is to evaluate the GHG emissions and primary energy consumption related to the production and the consumption of alternative fuels in the aviation sector on a "Well To Wake" (WTW) basis considering the current most promising alternative fuels. A WTW LCA study deals with the assessment of environmental impacts of a fuel all along its life cycle considering the raw material extraction, the fuel production step, the fuel distribution step and the fuel combustion step in a vehicle.

The methodology used is based on LCA principles as defined in the ISO standards [3, 4] and on the recommendations of the Renewable Energy Directive (RED) [8].

Two environmental indicators are assessed in this study: GHG emissions and Non Renewable Expended Primary (NREP) energy, defined as fossil energy by [9]. Non renewable primary energy represents the sum of all the fossil and mineral energy sources directly drawn from natural reserves such as crude oil, natural gas, coal and uranium.

For the GHG emissions indicator, only the 3 main GHG are taken into account – *i.e.* CO_2, CH_4 and N_2O – and converted into a CO_2 equivalent (CO_2eq) using coefficients called Global Warming Potentials (GWP at 100 years time horizon), presented in Table 1, defined and estimated by IPCC [4] (IPCC 2007 values have been chosen for this study). The final results are expressed as CO_2eq.

This study does not include the shorter term impact of aviation on climate (radiative forcing) due to non-CO_2 emissions and effects, which include the emissions of water vapour, particles and nitrogen oxides (NO_x).

A WTW approach has been chosen including algae production, recovery and treatment, extraction of raw material, conversion into fuel, transport and distribution of fuel and fuel use. Infrastructure impacts (and aircraft cycle in particular) are excluded from the scope of study as well as land use impacts.

In this study, the net balance between CO_2 captured for algae growth and CO_2 released to the atmosphere during fuel combustion in the engine is assumed to be equal to zero. Assuming that CO_2 used for algae growth would have been emitted to the atmosphere regardless, calculations take into account neither CO_2 entering the system for algae growth nor CO_2 leaving the system during the combustion step. Cultivating algae using waste CO_2 from industrial plant is a way to recycle those emissions but not to stock them.

All impacts are expressed on the basis of the same functional unit that is the MJ of jet fuel produced.

Calculations are made with a LCA software GABI (v4.3). Data are extracted from literature and the EcoInvent v2.2 database [10].

Methodology for taking into account coproducts in the first pathway has been defined considering the RED recommendations. Impacts are allocated to co-products (excluding waste that has no impact and electricity that is taken into account differently) on the basis of their energy content (Lower Heating Value, LHV). As a consequence this allocation method is used to share the impacts between products and coproducts from hydrotreatment process in all pathways and for biomass residues in the first pathway.

Concerning excess electricity coproduction by cogeneration, we used the avoided impact methodology considering a credit equal to the amount of impacts that would be produced when the same amount of electricity is generated in a power plant using the same feedstock in a cogeneration unit (as recommended in the RED). So excess electricity produced in pathway 2 substitutes to electricity produced by cogeneration from biomass residues. Impacts induced by electricity production from anaerobic digestion of algae biomass residues are taken into account in this second pathway.

2 DEFINED PATHWAYS AND CORRESPONDING DATA

Process chains for jet biofuel production from algae studied in this publication are prospective and extrapolated from lab-scale data since no industrial pathway exists yet. Yields and process choice and process data are extracted from available literature and adapted for each considered pathways in order to minimize burdens (*Tab. 2*).

Two pathways are considered for HVO production from microalgae. Those pathways have similarities and differences in terms of biomass characteristics and in terms of considered processes (*Tab. 3*). The first pathway will be described in detail (*Fig. 1*); the other will be described on the basis of the first only highlighting the differences (*Fig. 2*).

The average European electricity mix provided by the EcoInvent database is considered in the calculation.

2.1 First Pathway: Base Case

Microalgae are grown in open raceways mixed with paddle wheels in a facility covering about 100 ha [11]. Open ponds have been chosen mainly because the alternative technology, PhotoBioReactors (PBR), seems to be too expensive in terms of capital costs [12]. Furthermore energy costs of PBR are claimed to be too high for microalgae biomass production [13]. Ponds are located near a power plant in order to recycle the flue gases with 15% mass of CO_2 [14]. A purification step of flue gases is required in order to extract CO_2 [15] and to inject it in the growth medium. The uptake efficiency of CO_2 is set to 90% according to [16]. Nutrients are supplied by sludge from waste water treatment plant in order to lighten the environmental burden in a best case perspective [17, 18]. The algae culture medium is freshwater that is recycled up to 90% after harvesting, *i.e.* all the water removed from the harvesting process is feeding back to the pond. The remaining 10% is processed in a waste water treatment plant. No N_2O emission at the pond level is taken into account, in accordance with most of the LCA studies on bioenergy production from microalgae. This assumption is confirmed by the experimental data from [19] suggesting that N_2O emissions at pond level are much lower than conventional terrestrial crops.

Microalgae productivity is 25 g/(m^2.day) with an algae concentration in ponds of 0.5 kg/m^3 [16]. The oil content is 34% of dry matter which is the average of values available in the literature.

Microalgae harvesting is carried out by a Dissolved Air Flotation system (DAF) with an addition of hydrophobic polymer to reach a concentration of 20 g/L of dry matter with an efficiency of 95% [14]. Recovered water during harvesting is filtered to remove polystyrene before going back into the ponds.

Microalgae are pretreated before oil extraction. First harvested algae are concentrated up to 200 g/L Total Solid Suspended (TSS) by centrifugation before being dried. The heat used for drying is waste heat from the power plant, assuming that this plant produces enough heat to support algae drying needs and achieve an algal paste concentration of 90% TSS [20].

Oil is then extracted from microalgae by solvent extraction with hexane. This type of extraction process on dried microalgae paste leads to a recovery of 95% of their oil content [14]. This step leads to the production of microalgae oilcakes, which are taken into account with an energetic allocation.

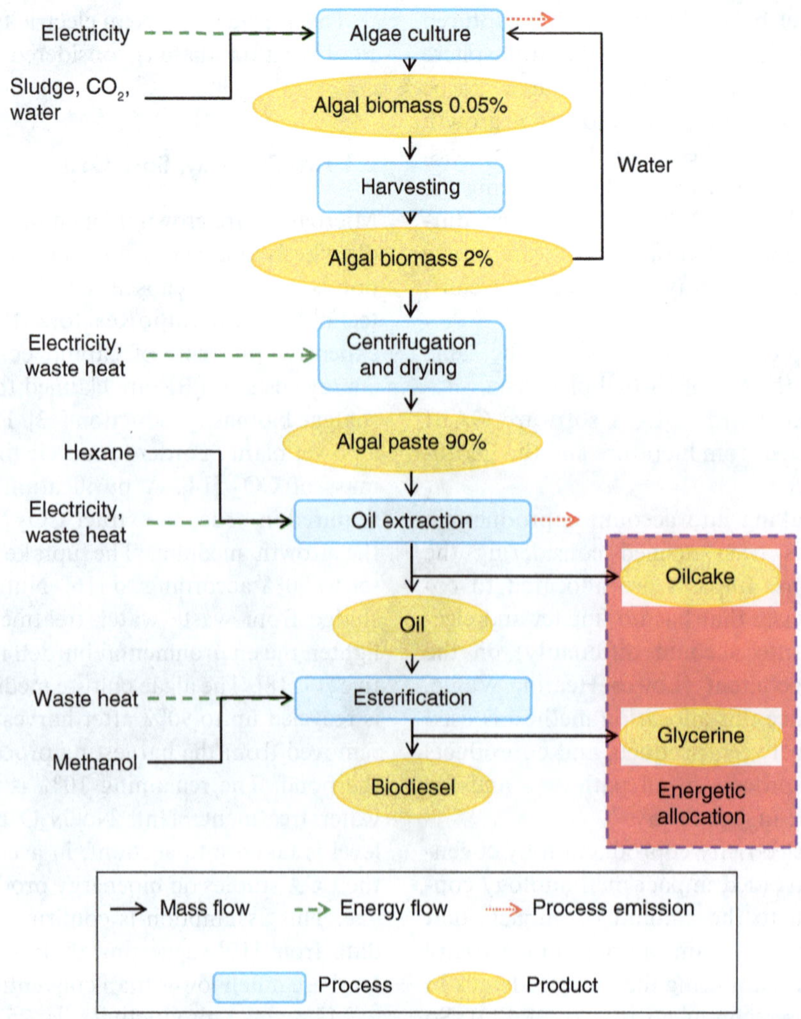

Figure 1
System boundaries of the first pathway.

Algal oil needs to be refined before being hydrotreated. Technical data on oil refining and hydrotreatment processes considered in this study correspond to the NexBTL process commercialized by *Neste Oil.* NexBTL is a HVO process that coproduced biogasoline and fuel gas converted into electricity and steam, part of the electricity production is used as utility for the process [21]. Data are presented for one MJ of HRJ even if others fuels (gasoline, naphtha) are produced *i.e.* an energetic allocation has already been made between fuels leaving processes. Other coproducts are taken into account with an energetic allocation in our calculations.

HRJ is supposed to be transported from the production site to a depot and then to a filling station in a 40 tons truck over a distance of 150 km. The corresponding Diesel oil consumption – including return trip (empty) – is 0.00344 MJ per MJ of HVO. The fuel is then stored in a depot. Corresponding electricity consumption is 0.00084 MJ per MJ of fuel. Electricity consumption associated with fuel distribution in a filling station is supposed to be 0.0034 MJ par MJ of fuel. These assumptions are in line with the ones from JEC 2008 study [22].

2.2 Second Pathway

The second pathway is quite similar to the first one. Main differences are process choices for nutrient supply, harvesting, and algal residues valorisation steps.

Harvesting is done thanks to settling ponds. Algae culture medium is fed into these settlers that allow

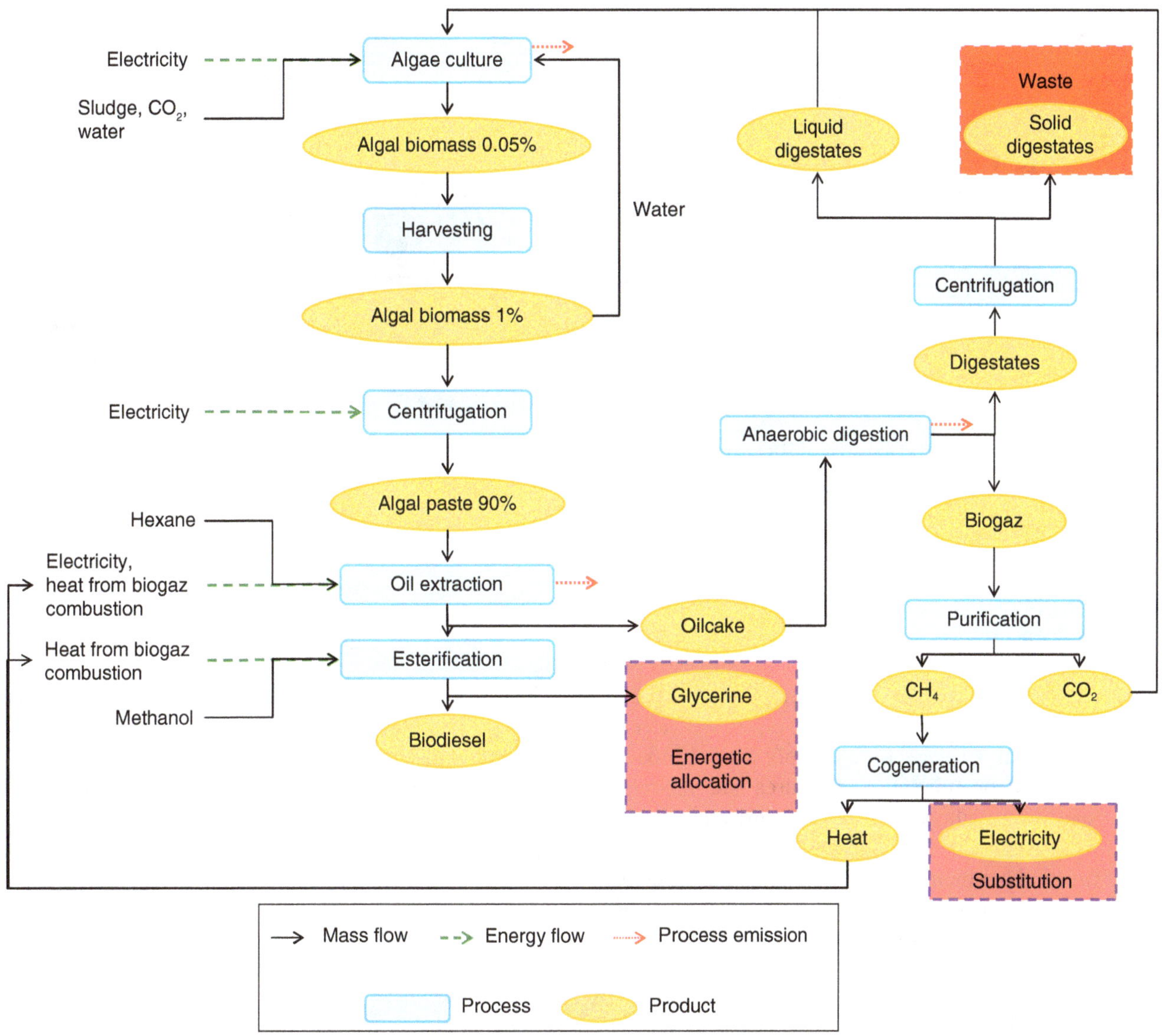

Figure 2

System boundaries of the second pathway.

---

to recover 65% of microalgae with a concentration of 10 g/L [16]. Electricity consumptions are lower compared to DAF system but final algae concentration is reduced. A rotary press is used to concentrate algae instead of centrifugation. Electric consumption associated to rotary press is smaller than centrifugation for almost the same efficiency [23]. The oil is extracted by press, with a lower efficiency than the hexane extraction (75%).

Algal biomass residues are converted into biogas for electricity and heat production by cogeneration. Algal biomass residues are fed into anaerobic digester for biogas conversion containing 70%vol of methane and 30%vol of carbon dioxide. The methane potential is equal to 209 mL CH_4/gTSS, which is a conservative value in the range of the experimental data of [24]. Heat required for digester operation is provided by steam from cogeneration. Biogas is then purified into methane by using water purification process. It consists in bubbling biogas in pressurized water in order to dissolve CO_2 in it and recover CH_4. Methane is converted into electricity with an efficiency of 32% [14] by cogeneration and electricity is fed into the grid. In this pathway, 1 kg (dry mass) of cultivated algae leads to an electricity production of 2.28 MJ.

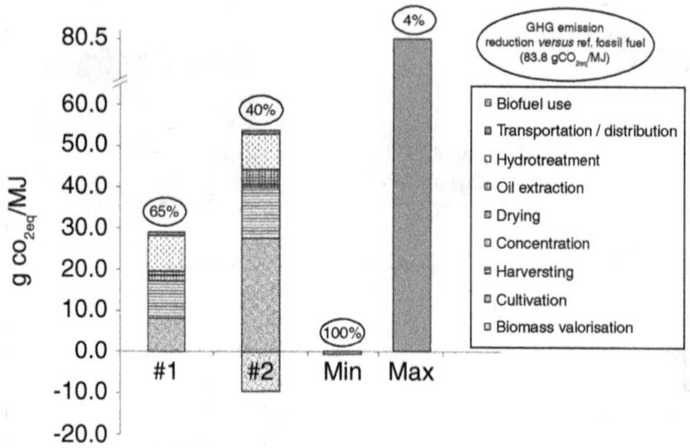

Figure 3

WTW GHG emissions in gCO_{2eq}/MJ of HRJ from microalgae.

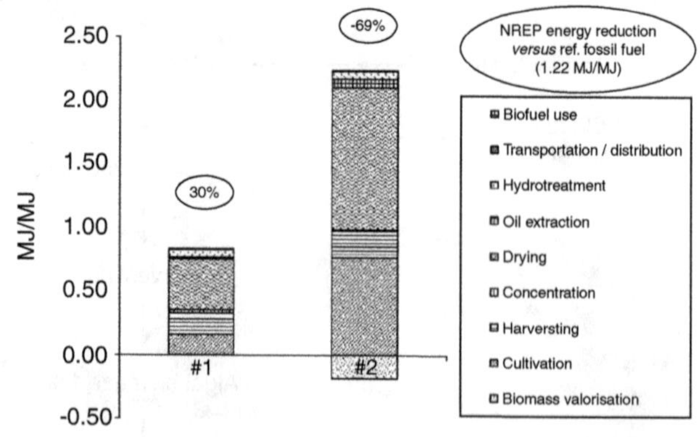

Figure 4

WTW non renewable expended primary energy in MJ/MJ of HRJ from microalgae.

Liquid digestates are recovered and fed into cultivation ponds. It allows to recycle 75% of nitrogen and 50% of phosphorus fed into cultivation ponds and to use it as nutrients for microalgae [25]. An addition of chemical nutrient is necessary to comply with photosynthesis needs. Nitrogen is supplied by urea and phosphorus by single superphosphate. In this pathway, CO_2 is provided by two ways: CO_2 recovered from the purification of the biogas which is dissolved in water and the rest from purified flue gas as in the first pathway. Solid digestates are considered as wastes.

3 RESULTS

3.1 GHG Emissions and Expended Primary Energy for Defined Pathways

Results presented below are relative to the production of HRJ from microalgae for the two pathways defined in the previous section. The two graphs display GHG emissions (*Fig. 3*) and NREP energy (*Fig. 4*) with the relative contribution of every life cycle steps for each pathway.

The GHG emission reduction of the two pathways compared to the RED fossil fuel reference (83.8 gCO_{2eq}/MJ) is mentioned. The minimum [14] and maximum [31] values of literature for similar systems (production of biodiesel in open raceways) have also been added to give some idea about the variability that exists in the literature for this pathway. We could not perform a full Monte-Carlo analysis to estimate uncertainties related to our model as we were not able to estimate a distribution for all parameters. Nevertheless we conduct a Monte-Carlo analysis for 3 main parameters (Sect. 3.2.2).

The NREP energy can be compared with the corresponding for fossil fuel which is equal to 1.22 MJ/MJ of Diesel according to EcoInvent 2.2. Comparison with the references used in Figure 4 [14, 31] was not possible because the results of the energy consumption were not expressed in the same metric. In this section, we also propose to analyse the influence of the coproduct management on the results.

3.1.1 Management of the Coproducts

Following the methodology for LCA of biofuel defined in the RED, an allocation based on energy content is applied in the first pathway to allocate the impacts between algae oil and algal residues. For the second pathway, the avoided impact method is used for taking into account the coproduction of electricity from algal residues by cogeneration as recommended by the RED methodology. In this case, all the impacts are only allocated to the main product (here one MJ of HRJ fuel) and a credit is given to the electricity coproduced from algal residues. This credit is equal to the amount of impacts that would be generated by the production of the same amount of electricity produced in a conventional way. These different methodological choices induce lower impacts in the pathway 1 than in the pathway 2.

Indeed with the energetic allocation in the pathway 1, 45% of the impacts from cultivation to oil extraction are allocated to algae oil. But all these impacts are allocated to the oil when using the avoided impact method in the

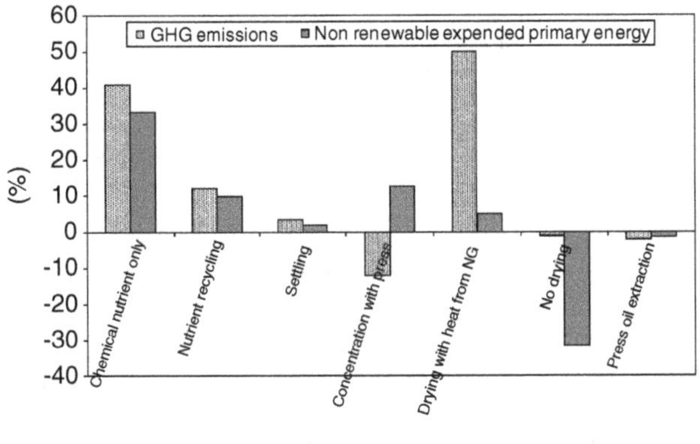

Figure 5

Sensitivity to technical choices compared to base case.

pathway 2. In this pathway, impact credit induced by the production of electricity from biomass residues just offset about 15% of the impacts produced by pathway while 55% of "real" impacts are allocated to biomass residues in pathway 1. This high sensibility of the results to the chosen allocation method is sharpened by the high energy content of the oilcake (21 MJ/kg) contrary to other coproducts with low energy content like straw in other biofuel pathways.

It is important to remind that the choice of a method for taking into account coproduct is determined by technical choice for the valorisation of biomass. Indeed, pathways with electricity coproduction by cogeneration cannot be studied with the same method as pathways coproducing oilcakes for animal feed for instance when following RED recommendations. Therefore both pathways presented here do not have to be compared one to each other but help for the understanding of algae pathway mechanisms.

3.1.2 Climate Change Impact

GHG emissions related to the production of one MJ of HRJ from microalgae are equal to 29.4 and 44.2 gCO_{2eq} for the first and the second pathway. It corresponds to a GHG emission reduction of respectively 65% and 40% compared to the RED fossil fuel reference. Consequently, results for the first pathway comply with the GHG emission reduction threshold of 60% set by the RED from 2017 [8].

For GHG emissions, the most impacting steps are the cultivation step (28% for pathway 1 and 51% for the pathway 2), the harvesting step (respectively 31% and 23%) and the hydrotreatment of algal oil for the first pathway (30% of total GHG emissions). At cultivation step, more than 90% of impacts for this step are due to electricity consumption. This electricity is used for CO_2 supply, water pumping and for water mixing in ponds. GHG emissions for harvesting step are due to electricity consumption for DAF operation in pathway 1 and water pumping for both pathways. GHG emissions associated with hydrotreatment step are linked with the hydrogen supply that comes from on site natural gas steam reforming. In the second pathway, 18% of GHG emissions are compensated by substituting the co-produced electricity. It should be underlined that there is no CO_2 emission linked with the drying step in both pathways. This is due to the use of waste heat with no environmental burden.

3.1.3 Energetic Balance

Focusing on the NREP energy, the first pathway leads to the consumption of less than 1 (0.85) MJ of fossil fuel for the production of 1 MJ of HRJ. In the second pathway, the ratio is higher than one (2.06 MJ/MJ), which underlines the higher dependence to fossil energy consumption of this pathway.

For NREP energy, most impacting steps are cultivation, harvesting and drying steps for the first pathway (respectively 19%, 21% and 47% of the NREP energy), and cultivation and drying steps for the second pathway (34% and 50% of the total NREP energy respectively). As for GHG emissions, NREP energy at cultivation and harvesting steps is due to electricity consumption. Drying step consumes a large amount of heat for drying algae (pathway 1 and 2). Here the heat consumed is waste heat coming from an industrial plant. This wasted heat is assumed to be an elementary flow of primary energy, meaning that the consumption of one MJ of this heat induces the consumption of one MJ of non renewable primary energy. It means that the consumption of this wasted heat does not induce any environmental burden. Hence, non renewable primary energy consumption associated with heat supply for drying algae represents 50% of non renewable primary energy consumption of pathways 1 and 2.

Next sections will analyse different aspects of the model (technical choices and main parameter values) in order to test their sensitivity.

3.2 Sensitivity Analyses

Sensitivity analyses are performed both on technical choices and on main parameters affecting GHG emissions. Technical sensitivity is done through scenarios comparison, and sensitivity analysis on the relevant parameters through Monte-Carlo analysis.

TABLE 2
Process options for each pathway to be assessed in SWAFEA project

Step	Input/process	Available options	Pathway 1	Pathway 2
Cultivation in open pounds	CO_2	From flue gases	x	x (in complement)
		From CO_2 recycling from anaerobic digestion		x
	Nutrients	From chemical only		x (in complement)
		From sludge	x	
		From recycling		x
	Light	Sunlight	x	x
	Water	Freshwater recycling	x	x
Harvesting		Settling		x
		Dissolved Air Flotation (DAF)	x	
Algae pretreatment	Concentration	Centrifugation	x	
		Mechanical press		x
	Drying	Heat from waste heat of industrial plant	x	x
		Heat from natural gas		
		No drying		
Extraction of raw material	Oil extraction	Hexane extraction	x	
		Mechanical press		x
Conversion into fuel	Hydrotreatment of vegetable oil	No propane recycling for utility production	x	x
Transport and distribution of fuel			x	x
Coproduct valorisation	Valorisation of biomass residues	Methanisation for electricity production (substitution with electricity from biomass cogeneration in Europe)		x
		Energetic allocation	x	
	Other fuels from hydrotreatment plant	Energetic allocation	x	x

3.2.1 Technical Sensitivity

A sensitivity analysis on technological choices on various steps of the pathway is done to compare each alternative scenario with the base case, defined as the pathway 1. Every alternative technology is presented below in Table 4.

Sensitivity results are presented below (*Fig. 5*). A positive percentage traduces an increase of the impact compared to the base case.

The use of chemical nutrients for algae cultivation increases by more than 40% the amount of GHG emissions compared to the use of sludge as nutrient (considered as a waste *i.e.* with no impact associated with its supply). The rise of GHG emissions and non renewable primary energy corresponds to the manufacture and the supply of chemical nutrient. In the nutrient recycling scenario, 75% of N and 50% of P are recycled, and the missing part is supplied by chemicals. This supply of

TABLE 3
Main data for pathways 1 and 2. Most of data is extracted from literature [10, 11, 14, 16, 20, 22, 24-30]

Data	Pathway 1	Pathway 2	Unit	Sources
Cultivation	In open pounds			
Productivty	21		$g/(m^2.day)$	[16]
Algae concentration	0.5		kg/m^3	[16]
Pond area	100		ha	[16]
Water depth	0.3		m	[16]
CO_2 content in flue gas	15%		(mass)	[14]
CO_2 needs	1.35		kg/kg of dry algae	[24]
N needs	0.061		kg/kg of dry algae	[24]
P needs	0.0081		kg/kg of dry algae	[24]
CO_2 recycle rate	-	33%	(mass) of original input	[25]
N recycle rate	-	75%	(mass) of original input	[25]
P recycle rate	-	50%	(mass) of original input	[25]
Electricity for CO_2 transportation	0.022		kWh/kg of injected CO_2	[26]
Electricity for paddle wheel	0.123		kWh/ha of pound	[27]
Electricity for water pumping	0.087		kWh/m^3 of water pumped	[27]
Harvesting	DAF	Settling		
Algae concentration out	20	10	kg/m^3	[14, 16]
Harvesting rate	95%	65%	(mass)	[14, 16]
Electricity	0.205	0.05	kWh/m^3 of input water	[14, 16]
Polystyrene	0.00242	-	kg/kg of dry harvested algae	[14]
Concentration of biomass	Centrifugation	Rotary press		
Algae concentration out	200	200	kg/m^3	[28] Fournier industries
Electricity	1.25	0.1	kWh/m^3 of input wet algae	[28] Fournier industries
Drying	Drying with waste heat from			
Algae concentration out	90		% of dry matter	[11]
Heat	2570		MJ/m^3 of water to be evaporate	[21]

(continued)

TABLE 3 *(continued)*

Data	Pathway 1	Pathway 2	Unit	Sources
Oil extraction	Hexane extraction	Oil press		
Oil content	34%		(mass) dry matter	Average from literature
LHV algae oil	36		MJ/kg	[29]
LHV algae before oil extraction	25.83		MJ/kg (dry matter)	[26]
Oil capture rate	95%	75%	(mass)	[14] constructor data from algaefuelsystems.com
Hexane losses	1.5	-	kg/t of dry algae	[14]
Heat	729	-	kg/t of dry algae	[14]
Electricity	70.18	125	kWh/kg of dry algae	[14] constructor data from algaefuelsystems.com
Hydrotreatment	NexBtL process			
Vegetable oil input	1.03		MJ/MJ of HRJ	[22]
Hydrogen input	0.09		MJ/MJ of HRJ	[22]
Electricity coproduced	0.002		MJ/MJ of HRJ	[22]
Steam coproduced	0.008		MJ/MJ of HRJ	[22]
LHV of HRJ	44.1		MJ/kg	[30]
Storage, transport and distribution of HRJ	150 km			
Diesel oil	0.00344		MJ/MJ of HRJ	[22]
Electricity	0.0042		MJ/MJ of HRJ	[22]
Valorisation of biomass coproduct	Animal meal	Methane to electricity		
LHV of biomass residues	21		MJ/kg of dry matter	Calculated
Conversion rate of biomass into CH_4	-	16.5	% of dry biomass converted in CH_4	Extrapolated from [22]
Electricity for mixing	-	5 388	kWh/m^3 of input biomass	[24]
Ekectricty for pumping	-	0.301	kWh/m^3 of input biogas	[16]
Autoconsumption of biogas	-	26%	of produced biogas	[16]
Yield of electricity production from CH_4	-	32	% energy	[14]
CO_2 credit for electricity coproduction	-	48	gCO_{2eq}/MJ of electricity	[10]

chemical nutrients explains the increase of the impacts for this scenario compared to scenario with the use of sludge as nutrient.

The choice of process option for the harvesting step impacts both the NREP energy and the GHG emissions. DAF system concentrates the algae to 20 kg/m^3 with a harvesting rate of 95% whereas settling system concentrates algae to 10 kg/m^3 with a harvesting rate of 65%. Consequently, more algae has to be cultivated with settling option in order to produce the same amount of algae harvested because there is no recycling of the overflow in the scope of this study. So these assumptions lead to an increase of the impacts linked with cultivation step for settling scenario.

Choosing a mechanical press instead of centrifugation for algae concentration reduces GHG emissions of the pathway but increases the NREP energy consumption. Indeed electricity consumption of the press is about ten times lower than the one for centrifugation. Even if the output algae concentration is lower, the amount of electricity consumed is smaller so as associated GHG emissions. Nevertheless more heat is consumed at drying step in order to reach the algae concentration of 90% TSS needed for dry hexane extraction. Therefore, NREP energy increases at drying step but GHG emissions do not increase because consumed heat is considered as waste with no CO_2 emissions associated.

Impacts associated with drying step could be significantly different depending on the process option. First the origin of heat used to dry algae is critical. The use of heat from natural gas leads to an increase of GHG emissions and NREP energy compared with the use of waste heat from industrial plant. Then the fact of drying or not algae before oil extraction modifies the energy balance of the system. Oil can be extracted from wet biomass by supercritical CO_2 but it lowers the efficiency of the oil extraction of 33% [32] so the oil recovering rate is equal to 62% instead of 95%. Hence upstream impacts

TABLE 4
Process options for the technical sensitivity

Step	Sensitive parameter	Base case	Alternative scenario
Cultivation	Nutrient supply	Sludge	Chemical nutrient only
Cultivation	Nutrient supply	Sludge	Nutrient recycling
Harvesting	Process option	DAF	Settling
Concentration	Process option	Concentration by centrifugation	Concentration with press
Drying	Process option	Drying with waste heat from industrial plant	Drying with heat from NG
Drying	Process option	Drying with waste heat from industrial plant	No drying
Oil extraction	Process option	Hexane extraction	Press for oil extraction

TABLE 5
Monte-Carlo analysis testing the sensitivity of each main parameter on GHG emission results for the first pathway

	Parameter Unit	Algae concentration kg/m^3	Algae productivity $g/(m^2.day)$	Oil content %
Parameter distribution	Mean	0.76	28.04	0.34
	Std deviation	65.72%	32.16%	39.70%
	Min	0.13	10.5	0.1
	Max	1.67	48	0.7
GHG emission results distribution (gCO_{2eq}/MJ)	Mean	18.2	19.9	20.1
	Std deviation	32.8%	0.3%	1.9%
	10th percentile	13.3	19.9	19.5
	25th percentile	14.3	19.9	19.8
	Median	16.2	19.9	20.1
	75th percentile	19.7	20.0	20.3
	90th percentile	25.8	20.0	20.6

associated with algae production are higher to produce the same amount of oil. Nevertheless, results for assessed impacts are lower than for the base case, which is in line with [31]. Indeed the amount and the LHV of biomass residues coproduced are higher and it represents 71% of the energy output at extraction step. Consequently the use of the energetic allocation for sharing the impacts between products and coproducts minimises the impacts allocated to the oil because of the high energy content of the oilcakes (see previous section about methodological sensibility).

The choice between an oil press instead of the hexane extraction induces a slight reduction of GHG emissions and NREP energy because no chemical inputs is needed for oil press extraction.

3.2.2 Sensitivity of Main Parameters for GHG Emissions

Several algae parameters are often mentioned as key parameters for algae production in the literature. These parameters are algae productivity and oil content on dry mass of algae [33], and algae concentration in ponds. Here, we conduct a Monte-Carlo analysis testing the distribution of each parameter one by one in order to evaluate its influence on GHG emissions result of the entire system. The same approach has been used by Benoist *et al.* to compare the GHG emissions of three first generation biofuels [34]. Parameter distributions correspond to distributions of the values found in the literature [1, 11, 13, 14, 16-18, 26, 31, 33, 35-43].

The Monte-Carlo analysis results show a weak system sensibility to algae productivity and oil content but a high system sensibility to algae concentration during cultivation that corresponds to a standard deviation of about 33% for the GHG emission results.

The high sensibility of the system to algae concentration value could be due to the amount of water to manage. Indeed the less algae are concentrated in water the higher the amount of water to be pumped, mixed and then to be removed to obtained the desired amount of algae will be. This water management is associated to electrical consumption that represents 45% of the electrical consumption of the culture step. GHG emissions associated with electricity consumption for water management (pumping and mixing at cultivation step and water removal at harvesting step) represent 36% of GHG emissions for the base case. Consequently, the less algae are concentrated the more GHG are emitted for culture and harvesting step.

The oil content influences both in opposite ways the amount of HRJ by kilogram of algae and the LHV of the biomass residues. When algae contain more oil, more biofuel is produced but the LHV of algal residues is lower. As the impacts allocated to HRJ are based on energy content between oil and algal residues, a higher oil content leads to a higher amount of impacts allocated to the production of HRJ. In the end, the sensibility to oil content is weak as indicated by the value of the standard deviation.

In our system, the value of algae productivity hardly affects the GHG emissions. In fact, impacts are expressed for one MJ of fuel produced. This production needs a certain amount of algae. When the productivity of algae increases, this needed amount of algae to produce 1 MJ of biofuel remains the same, as well as cultivation inputs (expressed per kg of algae produced). The only impacted processes are the ones which are independent from the growth rate: electricity for the paddlewheel and building of the infrastructures. However algae productivity strongly influences the land use impact that is not taken into account in our study.

DISCUSSION AND CONCLUSIONS

Since a couple of years LCA method is more and more used to assess the environmental impacts of bioenergy production from microalgae. Most studies are focused on the production of biodiesel, and some of them on biogas. The presented work is focused on the production on a type of biodiesel, the biojetfuel. The production of biogas by anaerobic digestion of the oilcakes and its use have also been analyzed. The main goal of this study is to perform sensitivity analyses on technical choices and on the main parameter values affecting energy consumption and GHG emissions. The sensitivity analysis on technological choices allows us to identify the best system configuration in terms of reduction of energy consumption and GHG emissions. The best scenarios are the ones with the use of nutrients from wastewater, and with extraction of oil from wet biomass by supercritical CO_2. Finally Monte-Carlo approach has been undertaken to link the variations of important parameters (oil content, algae productivity and algae concentration in the ponds) with the variations of GHG emissions. This analysis shows that the main parameter influencing GHG emissions is the algae concentration in the ponds. However, this result is strongly affected by the assumptions of the study, like the allocation rule chosen to account for coproducts impacts. Furthermore, nitrogen deprivation and its implication on productivity and oil content have not been taken into account, and could be the objects of future developments. In fact, given that growth rate and lipid content conflict with each other, a trade-off must be made between these two parameters when producing biofuel. The use of Monte-Carlo analysis allows us to incorporate probabilistic uncertainty analysis into LCA, and therefore to better understand the reliability of the results, even if we were not able to perform it taking into account all parameters distribution. The direct comparison with other studies is most of the time not straightforward, because of the different assumptions, data and system boundary choices set by each micro-algae or algae production chain. Nevertheless our results are in line with results from the literature [44, 45], both for energy consumption and GHG emissions.

ACKNOWLEDGMENTS

This work was done as an additional contribution to the SWAFEA Project with IFP Energies nouvelles as contributor.

REFERENCES

1 Chisti Y. (2007) Biodiesel from microalgae, *Biotechnology Advances* **25**, 3, 294-306. doi:10.1016/j.biotechadv.2007.02.001.

2 Brennan L., Owende P. (2010) Biofuels from microalgae — A review of technologies for production, processing, and extractions of biofuels and co-products, *Renewable and Sustainable Energy Reviews* **14**, 2, 557-577. doi:10.1016/j.rser.2009.10.009.

3 International Organization for Standardization (ISO) (2006) ISO14040:2006 Environmental management-life cycle assessment-principles and framework.

4 International Organization for Standardization (ISO) (2006) ISO14044:2006 Environmental management - Life cycle assessment - Requirements and guidelines

5 Luo L., Voet E., Huppes G., Udo de Haes H.A. (2009) Allocation issues in LCA methodology: a case study of corn stover-based fuel ethanol, *The International Journal of Life Cycle Assessment* **14**, 6, 529-539. doi:10.1007/s11367-009-0112-6.

6 González-García S., Gasol C.M., Gabarrell X., Rieradevall J., Moreira M.T., Feijoo G. (2010) Environmental profile of ethanol from poplar biomass as transport fuel in Southern Europe, *Renewable Energy* **35**, 5, 1014-1023. doi:10.1016/j.renene.2009.10.029.

7 Novelli P., Marizy C., Bringtown S. Tamboer B., Cariolle D., Paoli R., Roetger T., Prieur A., Patouillard L., Thellier L., Bessagnet B., Metz S., Novelli P., Vancassel X., Jongschaap R., Conijn J.G., Rutgers B., Courtet C., Rollin G. (2011) SWAFEA project - Environmental impact analysis report, **Vol. 2**.

8 European Parliament (2009) Directive 2009/28/EC of the European Parliament and of the Council of 23 April 2009 on the promotion of the use of energy from renewable sources and amending and subsequently repealing Directives 2001/77/EC and 2003/30/EC, *Official Journal of the European Union* **L 140/16**, 16-62.

9 Arvidsson R., Fransson K., Fröling M., Svanström M., Molander S. (2012) Energy use indicators in energy and life cycle assessments of biofuels: review and recommendations, *Journal of Cleaner Production* **31**, 54-61. doi:10.1016/j.jclepro.2012.03.001.

10 Dones R., Bauer C., Bolliger R., Burger B., Heck T., Röder A., Emmenegger M.F., Frischknecht R., Jungbluth N., Tuchschmid M. (2007) Life Cycle Inventories of Energy Systems: Results for Current Systems in Switzerland and other UCTE Countries Data v2.0, *Ecoinvent report No. 5.*

11 Lardon L., Hélias A., Sialve B., Steyer J.-P., Bernard O. (2009) Life-Cycle Assessment of Biodiesel Production from Microalgae, *Environmental Science & Technology* **43**, 17, 6475-6481. doi:10.1021/es900705j.

12 Del Campo J.A., García-González M., Guerrero M.G., (2007) Outdoor cultivation of microalgae for carotenoid production: current state and perspectives, *Applied Microbiology and Biotechnology* **74**, 6, 1163-1174. doi:10.1007/s00253-007-0844-9.

13 Jorquera O., Kiperstok A., Sales E.A., Embiruçu M., Ghirardi M.L. (2010) Comparative energy life-cycle analyses of microalgal biomass production in open ponds and photobioreactors, *Bioresource Technology* **101**, 4, 1406-1413. doi:10.1016/j.biortech.2009.09.038.

14 Campbell P.K., Beer T., Batten D. (2010) Life cycle assessment of biodiesel production from microalgae in ponds, *Bioresource Technology* **102**, 1, 50-56. doi:10.1016/j.biortech.2010.06.048.

15 Singhal A.K., Mehrotra R.K. (2000) *Environmental Issues and Management of Waste in Energy and Mineral Production*, A.A.Balkema.

16 Collet P., Hélias A., Lardon L., Ras M., Goy R.-A., Steyer J.-P. (2011) Life-cycle assessment of microalgae culture coupled to biogas production, *Bioresource Technology* **102**, 1, 207-214. doi:10.1016/j.biortech.2010.06.154.

17 Sander K., Murthy G.S. (2010) Life cycle analysis of algae biodiesel, *The International Journal of Life Cycle Assessment* **15**, 7, 704-714. doi:10.1007/s11367-010-0194-1.

18 Clarens A.F., Resurreccion E.P., White M.A., Colosi L.M. (2010) Environmental life cycle comparison of algae to other bioenergy feedstocks, *Environmental Science & Technology* **44**, 5, 1813-1819. doi:10.1021/es902838n.

19 Fagerstone K.D., Quinn J.C., Bradley T.H., De Long S.K., Marchese A.J. (2011) Quantitative measurement of direct nitrous oxide emissions from microalgae cultivation, *Environmental Science & Technology* **45**, 21, 9449-9456. doi:10.1021/es202573f.

20 Reinhardt G., Gärtner S., Helms H., Rettenmaier N. (2006) An assessment of energy and greenhouse gases of NExBTL, *Ifeu - Inst. Energy Environ. Res Heidelb.*

21 Stasta P., Boran J., Bebar L., Stehlik P., Oral J. (2006) Thermal processing of sewage sludge, *Applied Thermal Engineering* **26**, 13, 1420-1426. doi:10.1016/j.applthermaleng.2005.05.030.

22 JEC - Joint Research Centre-EUCAR-CONCAWE collaboration (2008) Well-to-Wheels Analysis of Future Automotive Fuels and Powertrains in the European Context (Version 3) (p. 88) *JRC/EUCAR/CONCAWE European Commission Joint Research Center*. doi:10.2788/79018.

23 Earle J.K.J. (2005) Wheels of Progress: Rotary Press Selection for Plum Island, *Florida Water Resources Journal*, April, 48-52. http://www.fwrj.com/techarticle05/0405%20fwrj%20tech%202.pdf.

24 Ras M., Lardon L., Bruno S., Bernet N., Steyer J.-P. (2011) Experimental study on a coupled process of production and anaerobic digestion of Chlorella vulgaris, *Bioresource Technology* **102**, 1, 200-206. doi:10.1016/j.biortech.2010.06.146.

25 Weissman J.C., Goebel P.R. (1987) *Design and Analysis of Microalgal Open Pond Systems for the Purpose of Producing Fuels*, Golden, Colorado. http://www.nrel.gov/docs/legosti/old/2840.pdf.

26 Kadam K.L. (2002) Environmental implications of power generation *via* coal-microalgae cofiring, *Energy* **27**, 10, 905-922. doi:10.1016/S0360-5442(02)00025-7.

27 Sazdanoff N. (2006) *Modeling and Simulation of the Algae to Biodiesel Fuel Cycle*, The Ohio State University. https://kb.osu.edu/dspace/bitstream/handle/1811/5981/?sequence=1.

28 Fournier Industries at http://www.rotary-press.com/ (n.d). Retrieved from http://www.rotary-press.com/.

29 Minowa T., Yokoyama S., Kishimoto M., Okakura T. (1995) Oil production from algal cells of Dunaliella tertiolecta by direct thermochemical liquefaction, *Fuel* **74**, 12, 1735-1738. doi:10.1016/0016-2361(95)80001-X.

30 Hileman J.I., Stratton R.W., Donohoo P.E. (2010) Energy Content and Alternative Jet Fuel Viability, *Journal of propulsion and Power* **26**, 6, 1184-1195. American Institute of Aeronautics and Astronautics. Retrieved from http://cat.inist.fr/?aModele=afficheN&cpsidt=23434811.

31 Brentner L.B., Eckelman M.J., Zimmerman J.B. (2011) Combinatorial life cycle assessment to inform process design of industrial production of algal biodiesel, *Environmental Science & Technology* **45**, 16, 7060-7067. doi:10.1021/es2006995.

32 Halim R., Gladman B., Danquah M.K., Webley P.A. (2011) Oil extraction from microalgae for biodiesel production, *Bioresource Technology* **102**, 1, 178-185. doi:10.1016/j.biortech.2010.06.136.

33 Hou J., Zhang P., Yuan X., Zheng Y. (2011) Life cycle assessment of biodiesel from soybean, jatropha and microalgae in China conditions, *Renewable and Sustainable Energy Reviews* **15**, 9, 5081-5091. doi:10.1016/j.rser.2011.07.048.

34 Benoist A., Dron D., Zoughaib A. (2012) Origins of the debate on the life-cycle greenhouse gas emissions and energy consumption of first-generation biofuels – A sensitivity analysis approach, *Biomass and Bioenergy* **40**, 133-142. doi:10.1016/j.biombioe.2012.02.011.

35 Vera-morales M., Schäfer A. (2009) *Final Report: Fuel-Cycle Assessment of Alternative Aviation Fuels*, pp. 1-39, http://www.cate.mmu.ac.uk/wp-content/uploads/2012/06/10-Final-Report-Sustainable-Fuels.pdf.

36 Stephenson A.L., Dupree P., Scott S.A., Dennis J.S. (2010) The environmental and economic sustainability of potential bioethanol from willow in the UK, *Bioresource Technology* **101**, 24, 9612-9623. doi:10.1016/j.biortech.2010.07.104.

37 Stratton R.W., Wong H.M., Hileman J.I. (2010) *Life Cycle Greenhouse Gas Emissions from Alternative Jet Fuels, PARTNER Project 28 report Version 1.1.*, Cambridge, Massachusetts, http://web.mit.edu/aeroastro/partner/reports/proj28/partner-proj28-2010-001.pdf.

38 Khoo H.H., Sharratt P.N., Das P., Balasubramanian R.K., Naraharisetti P.K., Shaik S. (2011) Life cycle energy and CO_2 analysis of microalgae-to-biodiesel: preliminary results and comparisons, *Bioresource Technology* **102**, 10, 5800-5807. doi:10.1016/j.biortech.2011.02.055.

39 Razon L.F., Tan R.R. (2011) Net energy analysis of the production of biodiesel and biogas from the microalgae: Haematococcus pluvialis and Nannochloropsis, *Applied Energy* **88**, 10, 3507-3514. doi:10.1016/j.apenergy.2010.12.052.

40 Batan L., Quinn J., Willson B., Bradley T. (2010) Net energy and greenhouse gas emission evaluation of biodiesel derived from microalgae, *Environmental Science & Technology* **44**, 20, 7975-7980. doi:10.1021/es102052y.

41 Clarens A.F., Nassau H., Resurreccion E.P., White M.A., Colosi L.M. (2011) Environmental impacts of algae-derived biodiesel and bioelectricity for transportation, *Environmental Science & Technology* **45**, 17, 7554-7560. doi:10.1021/es200760n.

42 Yang J., Xu M., Zhang X., Hu Q., Sommerfeld M., Chen Y. (2011) Life-cycle analysis on biodiesel production from microalgae: water footprint and nutrients balance, *Bioresource Technology* **102**, 1, 159-165. doi:10.1016/j.biortech.2010.07.017.

43 Sturm B.S.M., Lamer S.L. (2011) An energy evaluation of coupling nutrient removal from wastewater with algal biomass production, *Applied Energy* **88**, 10, 3499-3506. doi:10.1016/j.apenergy.2010.12.056.

44 Sills D.L., Paramita V., Franke M.J., Johnson M.C., Akabas T.M., Greene C.H., Tester J.W. (2013) Quantitative uncertainty analysis of Life Cycle Assessment for algal biofuel production, *Environmental Science & Technology* **47**, 2, 687-694. doi:10.1021/es3029236.

45 Slade R., Bauen A. (2013) Micro-algae cultivation for biofuels: Cost, energy balance, environmental impacts and future prospects, *Biomass and Bioenergy* **53**, 0, 29-38. doi:10.1016/j.biombioe.2012.12.019.

RHEVE 2011: International Scientific Conference on Hybrid and Electric Vehicles
RHEVE 2011 : Conférence scientifique internationale sur les véhicules hybrides et électriques

Effective Viscosity in Porous Media and Applicable Limitations for Polymer Flooding of an Associative Polymer

Peng Zhang[1,2*], Yefei Wang[2], Yan Yang[2], Wuhua Chen[2] and Shixun Bai[2]

[1] Chongqing University of Science & Technology, School of Chemistry and Chemical Engineering, Chongqing 266555 - China
[2] China University of Petroleum (East China), School of Petroleum Engineering, Qingdao 266555 - China
e-mail: zhangpengchina@foxmail.com - wangyf@upc.edu.cn - 491095127@qq.com - 578974157@qq.com - orinocoflow@qq.com

* Corresponding author

Résumé — Viscosité effective dans des médias poreux et limites d'application de l'injection de polymères associatifs — Le polyacrylamide associatif hydrophobe (HAPAM) est considéré comme un candidat prometteur pour l'injection de polymère, de par son excellente capacité apparente de viscosification. Par rapport au polyacrylamide partiellement hydrolysé (HPAM), les facteurs de résistance et de résistance résiduelle causés par le HAPAM sont généralement plus élevés. Cependant, la viscosité effective du HAPAM est plus faible que celle d'un polymère conventionnel pour une concentration de 2 000 mg/L. La capacité de rétention dynamique de HAPAM est environ 2,3 fois celle du HPAM. L'efficacité pour déplacer de l'huile avec du HAPAM est plus faible que celle obtenue avec un polymère conventionnel pour une concentration de 2 000 mg/L et pour un modèle de type sandpack homogène. L'efficacité de déplacement de l'huile avec du HAPAM est plus élevée que celle de HPAM uniquement pour un modèle hétérogène (ratio de perméabilité 2,8). Les ratios de perméabilité élevés et faibles ne sont pas bons pour obtenir des déplacements efficaces de l'huile avec du HAPAM.

Abstract — Effective Viscosity in Porous Media and Applicable Limitations for Polymer Flooding of an Associative Polymer — *Hydrophobically associating polyacrylamide (HAPAM) is considered to be a promising candidate for polymer flooding because of its excellent apparent viscosifying capability. Compared with partially hydrolyzed polyacrylamide (HPAM), the resistance factor and residual resistance factor caused by HAPAM tend to be higher. However, the effective viscosity of HAPAM is lower than that of conventional polymer at a concentration of 2 000 mg/L. The dynamic retention capacity of HAPAM is about 2.3 times that of HPAM. The oil displacement efficiency of HAPAM is lower than that of conventional polymer at a concentration of 2 000 mg/L in the homogeneous sandpack model. The oil displacement efficiency of HAPAM is higher than that of HPAM only in the heterogeneous model (permeability ratio 2.8). Neither high nor low permeability ratios are good for the oil displacement efficiency of HAPAM.*

INTRODUCTION

Regarding the use of partially hydrolyzed polyacrylamide (HPAM) for high-temperature and high-salinity reservoirs, there are many unfavorable factors [1] such as thermal oxidation degradation, salt sensitivity and mechanical degradation.

To overcome these weaknesses of HPAM, hydrophobically associating polyacrylamide (HAPAM) has been paid more and more attention [2]. In aqueous solutions, above a critical association concentration, HAPAM has excellent apparent viscosity due to its unique structure. In addition, many research results show HAPAM has some excellent properties such as anti-shear, salt tolerance and temperature resistance [3-5].

Besides the solution properties, many experimental studies have been carried out to evaluate the performance of HAPAM in porous media [6, 7]. The main research objectives were usually to determine the resistance factors, retention, residual resistance factors and oil recovery. Many research results [8-10] indicate significantly higher resistance factors and residual resistance factors of HAPAM than those of corresponding HPAM. Maia *et al.* [10] believe that HAPAM may form agglomerates or associations in porous media. In addition to significant adsorption (multilayer adsorption [11, 12]) and enhanced viscosity, the residual resistance factor of HAPAM is several times greater than that of HPAM. Zhang *et al.* [13] and Ye *et al.* [14] reported that HAPAM can generally form aggregates with super-molecular and network structures even in porous media. Based on these results, HAPAM can establish a higher resistance factor and residual resistance factor in porous media. Wang *et al.* [15] proposed that the resistance factor was used to describe the total effect for reducing aqueous phase fluidity. So, in a sense, the resistance factor reveals a kind of general result. To maintain the effectiveness and persistence of fluid displacement, the residual resistance factor should be reduced in order to increase the effective viscosity in porous media, because the retention should be kept low in order to ensure an in-depth propagation of the viscous polymer slug. The retention will inevitably cause a reduction of the concentration of movable polymer through porous media, which can cause a decrease in effective viscosity, and eventually leads to a decrease in the swept area [15]. However, effective viscosity in porous media is also very important for tertiary oil recovery [16].

Previous experimental work from Buchgraber *et al.* [17] showed that associative polymer concentrations from 1 250 mg/L to 1 500 mg/L led to the best swept area and recovery results. However, 2 500 mg/L solution led to poor recovery and low sweep efficiency due to very serious pore plugging by multi-layer polymer adsorption. Taylor and Nasr-El-Din [7] pointed out that adsorption was a concern with polymer flooding and has limited this application. On the other hand, the significant adsorption and physical gel properties of HAPAM have been proved to be industrially useful when applied to water shutoff and conformance control [18].

The effective viscosity and dynamic retention capacity of HPAM and HAPAM were studied in this paper. Moreover, the ability to enhance oil recovery of the two types of polymers was also studied in homogeneous and heterogeneous models. The results show that the application of HAPAM in polymer flooding has some limitations.

1 EXPERIMENT

1.1 Materials

Two types of polyacrylamides were studied in this paper: HPAM (HJ, degree of hydrolysis: 21.2% [η]: 2 510 mL/g) and HAPAM (SHDH5, degree of hydrolysis: 25.4% [η]: 2 045 mL/g). The synthetic formation water for all the tests was a brine with TSD = 20 000 mg/L (deionized water with 1.861 wt% sodium chloride and 0.139 wt% calcium chloride) and viscosity of 1.22 mPa.s at 25°C. The oil sample used in this study was dehydrated oil (asphaltenes 7.8%, resins 33.9%, oily constituents 58.3%) from *Shengli Oilfield*, with viscosity of 56 mPa.s at 80°C. The displacement tests were conducted by using artificial cores (Φ2.5 × 10 cm) and sandpacks (Φ2.5 × 20 cm) at 80°C. The petrophysical properties of the artificial cores are summarized in Table 1.

1.2 Experimental Procedures

The seepage experiment to determine the effective viscosity was carried out as follows. The artificial core was dried, weighed and saturated by synthetic water.

TABLE 1
Petrophysical properties of artificial cores

No.	Porous volume (mL)	Porosity (%)	Brine permeability (μm^2)
6-1	7.17	29.65	0.562
6-2	7.37	30.48	0.562
7-1	7.76	32.09	0.836
7-2	7.98	33.00	0.836

Then, synthetic water was injected into the core at a constant rate of 0.34 mL/min until a stable pressure was established. After that, 10 Pore Volumes (PV) of polymer slug were injected. Next, synthetic water was injected until a plateau pressure was obtained again.

Previous research mainly focused on the apparent viscosity of polymer in solution. Actually, effective viscosity can provide more accurate viscosity when polymer solution flows in porous media. Effective viscosity (also called Darcy viscosity) is calculated by Darcy's law (*Eq. 1*), where ΔP is the pressure drop of 10 PV polymer solution, and K_p is the permeability of the polymer solution in the core. However, it should be pointed out that it is difficult to obtain accurate permeability of polymer solution. K_p can be approximately replaced by brine permeability (K_f) after polymer injection [15]. So, μ_{ef} has another expression form (*Eq. 2*) [15], where F_R is the resistance factor, F_{RR} is the residual resistance factor, and μ_w is the water viscosity at the test temperature. F_R can be calculated by Equation (3) by the pressure drop of the polymer solution (ΔP_P) and brine (ΔP_B) through porous media at the same flow rate. F_{RR} can be obtained by Equation (4) based on ΔP_B and the pressure drop (ΔP_F) of the brine after polymer injection at the same flow rate:

$$\mu_{ef} = \frac{K_p \Delta P}{v_D L} \tag{1}$$

$$\mu_{ef} = \frac{F_R}{F_{RR}} \mu_w \tag{2}$$

$$F_R = \frac{\Delta P_p}{\Delta P_B} \tag{3}$$

$$F_{RR} = \frac{\Delta P_F}{\Delta P_B} \tag{4}$$

The experiment to determine the dynamic retention capacity was carried out as follows. The artificial core was dried, weighed and saturated by synthetic water. Then, the polymer solutions (2 000 mg/L) of HJ and SHDH5 were injected into the cores at a constant rate of 0.34 mL/min, respectively.

The concentration of polymer solution at the outlet was monitored by the starch-cadmium iodide method until the concentration reached a stable value. Using the material balance method, the retention capacity in porous media can be calculated by Equation (5) [19], where Q is the retention capacity (mg/g), ρ_0 is the concentration of injected polymer solution (mg/L), V_0 is the accumulated injection volume (mL), ρ_i is the outflow concentration of the i'th sample (mg/L). V_i is the outflow volume of the i'th sample (mL), n is the number of outflow samples, and W is the mass of the dry core:

$$Q = \left(\rho_0 V_0 - \sum_{i=1}^{n} \rho_i V_i \right) / W \tag{5}$$

The laboratory oil displacement experiment was carried out as follows. The wet-packed sandpack was flooded with the oil sample until water production ceased. Thereafter, brine flooding continued until the water cut was greater than 98%. After that, 0.3 PV of polymer slug was injected (2 000 mg/L), followed by water flood until the water cut of the efflux reached 98% again.

Lyophilization and microscopic observations of polymer solution in porous media were carried out as follows. About 1 PV of polymer solution was injected into the artificial core (Φ25 × 90 mm). Next, the core was quickly transferred to a vacuum cup of liquid nitrogen and the vacuum cup was placed in a freezer for 24 h at −20°C. Then, the core was put into a lyophilizer for 24 h with a vacuum pressure of 7-8 Pa. After being freeze-dried, the core was crushed at the inlet and 30-40 mm away from the inlet and core fragments were examined with a scanning electron microscope.

2. RESULTS AND DISCUSSION

2.1 Seepage Properties and Effective Viscosity

The artificial cores were injected with polymer solution of HJ and SHDH5, respectively. As shown in Figure 1,

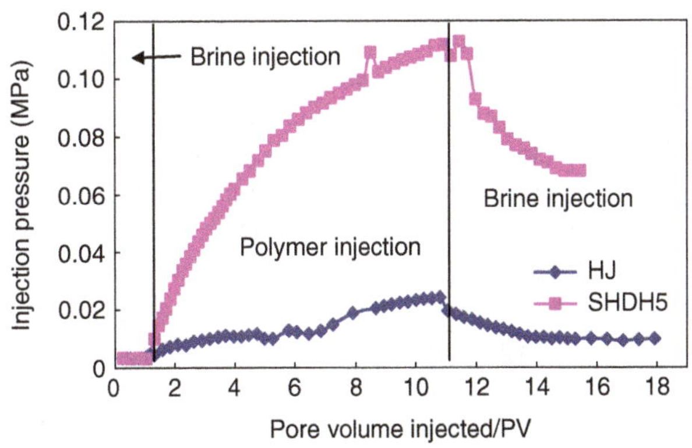

Figure 1

Variation of injection pressures during the injection of the two 2 000 mg/L polymer solutions.

TABLE 2
The seepage parameters and effective viscosity of the two polymers

Polymer	Initial permeability (μm^2)	Core code	Injection pressure of polymer flooding (MPa)	Injection pressure of subsequent water flooding (MPa)	F_R	F_{RR}	Effective viscosity (mPa.s)
HJ	0.562	6-1	0.062	0.0201	5.743	1.875	1.11
SHDH5		6-2	0.1918	0.108	19.636	11.157	0.64
HJ	0.836	7-1	0.024	0.0092	6.67	2.56	0.95
SHDH5		7-2	0.1119	0.068	31.97	19.43	0.59

the injection pressure of SHDH5 is much higher than that of HJ. Also, the injection pressure of the subsequent water flooding of HAPAM is higher than that of HPAM. The seepage parameters and effective viscosity of the two polymers are shown in Table 2. The F_{RR} values indicate that the ability to reduce permeability of SHDH5 is higher than that of HJ. However, the strong ability to reduce permeability means an increase in retention loss. The retention capacity of HAPAM may be higher than that of HPAM. It will be demonstrated in the next section. The retention loss would result in a reduction in the amount of movable polymer flowing in porous media, which can cause a decrease in effective viscosity. As shown in Table 2, the effective viscosity of SHDH5 is lower than that of HJ. Under the same conditions, the bulk viscosity of SHDH5 (49.2 mPa.s) is higher than that of HJ (19.7 mPa.s).

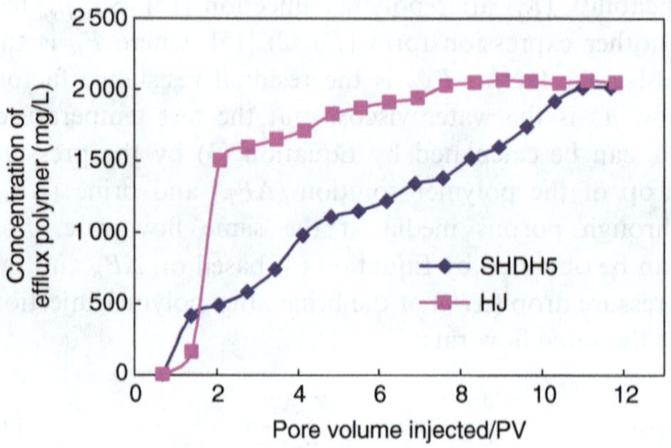

Figure 2

Variation of efflux concentration during the injection of the two 2 000 mg/L polymer solutions (permeability: 1.121 μm^2).

2.2 Dynamic Retention Capacity

In order to reveal the reason that HAPAM can build excellent flow resistance in porous media, experiments of dynamic retention capacity were carried out. The relationship between efflux concentration and the pore volume injected is shown in Figure 2. For HPAM, the polymer retention can reach a balance in a short period of time. However, for HAPAM, the injected PV needs to exceed at least 11 before the polymer achieves the concentration balance. According to Equation (5), the result indicates that the dynamic retention capacity of SHDH5 (1 450.98 μg/g) is 2.3 times that of HJ (626.70 μg/g). Thus, it can be seen that the high flow resistance of HAPAM is mainly due to high retention. However, could this high flow resistance become an advantage for polymer flooding? The next section may give an answer.

2.3 Ability to Enhance Oil Recovery

A series of oil displacement experiments were carried out in homogeneous sandpack models of different permeabilities (0.5-3.0 μm^2). The tertiary recovery of polymer flooding and the final oil recovery of each flood test are shown in Table 3. The oil displacement efficiency of SHDH5 is always lower than that of HJ for different permeabilities at the same concentration of 2 000 mg/L. Only when permeability was approximately 1 μm^2 was the tertiary recovery of the two kinds of polymers approximately equivalent, and higher than those with other permeabilities. Low permeability is not favorable for injecting polymer and high permeability may result in fingering of subsequent water flooding. Therefore, there is an optimum permeability for polymer flooding. With permeability of approximately 1.7 μm^2, the curves of polymer flooding of the two kinds of polymers are

TABLE 3

Summary of polymer flood tests in the homogeneous sandpack model

Permeability (μm^2)	Polymer	Recovery of water flood (%)	Final oil recovery (%)	Tertiary recovery (%)
0.503	HJ	45.5	62.3	16.8
0.543	SHDH5	41.9	55.3	13.4
0.905	HJ	49.5	72.1	22.6
0.970	SHDH5	43.3	65.5	22.2
1.697	HJ	51.9	70.3	18.4
1.835	SHDH5	48.4	58.1	9.7
2.425	HJ	47.3	64.6	17.3
2.514	SHDH5	38.4	53.3	14.9
3.017	HJ	60.8	77.3	16.5
3.086	SHDH5	53.8	66.3	12.5

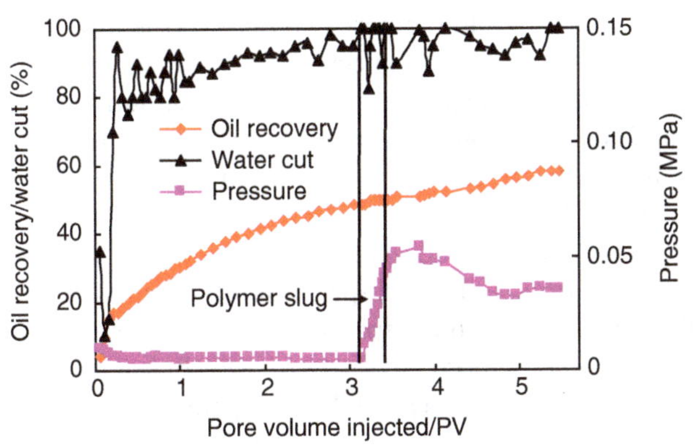

Figure 3

Polymer flooding curve of SHDH5 in the homogeneous sandpack model. (Permeability: 1.835 μm^2).

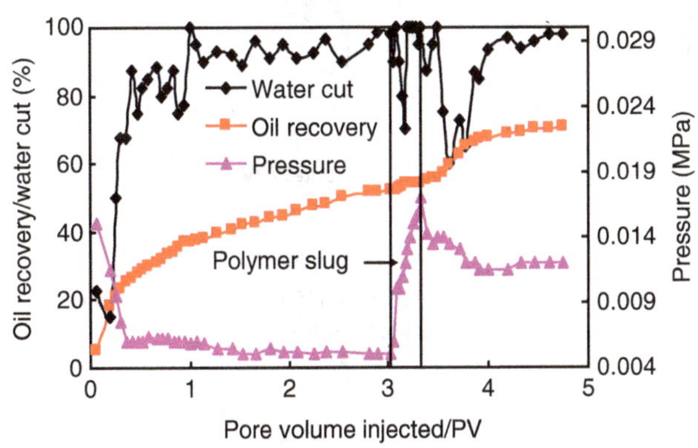

Figure 4

Polymer flooding curve of HJ in the homogeneous sandpack model. (Permeability: 1.697 μm^2).

shown in Figures 3 and 4. Although the pressures of SHDH5 polymer flooding and subsequent water flooding are higher than those of HJ, the decreased degree of water cut and tertiary oil recovery of HJ are high than those of SHDH5. In the homogeneous model, significant improvement of the water injection profile is not required. So, compared with HPAM, HAPAM has no advantage for polymer flooding. The great retention loss of HAPAM can cause a reduction in the amount of movable polymer through porous media, which will cause a decrease in effective viscosity and lead to low tertiary oil recovery.

However, in the heterogeneous model, the oil displacement effect of SHDH5 has some improvement. The heterogeneous model was simulated by two parallel sandpacks ($\Phi 2.5 \times 20$ cm) with different permeabilities. A summary of the polymer flood tests in the heterogeneous model is shown in Table 4. The tertiary oil recovery of SHDH5 (23.8%) is higher than that of HJ (15.7%) when the permeability ratio is approximately 2.8. In these conditions, the curves of polymer flooding are shown in Figures 5 and 6. In the conditions of a certain permeability ratio, the strong retention properties of HAPAM can only adjust the interlayer heterogeneity.

TABLE 4

Summary of polymer flood tests in the heterogeneous sandpack model

Polymer	Permeability (μm^2)	Permeability ratio	Recovery of water flood (%)	Final oil recovery (%)	Tertiary recovery (%)
SHDH5	3.57/1.28	2.79	24.6	48.4	23.8
HJ	3.39/1.17	2.90	32.8	48.5	15.7
SHDH5	2.26/1.26	1.79	40.2	54.1	13.9
HJ	2.34/1.28	1.83	38.5	58.9	20.4
SHDH5	5.43/1.13	4.81	40.9	53.4	12.5
HJ	5.43/1.13	4.81	35.4	52.6	17.2

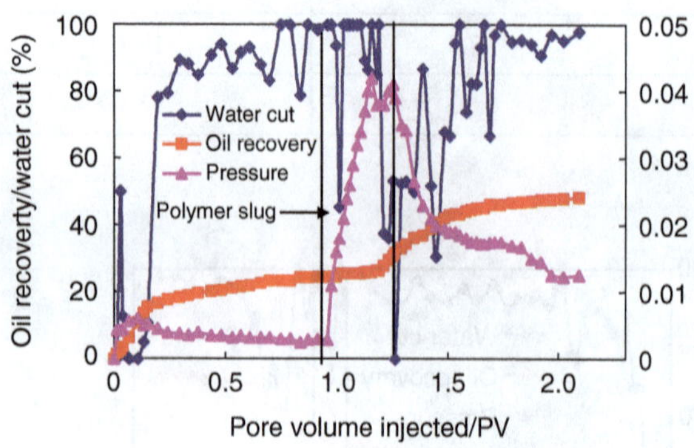

Figure 5

Polymer flooding curve of SHDH5 in the heterogeneous sandpack model. (Permeability ratio: 2.79 μm^2).

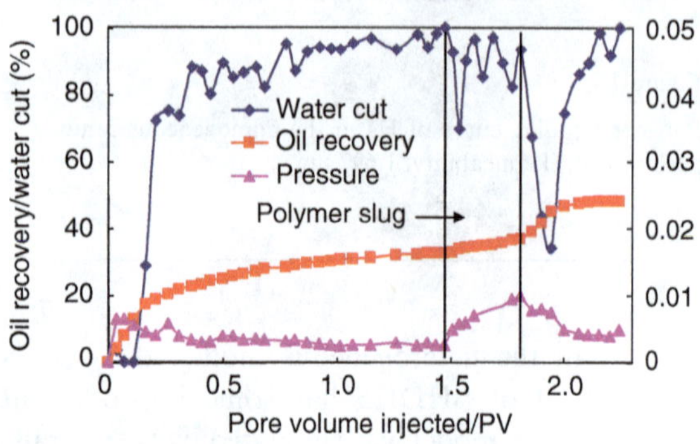

Figure 6

Polymer flooding curve of HJ in the heterogeneous sandpack model. (Permeability ratio: 2.90 μm^2).

It is similar to profile control, which diverts the fluids to improve oil recovery.

However, when the permeability ratio is approximately 1.8 or 4.8, the tertiary oil recovery of SHDH5 is lower than that of HJ. While the permeability ratio is small (about 1.8), it is similar to the homogeneous model. The great retention loss of HAPAM causes a decrease in effective viscosity, which leads to low tertiary oil recovery. While the permeability ratio is low, more associative polymer are injected into the low-permeability sandpack. Plugging due to strong retention tends to inhibit flow through low-permeability regions, therefore reducing the sweep of the injectant and tertiary oil recovery. While the permeability ratio is high (about 4.8), the polymer and subsequent water flooding tend to flow through the high-permeability sandpack. The displacing fluid cannot spread to the low-permeability sandpack with more residual oil, which leads to low tertiary oil recovery. The ability to enhance oil recovery of the two kinds of polymers in the heterogeneous model of high and low permeability is shown in Table 5. While the permeability ratio is approximately 1.8 or 4.8, the results indicate that the contribution of HAPAM to tertiary recovery of the low-permeability sandpack is very small (4.6-5.6%). Therefore, HAPAM may damage the low-permeability layer due to its strong retention properties.

2.4 Microstructures in Porous Media

The microstructures of the two kinds of polymers in porous media are shown in Figures 7-10. Figures 7 and 8 show micrographs of the conventional polymer HJ and associative polymer SHDH5 at the inlet of the core, respectively. The micrograph of HPAM has no significant difference compared with that of HAPAM. It is because both kinds of polymers have a large retention capacity at the inlet of the core. The polymer flakily adheres to the surface of the rock. The network structure

TABLE 5
Ability to enhance oil recovery in the heterogeneous sandpack model of high and low permeability

Polymer	Permeability (μm^2)	Permeability ratio	Recovery of water flood (%)	Final oil recovery (%)	Tertiary recovery (%)
SHDH5	High 3.57	2.79	46.6	66.9	20.3
	Low 1.28		0.5	28.1	27.6
HJ	High 3.39	2.90	57.0	77.9	20.9
	Low 1.17		7.8	18.3	10.5
SHDH5	High 2.26	1.79	40.3	63.6	23.3
	Low 1.26		40.0	44.6	4.6
HJ	High 2.34	1.83	38.2	68.4	30.2
	Low 1.28		38.8	48.9	10.1
SHDH5	High 5.43	4.81	46.6	63.2	16.6
	Low 1.13		27.8	33.4	5.6
HJ	High 5.43	4.81	58.7	76.0	17.3
	Low 1.13		13.0	30.9	17.9

Figure 7

The microstructure of HJ in the inlet of the core.

Figure 8

The microstructure of SHDH5 in the inlet of the core.

of the polymer cannot be observed at the inlet of the core. However, in the middle of the core, the micrographs of the different polymers are very different, as shown in Figures 9 and 10.

It can be seen that both polymers can form networks in the middle of the core. However, the networks of HPAM are larger and more integrated than those of HAPAM. The diameter of the HPAM skeleton is significantly larger than that of HAPAM. This is likely because the retention loss of HAPAM is greater than that of HPAM, causing a reduction in the amount of movable polymer chains in the pores, which produces a defective and small network and leads to a decrease in effective viscosity and ability to enhance oil recovery.

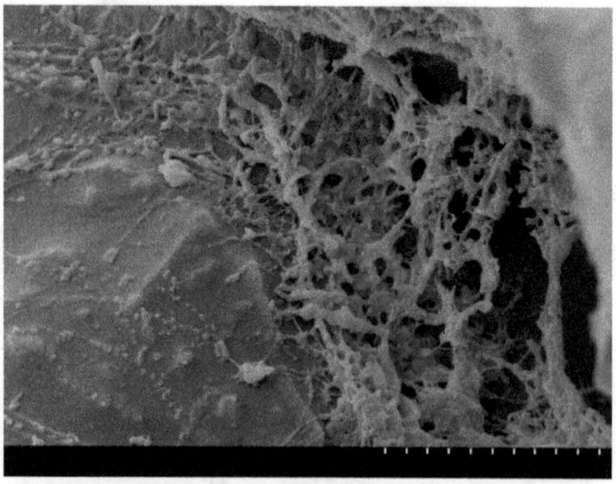

Figure 9

The microstructure of HJ in the middle of the core.

Figure 10

The microstructure of SHDH5 in the middle of the core.

CONCLUSIONS

Under certain conditions, HAPAM can establish a higher resistance factor and residual resistance factor. This is because the dynamic retention capacity of HAPAM is greater than that of HPAM. For the same reason, the effective viscosity of HAPAM is lower than that of HPAM. However, high flow resistance is not an advantage for polymer flooding in the homogeneous model. The oil displacement efficiency of HAPAM is always lower than that of HPAM for different permeabilities (0.5-3.0 μm^2) at the same concentration of 2 000 mg/L. Only in the heterogeneous model and at a suitable permeability ratio (about 2.8) is the tertiary oil recovery of HAPAM higher than that of HPAM. Thus, the application of HAPAM in polymer flooding has some limitations.

ACKNOWLEDGMENTS

This work was sponsored by the National S&T Major Project (Grant No.: 2011ZX05024-004) on chemical flooding for offshore heavy oil fields, and "Taishan Scholars" Construction Engineering (No. ts20070704).

REFERENCES

1 Zhong C.R., Huang R.H., Xu J.Y. (2008) Characterization, solution behavior, and microstructure of a hydrophobically associating nonionic copolymer, *J. Solution Chem.* **37**, 1227-1243.

2 Shashkina Y.A., Zaroslov Y.D., Smirnov V.A., Philippova O.E., Khokhlov A.R., Pryakhina T.A., Churochkina N.A. (2003) Hydrophobic aggregation in aqueous solutions of hydrophobically modified polyacrylamide in the vicinity of overlap concentration, *Polymer* **44**, 2289-2293.

3 Feng Y., Luo P., Luo C., Yan Q. (2002) Direct visualization of microstructures in hydrophobically modified polyacrylamide aqueous solution by environmental scanning electron microscopy, *Polym. Int.* **51**, 931-938.

4 Gao B.J., Wu N., Li Y.B. (2005) Interaction between the strong anionic character of strong anions and the hydrophobic association property of hydrophobic blocks in macromolecular chains of a water-soluble copolymer, *J. Appl. Polym. Sci.* **96**, 714-722.

5 Gouveia L.M., Paillet S., Khoukh A., Grassl B., Müller A. J. (2008) The effect of the ionic strength on the rheological behavior of hydrophobically modified polyacrylamide aqueous solutions mixed with sodium dodecyl sulfate (SDS) or cetyltrimethylammonium *p*-toluenesulfonate (CTAT), *Colloids Surf. A. Physicochem. Eng. Aspects* **322**, 211-218.

6 Taylor K.C., Nasr-El-Din H.A. (1998) Water-Soluble Hydrophobically Associating Polymers for Improved Oil Recovery: A literature review, *J. Petrol. Sci. Eng.* **19**, 265-280.

7 Taylor K.C., Nasr-El-Din H.A. (2007) Hydrophobically Associating Polymers for Oil Field Applications, *Petroleum Society's 8th Canadian International Petroleum Conference (58th Annual Technical Meeting)*, Calgary, 12-14 June.

8 Kun R., Jinlan W., Meiqin L., Chunming X. (2006) Rheological Properties of Hydrophobically Associating Polyacrylamide Solution, *Petrol. Sci.* **3**, 66-72.

9 Lu H., Feng Y., Huang Z. (2008) Association and Effective Hydro-dynamic Thickness of Hydrophobically Associating Polyacrylamide Through Porous Media, *J. Appl. Polym. Sci.* **110**, 1837-1843.

10 Maia A.M.S., Borsali R., Balaban R.C. (2009) Comparison between a Polyacrylamide and a Hydrophobically Modified Polyacrylamide Flood in a Sandstone Core, *Mater. Sci. Eng.* **29**, 505-509.

11 Argillier J.F., Audibert A., Lecourtier J., Moan M., Rousseau L. (1996) Solution and adsorption properties of hydrophobically associating water-soluble polyacrylamides, *Colloids Surf. A. Physicochem. Eng. Aspects* **113**, 247-257.

12 Volpert E., Selb J., Candau F., Green N., Argillier J.F., Audibert A. (1998) Adsorption of hydrophobically associating polyacrylamides on clay, *Langmuir* **14**, 1870-1879.

13 Zhang C., Ye Z., Shi L., Zhao W., Zhang Z. (2007) A study on the mechanism to build flow resistance by hydrophobic associated polymer, *China Offshore Oil Gas.* **19**, 4, 251-253 (China).

14 Ye Z., Jia T., Shi L., Chu Y., Shu Z. (2007) The mobility control capacity of hydrophobic associating polymer in high permeability porous media, *J. Southwest Petroleum University* **29**, 100-104 (China).

15 Wang J., Luo P.Y., Zhang G.Q. (2001) The flow properties of aqueous solution of hydrophobically associating amphoteric polymer NAPs through porous media, *Oilfield Chem.* **18**, 152-154 (China).

16 Wang J., Dong M. (2009) Optimum effective viscosity of polymer solution for improving heavy oil recovery, *J. Petrol. Sci. Eng.* **67**, 155-158.

17 Buchgraber M., Clemens T., Castanier L.M., Kovscek A. R. (2009) The Displacement of Viscous Oil by Associative Polymer Solutions, *SPE ATCE Annual Conference and Exhibition*, New Orleans, 4-7 Oct.

18 Dupuis G., Rousseau D., Tabary R., Grassl B. (2011) Flow of Hydrophobically Modified Water-Soluble-Polymer Solutions in Porous Media: New Experimental Insights in the Diluted Regime, *SPE J.* **16**, 43-54.

19 Lv K., Dai C.L., Zhang R.R., Liu X.Z., Zhang J., Tang E.G. (2010) Investigations of Adsorption Rules of Hydrophobically Associating Polymer in the Porous Medium, *Oilfield Chem.* **27**, 391-394 (China).

12

Pressure Transients in a Fractal-Cluster Model of Porous Media

Abraham de Swaan*

Chilpa 16-4, Mexico D.F. cp 03910 - Mexico
e-mail: adeswaan@hotmail.com

* Corresponding author

Abstract — *The reservoir is described as a "supercritical cluster"; that is, an aggregate of conductive elements that comprises a "backbone" of connected pores or fractures that span the zone of interest, and also a collection of "sub-critical clusters" or "dangling ends" joined to the backbone to a limited extent. The scheme resembles the usual fracture and matrix-blocks setting but both backbone and sub-clusters are of the same material and share similar petrophysical properties. Whereas the backbone is a homogeneous porous medium, the sub-critical clusters behave as fractal porous media. The backbone-cluster type of flow has been observed in laboratory experiments. The sub-critical clusters were approximated as linear fractal media characterized by static and dynamic fractal exponents and also by porosity and permeability of the compound medium. One of the ends of the linear clusters is closed and the other is joined to the backbone, where the mainstream occurs. A new solution was developed for that problem. The Laplace transform in time and space was used in the mathematical scheme. The theory developed was applied to field cases of interference between wells in aquifers. The matches of computed and observed dynamic pressures show fair fits.*

Résumé — **Transitoires de pression dans un modèle d'agrégat fractal de milieux poreux** — Le réservoir est décrit comme un « agrégat supercritique » ; à savoir un ensemble d'éléments conducteurs comprenant un « squelette » de pores connectés ou de fractures qui s'étendent sur la zone d'intérêt et également une collection « d'agrégats sous-critiques » ou « d'extrémités non saturées » jointes au squelette sur une distance limitée. Le schéma ressemble à la fracturation et aux assemblages blocs-matrice habituels mais le squelette et les sous-agrégats sont du même matériau et partagent des propriétés pétro-physiques similaires. Tandis que le squelette est un milieu poreux homogène, les agrégats sous-critiques se comportent comme des milieux poreux fractals. Ce type d'écoulement a été observé dans des expériences de laboratoire. Les agrégats sous-critiques ont été représentés comme des milieux fractals linéaires caractérisés par des exposants fractals statiques et dynamiques et également par la porosité et la perméabilité du milieu du composé. Une des extrémités des agrégats linéaires est fermée et l'autre est reliée au squelette, où coule le flux principal. Une nouvelle solution a été développée pour ce problème. La transformée de Laplace dans le temps et l'espace a été utilisée dans le développement mathématique. La théorie développée a été appliquée aux cas d'interférence sur le terrain entre les puits dans des aquifères. Les comparaisons de pressions dynamiques calculées et observées présentent des résultats compatibles.

NOMENCLATURE

f_{SC}	Fraction of SC-BB connected length
A_{BB}	Well surface at the pay zone
C	Coefficient in Equation (5)
d	Fractal exponent
h	Thickness (m)
H	Heaviside step distribution
K_v	Modified Bessel function of order v
L	Laplace transform
L^{-1}	Inverse Laplace transform
m	Integer
p	Pressure (Pa)
q	Flow rate (m^3/s)
r	Radial distance (m)
t	Time (s), (h in Fig. 9)
x	Linear coordinate (m)
α	Fractal dimension exponent ($\equiv d_f$)
β	Dynamic exponent ($= 2/d_w$)
θ	Exponent in Equation (2) in O'Shaugnessy and Procaccia (1985) and Barker (1988)
η	Hydraulic diffusivity (m^2/s)
μ	Viscosity (Pa.s)
τ	Time-Laplace transform parameter
κ	Space-Laplace transform parameter
∇_r^2	Radial Laplacian operator

SUBSCRIPTS AND SUPERSCRIPTS

BB	Backbone
f	Fractal dimension
SC	Subcritical cluster
w	Brownian or random-walk
$\tilde{F}$	Time-Laplace transform of F
$\hat{F}$	Space-Laplace transform of F
v	$= 1 - d_f/d_w$, order of Bessel K function in (Metzler *et al.*, 1994)

INTRODUCTION

Characterization of Fractal Porous Media

The main characteristics of fractal media are two exponents that are defined in deterministic fractals. d_f is the fractal dimension: the exponent that links a scalar factor applied to a regular fractal object and the number of elements that result from that scale-up. For instance in the Sierpinsky gasket when the scale is doubled and the central triangle is discarded the number of new elements is three, Figure 1.

In actual porous or fractured media d_f is obtained from sampling and through box counting or sand box algorithms. See ben-Avraham and Havlin (2000) and Schroeder (1990). The extent of pore area or volume; or the number of fractures in the sampling space, increases as the d_f power of the augmented search radius.

The dynamic exponent, d_w, is derived considering Brownian stochastic motion of a particle in the medium. It relates the scalar factor to the number of possible steps between two extreme nodes in a fractal net, Figure 2.

Computation of d_w even in deterministic media is difficult, numerical enumerations are often used, but the Alexander and Orbach (1982) conjecture, the ratio of the exponents is a constant, $d_f/d_w = 2/3$, allows determination of the dynamic from the fractal exponent.

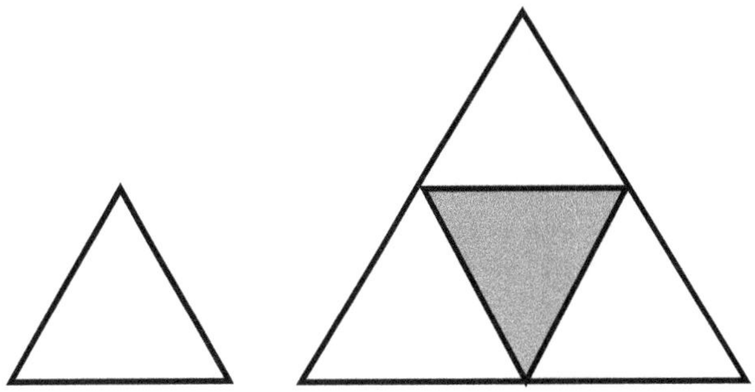

Figure 1

Upscaling in the Sierpinsky gasket.

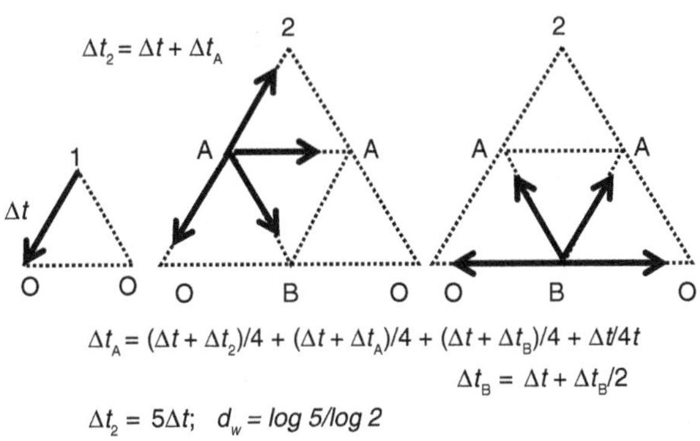

Figure 2

Brownian motion in the Sierpinsky gasket.

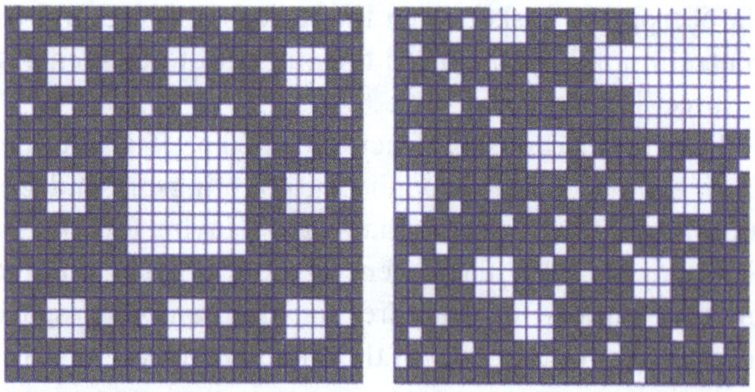

Figure 3

a) Deterministic and b) random fractals.

That ratio has been proven as accurate by a number of experimental numerical studies (Rammal *et al.*, 1984; Essam and Bhatti, 1985) although it is still a conjecture, *i.e.* with no analytical proof. In this paper, we adopt that link between static and dynamic exponents.

It is essential for the application of fractal principles to natural materials that random or stochastic fractals, representative of natural objects, preserve the properties of deterministic fractals, thus, the fractal in Figure 3b may behave similarly to the one in Figure 3a in actual flow processes, (Ben-Avraham and Havlin, 2000).

Generally, the values of the exponents are fractions not as simple as the ratios shown in Figures 1 and 2.

The fractal diffusion equation:

$$\nabla_r^{\alpha} p(r,t) - \eta \frac{\partial^{\beta} p(r,t)}{\partial t^{\beta}} = 0 \tag{1}$$

with $\alpha \equiv d_f$, $\beta = 2/d_w$, has a foundation on probabilistic considerations with Fourier and Laplace transforms. The step from diffusion in fractal media to the fractional-derivative representation was not immediate but required the formulation of probabilistic concepts, (Compte, 1996; Gorenflo and Mainardi, 1997; Metzler and Klafter, 2000; Mainardi *et al.*, 2001; Vlahos *et al.*, 2008).

The methods for porous or for fractured media are the same; "*Porous and fractured media appear to follow the same idealized scaling rule for both flow and transport*" (Neuman, 1990). By using scalar arguments O'Shaugnessy and Procaccia (1985) and Barker (1988) transformed Equation (1) into a differential equation with a space dependent diffusivity:

$$\eta \frac{\partial p}{\partial t} = \frac{1}{r^{d_f-1}} \frac{\partial}{\partial r}\left(r^{d_f-1-\theta} \frac{\partial p}{\partial r} \right) \tag{2}$$

θ includes d_w. Basically it results from passing the scaling exponential law of distances to the geometrical material balance. Its solution is:

$$p(r,t) = \frac{d_w}{d_f \Gamma\left(\frac{d_f}{d_w}\right)} \left(\frac{1}{a d_w{}^2} \right)^{\frac{d_f}{d_w}} exp\left(-\frac{r^{d_w}}{a d_w{}^2 t} \right) \tag{3}$$

this and similar solutions have been labeled "stretched exponentials", (Sahimi, 1995; Metzler and Klafter, 2000). Chang and Yortsos (1990) applied it to the analysis of well tests. The solution, Equation (3), does not account for interference tests (Sahimi, 1993; Camacho Velazquez *et al.*, 2008). It has been criticized for limiting diffusion as only space dependent but time independent (Sahimi, 1993; Leveinen, 2000).

A modification of Equation (2) was presented by Metzler *et al.* (1994) with a fractional time derivative dependent on d_w. (Raghavan, 2011).

$$\eta \frac{\partial^{\beta} p}{\partial t^{\beta}} = \frac{1}{r^{2d_f/d_w-1}} \frac{\partial}{\partial r}\left(r^{2d_f/d_w-1} \frac{\partial p}{\partial r} \right) \tag{4}$$

Its solution in the Laplace domain is:

$$\tilde{p}(r,\tau) = C\left(\frac{r}{\tau^{\frac{2}{d_w}}} \right)^{1-\frac{d}{2}} K_{\nu}\left(r\tau^{d_w} \right) \tag{5}$$

It has the advantage of describing interference, (de Swaan *et al.*, 2012), and it does reduce to the normal case for integer exponents.

Percolation Clusters

The concept of "percolation cluster" as representative of porous and fractured media has been advanced by several authors. (Sahimi, 1995; ben Avraham and Havlin, 2000). They have been called "*spatially periodic structures* [...] *homogeneous in the large scale but with a fractal microstructure*" by Adler and Thovert (1993).

It has been reckoned that in fractured reservoirs "*flow takes place in a relatively small number of fractures*" (Leveinen, 2000) and "*Only six fractures could be traced through nine sections* [in a rock block with 240 fractures]", (Adler and Thovert, 1999, page 213). Figure 4.

A cluster of bounding elements becomes connected across a region of interest when the probability of an element of area being a conducting pore or bond, is greater than a fraction that depends on the underlying type of lattice, (see page 11 table 2.1 in Sahimi (1994); ben Avraham and Havlin, 2000). A cluster connected throughout is a "super-critical" cluster, Figure 5a. It is

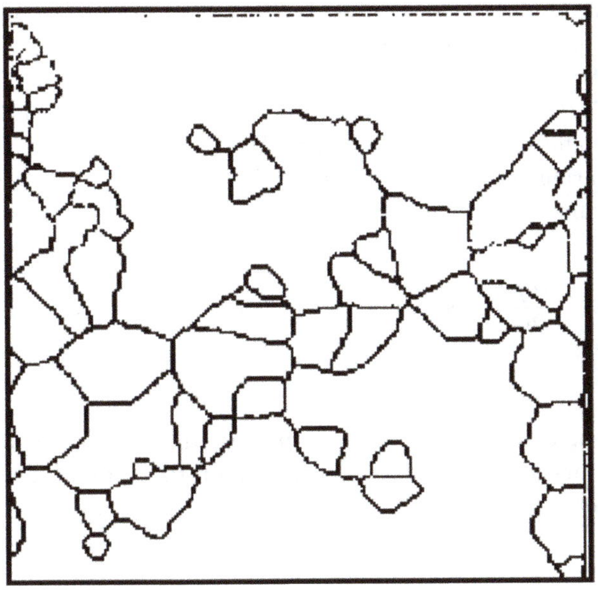

Figure 4

Fracture network, Adler and Thovert (1999).

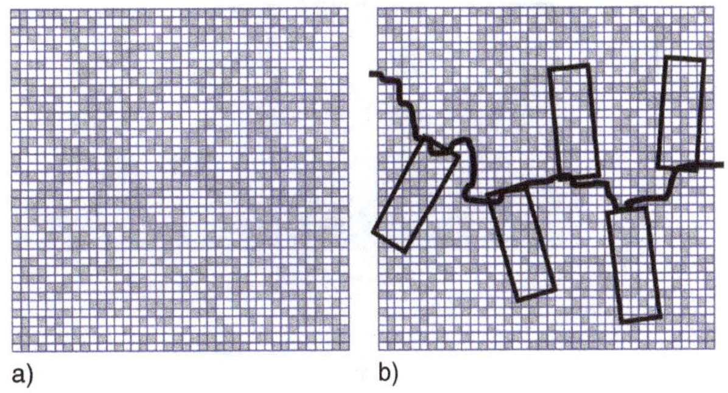

Figure 5

Supercritical cluster a) and its description as backbone and subcritical clusters b).

composed by one or several BackBones (BB) communicated to Sub-critical Clusters (SC); that is, aggregates of pores or fractures with a connection to the backbone but isolated among them, Figure 5b.

The backbone is a homogeneous medium whereas the sub-critical clusters, as randomly generated aggregates, behave as fractal media. Thus, in the backbone flow of fluids follows the normal diffusion law in a connected medium. The backbone-cluster type of flow has been observed in laboratory experiments (Klemm *et al.*, 1999). A turbiditic system (Chen *et al.*, 2012) with communicating channels and appended lobes may also be approximated by the BB-SC description – in this case with different properties.

1 THEORY

1.1 Diffusion in a Supercritical Cluster

The system BB plus SC is similar to the well known system of fracture and matrix-rock blocks (de Swaan, 1986), but in this case both media share the same petrophysical properties: local porosity, permeability and rock compressibility.

BB's generally have varying cross sections along their lengths. They have "hot spots" where the section reduces to a single communicating element (fracture or pore); in between, they may have wide openings or "blobs". In the present theory BB are assumed with a constant or homogeneous cross-section h_{BB}. Heterogeneous BB can be accounted for by composite linear, (Carslaw and Jaeger, 1959), radial or spherical regions, the last two with no variation in angular directions, (de Swaan, 1998).

We assumed that flow in the compound medium of backbones and fractal clusters is described by the following simplifying conditions:

1. The backbone is a uniform net of pores or fractures with average thickness h_{BB};
2. Tortuosity of the backbone was not considered;
3. The SC are prisms with an average length h_{SC} perpendicular to the backbone;
4. The SC contact a fraction f_{SC} of the backbones' length, Figure 5;
5. Values of local porosity, permeability and compressibility are common to both media;
6. A repetitive element within the pay zone has a thickness equal to $h_{BB} + h_{SC}$.

The system for a single extended backbone is depicted in Figure 6.

In radial – axial – symmetry the pressure equation is:

$$\nabla_r^2 p_{BB}(r,t) - \eta \frac{\partial p_{BB}(r,t)}{\partial t} - \frac{f_{SC}}{h_{BB}} \int_0^t \frac{\partial p_{BB}(x,t')}{\partial t'} q_{SC}(t')dt' = 0 \tag{6}$$

where q_{SC} is the inflow from a representative or average SC to the BB as the effect of a unit step pressure disturbance at the contacting surface. The SC contact a fraction f_{SC} along the radial BB's.

That outflow is convolved in time to account for the variable pressure in the backbone. The time-Laplace transform is:

$$\nabla_r^2 \tilde{p}_{BB}(r,t) - \eta\tau\tilde{p}_{BB} - \tau\frac{f_{SC}}{h_{BB}}\tilde{p}_{BB}\tilde{q}_{SC}(\tau) = 0 \tag{7}$$

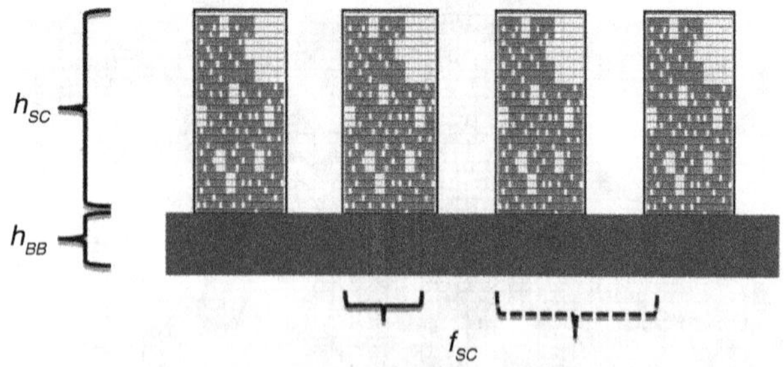

Figure 6

Schematic of backbone and subcritical clusters.

SC are considered with a linear finite geometry. One of the ends of the linear-fractal clusters is closed; the other extreme is connected to the BB. $\tilde{q}_{SC}$ is the transformed flow rate from a single SC after a unitary pressure disturbance – a heaviside step – in the BB.

The solution to Equation (7) in the time-Laplace domain is:

$$\tilde{p}_{BB} = \frac{-q\mu}{A_{BB}k} \frac{K_0\left(r\sqrt{\tau/\eta'}\right)}{\tau\sqrt{\tau/\eta'}\,K_1\left(r_w\sqrt{\tau/\eta'}\right)} \tag{8}$$

where,

$$\eta' = \eta + \frac{f_{SC}}{h_{BB}}\tilde{q}_{SC}(\tau) \tag{9}$$

Distributions of h_{SC} lengths can be incorporated in the formulation as in the multiply sized blocks and fractures formulation, de Swaan (2000). Different BB and SC properties, as in turbiditic channel-lobe systems can also be represented in that scheme.

1.2 Fractal Diffusion in a Finite Linear Medium

The fractional differential equation:

$$\frac{\partial^\alpha p}{\partial x^\alpha} = \eta\frac{\partial^\beta p}{\partial t^\beta} \tag{10}$$

has been solved in several papers for infinite linear media, (Metzler and Klafter, 2004), but not for finite media with the boundary conditions of the present problem.

The clusters-to-backbone flow term in Equation (6) is accounted for through the solution to the fractional diffusion of pressure equation in a finite linear medium, $\alpha \equiv d_f$, $\beta = \frac{2}{d_w}$. Fractional derivative definitions include Riemann–Liouville, Grunwald–Letnikov, Weyl, Caputo, Marchaud, and Riesz, (Agrawal, 2002; Oldham and Spanier, 1974; Miller and Ross, 1993; Podlubny, 1999; Butzer and Westphal, 2000). Of the various derivatives we chose Caputo's which Laplace transform is:

$$L_{x,\kappa}\left\{\frac{\partial^\alpha p}{\partial x^\alpha}\right\} = \kappa^\alpha \hat{p} - \sum_{j=0}^{m-1} \kappa^{\alpha-1-j} p^{(j)}(0+,t);\ m-1 < \alpha \leq m \tag{11}$$

where m is an integer; $p^{(j)}$ is the derivative of order j. Thus, the L transform of the Caputo time derivative is:

$$L_{t,\tau}\left\{\frac{\partial^\beta p}{\partial t^\beta}\right\} = \tau^\beta \tilde{p} - \tau p(x,0);\ \beta < 2 \tag{12}$$

The Laplace transform of the Caputo derivative has the convenience of including integer derivatives, in contrast with Sturm-Liouville and other definitions of fractal derivatives.

The boundary condition at the extreme length h_{SC} of the average sub-critical clusters is a Heaviside unit step function:

$$p(h_{SC}, t) = H(t) \tag{13}$$

with transform:

$$\tilde{p}(h_{SC}, \tau) = 1/\tau \tag{14}$$

As the space coordinate extends only in one direction from the origin, the Laplace transform can be used instead of the Fourier transformation. Circumflex means space-Laplace transform; a dash means the time-Laplace transform.

At $x = 0$ the boundary condition is no-flow, we use the gradient as the regular order one derivative:

$$\frac{\partial p(0,t)}{\partial x} = 0 \tag{15}$$

or,

$$\kappa\hat{p}(0,t) = 0 \tag{16}$$

Space and time transforms of Equation (10) result in:

$$\hat{\tilde{p}}(\kappa,\tau) = \frac{\kappa\tilde{p}(0,\tau)}{\kappa^\alpha - (\tau/\eta)^\beta} \tag{17}$$

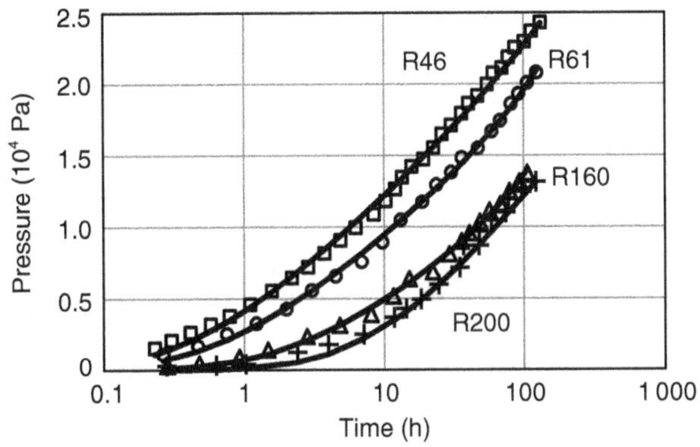

Figure 7

This theory (lines); observations (marks, Le Borgne *et al.*, 2004).

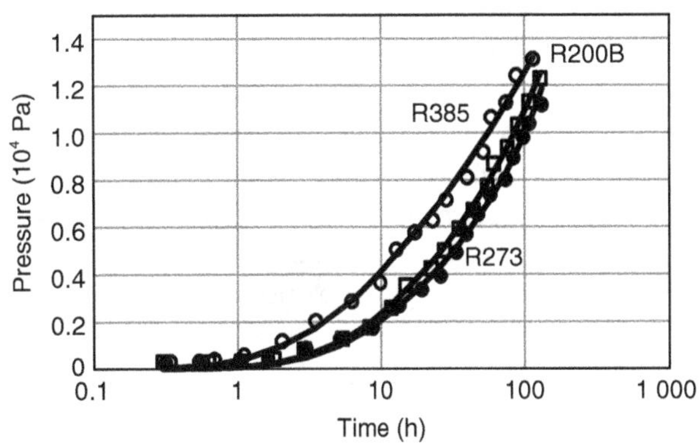

Figure 8

This theory (lines); observations (marks; Le Borgne *et al.*, 2004).

Notice that Equation (16) is space-transformed and the numerator in Equation (17) is time-transformed. The boundary condition at h_{SC}, Equation (14), requires:

$$\tilde{p}(0,\tau) = \frac{1}{\tau}\left[L^{-1}_{\kappa,x}\left\{\frac{\kappa}{\kappa^{\alpha} - (\tau/\eta)^{\beta}}\right\}\right]^{-1}_{x=h_{SC}} \tag{18}$$

For $\alpha = 2$ and $\beta = 1$, the diffusivity equation reduces to the normal or Gaussian case as:

$$L^{-1}\left\{\frac{\kappa}{\kappa^2 - a^2}\right\} = \cosh(ax) \tag{19}$$

and

$$\tilde{p}(x,\tau) = \frac{\cosh(\sqrt{\tau/\eta}x)}{\tau \cosh(\sqrt{\tau/\eta}h_{SC})} \tag{20}$$

See Carslaw and Jaeger (1959) page 309.

The flow into or from the fractal SC cluster is the derivative of pressure at the face $x = h_{SC}$ we use the regular, non-fractal, space derivative:

$$\widetilde{q_{SC}} = \frac{k}{\mu}\frac{\partial\tilde{p}(h_{SC},\tau)}{\partial x} \tag{21}$$

$$L_{x,\kappa}L_{t,\tau}\left\{\frac{\partial p(x,t)}{\partial x}\right\} = \frac{\kappa^2\tilde{p}(0,\tau)}{\tau\left[\kappa^{\alpha} - (\tau/\eta)^{\beta}\right]} \tag{22}$$

The L transform does not include fractional derivatives at $t = 0$, neither at $x = 0$, because we used the Caputo derivative.

The reduction to the normal case ($\alpha = 2$, $\beta = 1$) results in a hyperbolic sine divided by the hyperbolic cosine at the constant pressure face, (Carslaw and Jaeger, 1959).

2 RESULTS

The transformed Equation (8) with source term Equation (21) is back-transformed in time by means of the Crump (1976) and Schapery (1961) procedures. Validation of the formulation was made by comparing its results with the analytical normal results with exponents $\alpha = 2$, $\beta = 1$, (not shown).

The properties d_f, φ, k, h_{BB}, h_{SC}, f_{SC}, were adjusted to match observed interference pressure tests by an automatic match procedure, (Minpack, 1999).

Figures 7 and 8 show results matched to punctual data selected from Le Borgne *et al.* (2004), from tests in a highly heterogeneous fractured aquifer. The names of the wells include their distance to the pumping well. The pressure increments at the observation wells show up at intervals unrelated to those distances. Tidal effects were not filtered.

Figure 9 shows results matched to observations from Bernard *et al.* (2006).

Table 1 contains values of the properties derived from the matches. Table 2 shows values used in the model.

Some adjusted values show lack of consistency; in wells R200 and M5, the SC and BB thicknesses add significantly to more than the corresponding aquifer thickness.

Most of the d_f values in column 3 are between 1 and 2. That value means that the trajectory of flow is a crooked line that covers more than the linear dimension, one;

but less than the areal dimension (two), see the Koch line in Ben-Avraham and Havlin (2000) page 4. The exception is well M1 with a value of 2.69 which means a flow path close to be space-filling, (dimension 3).

By using a different technique, Bernard *et al.* (2006) give values of permeability of 4.5E-12 m^2; porosities from 0.005 to 0.08, from the pressure responses in the wells. Well M5 is in a perpendicular direction from, and farther from the injection well than, well M1. It shows an earlier response; therefore manifesting high anisotropy in the formation.

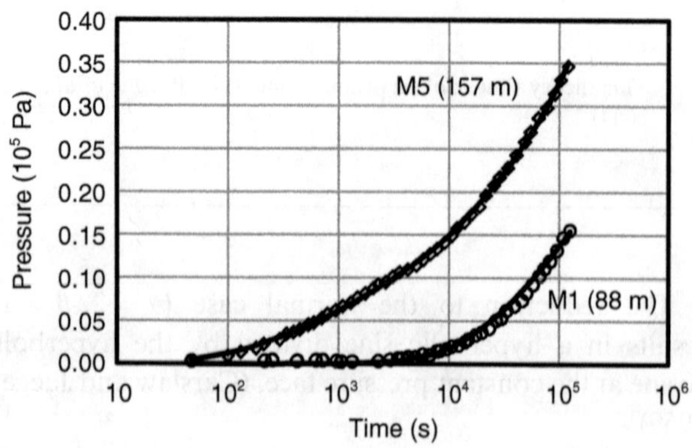

Figure 9

This theory (lines); observations (Bernard *et al.*, 2006).

3 COMMENTS

The presented matches are not unique; they depend on the starting values of the assumed properties; those presenting plausible values of the properties were selected and are shown in Figures 7 to 9 with the values in Table 1.

A full validation of the presented approach requires determination of fractal exponents and fracture space distribution besides the average petrophysical properties of the media where the well tests are conducted. Detailed numerical simulations of fractal media may be used to that avail.

A more realistic backbone with varying cross section can be represented as a linear medium with a collection of linear media (Carslaw and Jaeger 1959). In radial coordinates a collection of linked coronae can account for radial (not angular) variations in properties or thickness, (de Swaan, 1998).

Discrete distributions of varying properties and size of the dangling ends can be described in the manner of

TABLE 1

Adjusted properties, wells R46 to R385 from Le Borgne *et al.* (2004); wells M1 and M5 from Bernard *et al.* (2006)

Well	Distance (m)	d_f	d_w	φ	k (m^2)	f_{SC}	h_{SC} (m)	h_{BB} (m)
R46	46	1.92	2.88	0.0015	2.77E-13	0.035	55.18	53.13
R61	61	1.89	2.84	0.0021	2.53E-13	0.08	47.47	84.58
R160	160	1.01	1.51	0.092	7.50E-13	0.30	1.83	28.20
R200	200	1.68	2.52	0.0078	2.76E-13	0.11	36.64	100.9
R200B	200	1.87	2.80	0.072	4.55E-13	0.13	25.34	24.60
R273	273	1.83	2.75	0.044	3.53E13	0.42	27.53	67.07
R385	385	1.64	2.46	0.047	8.80E-13	0.0049	11.70	23.55
M1	88	2.69	4.04	0.00069	1.07E-15	0.68	10.94	3.40
M5	157	1.87	2.81	0.042	2.10E-13	0.19	242.11	34.05

TABLE 2

Aquifer interference tests and water properties

Aquifer reference	Thickness	Water injection rate	Water compressibility	Water viscosity
Le Borgne *et al.* (2004)	100 m	34 m^3/h	8.3E-10 Pa^{-1}	0.17E-3 Pa.s
Bernard *et al.* (2006)	120 m	0.0187 m^3/s	8.3E-10 Pa^{-1}	0.17E-3 Pa.s

distributions of matrix-rock blocks in fractured reservoirs, (de Swaan, 2000).

A full characterization of actual porous or fractured reservoirs in terms of backbone(s) and dangling-end sizes, as well as indicators of fractal exponents is not known to the author. That knowledge is important for secondary recovery practices since in dangling ends fluids displacement is not operative; they are stagnant unproductive zones.

CONCLUSIONS

1. The diffusion equation with fractional space and time derivatives, representing flow in a finite, fractal and linear medium, was formulated by using the Caputo fractional derivative both in the space and time terms. The presented approach was validated with comparisons to the normal case with integer exponents;
2. The resulting outflow function from that linear medium was included as a source term in the equation of pressure transient in an infinite radial connected medium. That global equation was back-transformed – in time and space – with approximate procedures and matched to actual pressure-interference data from tests in aquifers, showing the feasibility of the approach;
3. The present theory provides a mathematical description of transient flow in media that may be represented as clusters of pores or fractures formed by backbone(s) and subcritical clusters.

ACKNOWLEDGMENTS

The author acknowledges valuable help from Rodolfo Camacho-Velazquez, Mario Vásquez-Cruz and Gabriela Gomez-Ortiz in the development of this paper. Figure 4, from Adler and Thovert (1999), is reproduced with kind permission from Springer Science + Business Media B.V.
 This paper was presented at the EAGE Annual Conference & Exhibition Incorporating SPE Europec held in London, United Kingdom, 10-13 June 2013 as SPE Paper 164892.

REFERENCES

Adler P.M., Thovert J.F. (1993) Fractal Porous Media, *Transport in Porous Media* **13**, 41-78.

Adler P.M., Thovert J.F. (1999) *Fractures and Fracture Networks*, Kluwer, Dordrecht.

Agrawal O.P. (2002) Solution for a Fractional Diffusion-Wave Equation Defined in a Bounded Domain, *Nonlinear Dynamics* **29**, 145-155.

Alexander S., Orbach R. (1982) Density of states on fractals: 'fractons', *J. Phys. Lett.* **43**, L625-631.

Barker J.A. (1988) A generalized radial flow model for hydraulic tests in fractured rock, *Water Resour. Res.* **24**, 10, 1796-1804.

Ben-Avraham D., Havlin S. (2000) *Diffusion and Reactions in Fractals and Disordered Systems*, Cambridge, Massachusetts, Cambridge U Press.

Bernard S., Delay F., Porel G. (2006) A new method of data inversion for the identification of fractal characteristics and homogenization scale from hydraulic pumping tests in fractured aquifers, *J Hydrology* **328**, 647-658.

Butzer P.L., Westphal U. (2000) An introduction to fractional calculus, Hilfer R. (ed.), *Applications of Fractional Calculus in Physics*, World Scientific, Singapore, pp. 1-85.

Camacho Velazquez R., Fuentes-Cruz G., Vásquez-Cruz M. (2008) Decline Curve Analysis of Fractured Reservoirs with Fractal Geometry, *SPE Reservoir Evaluation & Engineering* **11**, 03, 606-619.

Carslaw H.S, Jaeger J.C. (1959) *Conduction of Heat in Solids*, 2nd ed., Oxford U. Press, New York.

Chang J., Yortsos Y.C. (1990) Pressure Transient Analysis of Fractal Reservoirs, *SPE Formation Evaluation* **5**, 01, 31-38.

Chen T., Noirot J.-C., Khandelwal A., Xue Guangri, Barton M.D., Alpak F.O. (2012) Estimating Stratigraphic Parameters from Well Test Data in Turbidite Reservoirs, *SPE Annual Technical Conference and Exhibition*, 8-10 Oct, San Antonio TX, USA, SPE Paper 159090.

Compte A. (1996) *Phys. Rev. E* **53**, 4, 4191-4193.

Crump K.S. (1976) Numerical Inversion of Laplace Transforms Using a Fourier Series Approximation, *Journal ACM* **23**, 1, 89-96.

de Swaan A. (1986) Influence of Shape and Skin of Matrix-Rock Blocks on Pressure Transients in Fractured Reservoirs, *61st Annual Technical Conference and Exhibition of the Society of Petroleum Engineers*, 5-8 Oct, New Orleans, LA, SPE Paper 15637.

de Swaan A. (1998) Transient Fluid Flow through Composite Geometries, *SPE International Petroleum Conference and Exhibition of Mexico*, 3-5 March, Villahermosa Mexico, SPE Paper 36777.

de Swaan A. (2000) Analysis of Well Tests in Multiple Fractured Reservoirs - Field Case Applications, *SPE International Petroleum Conference and Exhibition*, 1-3 Feb, Villahermosa, Mexico, SPE Paper 59015.

de Swaan A., Camacho-Velasquez R., Vasquez-Cruz M. (2012) Interference Tests Analysis in Fractured Formations with a Time-Fractional Equation, *SPE Latin American and Caribbean Petroleum Engineering Conference*, 16-18 April, Mexico-City, Mexico, SPE Paper 153615-PP.

Essam J.W., Bhatti F.M. (1985) Series expansion evidence supporting the Alexander-Orbach conjecture in two dimensions, *J. Phys. A* **18**, 3577-3584.

Gorenflo R., Mainardi F. (1997) Fractional calculus: integral and differential equations of fractional order, Carpinteri A., Mainardi F. (eds), *Fractals and Fractional Calculus in Continuum Mechanic*, Springer Verlag, Vienna, pp. 223-276.

Klemm A., Müller H.P., Kimmich R. (1999) Evaluation of fractal parameters of percolation model objects and natural porous media by means of NMR microscopy, *Physica A: Statistical Mechanics and its Applications* **266**, 1-4, 242-246.

Le Borgne T., Bour O., de Dreuzy J.R., Davy P., Touchard F. (2004) Equivalent mean flow models for fractured aquifers: Insights from a pumping tests scaling interpretation, *Water. Resour. Res.* **40**, W03512.

Leveinen J. (2000) Composite model with fractional flow dimensions for well test analysis in fractured rocks, *J. Hydrol.* **234**, 116-141.

Mainardi F., Luchko Y., Pagnini G. (2001) *Fractional Calculus and Applied Analysis* **4**, 2, 153-192.

Metzler R., Klafter J. (2000) The random walk's guide to anomalous diffusion: a fractional dynamics approach, *Physics Reports* **339**, 1-77.

Metzler R., Klafter J. (2004) The restaurant the end of the random walk: recent developments in the description of anomalous transport by fractional dynamics, *J. Phys. A: Math. Gen.* **37**, R161-R208.

Metzler R., Glockle W.G., Nonenmacher T.F. (1994) Fractional model equation for anomalous diffusion, *Physica A* **211**, 13-24.

Miller K.S., Ross B. (1993) *An Introduction to the Fractional Calculus and Fractional Differential Equations*, Wiley, New York.

Minpack (1999) www.netlib.org.

Neuman S.P. (1990) Universal scaling of hydraulic conductivities and dispersivities in geologic media, *Water Res. Research* **26**, 8, 1749-1758.

O'Shaugnessy B., Procaccia I. (1985) Analytical solutions for diffusion on fractal objects, *Phys. Rev. Lett.* **54**, 5, 455-458.

Oldham K.B., Spanier J. (1974) *The Fractional Calculus*, Academic Press, New York.

Podlubny I. (1999) *Fractional Differential Equations*, Academic Press, New York.

Raghavan R. (2011) Fractional derivatives: Application to transient flow, *Journal of Petroleum Science and Engineering* **80**, 1, 7-13.

Rammal R., Angles D'Auriac J.C., Benoit A. (1984) Universality of the spectral dimension of percolation clusters, *Phys. Rev. B* **30**, 4087.

Sahimi M. (1993) Fractal and Superdiffusive Transport and Hydrodynamic Dispersion in Heterogeneous Poroous Media, *Transport in porous Media* **13**, 3-40.

Sahimi M. (1994) *Applications of Percolation Theory*, Taylor and Francis, Bristol.

Sahimi M. (1995) *Flow and Transport in Porous Media and Fractured Rock: From Classical Methods to Modern Approaches*, VCH, Weinheim, pp. 204-205.

Schapery A. (1961) Approximate Methods of Transform Inversion for Viscoelastic Stress Analysis, Technical Report, *4th US National Congress Appl. Math.*

Schroeder M. (1990) *Fractals, Chaos, Power Laws*, Freeman, New York.

Vlahos L., Isliker H., Kominis Y., Hizanidis K. (2008) Normal and Anomalous Diffusion: A Tutorial, arXiv:0805.0419v1.

13

Thermodynamic Characterization of Undefined Petroleum Fractions of Gas Condensate using Group Contribution

Veronica Uribe-Vargas*, Bernardo Carreón-Calderón, Edgar Ramírez-Jaramillo and Mario Ramírez-de-Santiago

Mexican Institute of Petroleum, Mexico City - Mexico

e-mail: vuribe@imp.mx - bcarreon@imp.mx - eramirez@imp.mx - msantiag@imp.mx

* Corresponding author

Abstract — *A methodology proposed in a previous paper [Carreón-Calderón* et al. *(2012)* Ind. Eng. Chem. Res. ***51****, 14188-14198] for thermodynamic characterization of undefined petroleum fractions was applied to gas-condensate fluids. Using this methodology, input parameters of cubic equations of state and their mixing rules, critical properties and chemical pseudostructures are determined for undefined fractions by minimizing their Gibbs free energy. The results show the feasibility of applying this approach to gas-condensate fluids without making use of either cubic equations of state or mixing rules with specific adjusted parameters for petroleum fluids. Besides, it is shown that the phase equilibrium envelopes of gas-condensate fluids are highly dependent on the critical properties assigned to the undefined petroleum fractions of such fluid fractions and less dependent on the equation used for modeling gas-condensate fluids as a whole. The Absolute Average Error (AAE) considering the best arrangement is 1.79% in predicting the dew point.*

Résumé — Caractérisation thermodynamique des fractions indéfinies d'un condensat de gaz à l'aide de la méthode de contribution de groupes — Une méthodologie proposée dans un article précédent [Carreón-Calderón *et al.* (2012) *Ind. Eng. Chem. Res.* **51**, 14188-14198] pour la caractérisation thermodynamique des fractions de pétrole non définies a été appliquée à des gaz à condensats. En utilisant cette méthodologie, les paramètres d'entrée de l'équation d'état cubique et leurs règles de mélange, propriétés critiques et pseudo-structures chimiques, sont déterminés pour les fractions non définis en minimisant l'énergie libre de Gibbs. Les résultats montrent que l'application de cette approche est appropriée aux fluides provenant des condensats de gaz sans utiliser, pour l'équation d'état cubique ou les règles de mélange, de paramètres caractéristiques ajustés pour les fluides pétroliers. En plus, il est montré que les courbes d'équilibre de phases des fluides de condensat de gaz sont fortement dépendantes des propriétés critiques associées aux fractions de pétrole non définies et moins dépendantes de l'ensemble des équations modélisant les fluides de condensat de gaz. Pour la prédiction du point de rosée, l'erreur moyenne absolue est 1,79 % pour le meilleur arrangement.

NOTATION

AAD	Absolute Average Deviation
FG	Number of functional group types
JR	The Joback-Reid method
MW	Molecular weight
$MHV1$	Modified Huron-Vidal mixing rule of one parameter
P	Pressure (MPa)
P_C	Critical pressure (MPa)
p_c	Contribution to P_c by functional groups
$PSRK$	Predictive Soave-Redlich-Kwong mixing rule
PR	Peng-Robinson equation of state
R	Gas constant, 8.31441 J/mol.K
SRK	Soave-Redlich-Kwong equation of state
T	Temperature (K)
T_b	Boiling temperature at 0.101325 MPa (K)
T_c	Critical temperature (K)
t_c	Contribution to T_c by functional groups
V	Volume (m^3/mole)
V_c	Critical volume (m^3/mol)
VDW	van der Waals mixing rule
v_c	Contributions to V_c by functional groups
x	Mole fraction
Z_c	Critical compressibility factor

GREEK LETTERS

α	Coefficient in Equation (14)
β	Coefficient in Equation (14)
χ	Coefficient in Equation (14)
Δv	Group volume increment, Equation (13)
ϕ	Fugacity coefficient
γ	Activity coefficient
ρ	Mass density (kg/m^3)
υ	Number of functional groups
ω	Acentric factor

SUPERSCRIPTS

EOS	Equation Of State
EXP	Experimental value

SUBSCRIPTS

c	Critical property
i	Functional group i or pure component i
k	Undefined petroleum fraction k or pure component k

INTRODUCTION

Gas-condensate fluids from reservoirs are composed of hydrocarbon compounds from C_1 to C_6 and at least one undefined fraction, containing unidentified compounds ranging from C_7 to heavier hydrocarbon compounds. Similar to other petroleum fluids, gas-condensate fluids are very complex to model, not only because of the presence of unidentified compounds but also because of the presence of non-hydrocarbon compounds such as hydrogen sulfide, carbon dioxide and nitrogen.

In the same way as heavier petroleum fluids, the characterization of undefined fractions has a great effect on phase equilibrium simulations of gas-condensate fluids. Several methods have been proposed for characterizing undefined fractions, where experimental information in the form of molar distributions, the mole fraction against the Carbon Number (CN), is taken as starting point. For instance, it has been observed that a single straight-line relationship exists between the logarithm of the mole fraction and the molecular weight of the CN (Ahmed, 1989; Hosein and McCain, 2007). On the other hand, Whitson (1983, 1984) used a gamma distribution to establish the molar distribution. Soreide (1989) modified Whitson's method using a K factor in order to generate the distribution of specific gravities. When an experimental molar distribution of undefined fractions is absent, the methods commonly employed in the oil industry are those given by Whitson *et al.* (1990) and Pedersen *et al.* (1992).

Once the molar distribution has been defined, the characterization procedure requires the estimation of the critical properties for each CN or a mixture of them. Such mixtures are better known as pseudocomponents. The estimation of critical properties is usually performed using empirical correlations. Some examples of such correlations are those obtained by Kesler and Lee (1976) and Pedersen *et al.* (1980) among others. Sinayuc and Gumrah (2004) established that different sets of correlations provide very dissimilar results. These differences may be explained by the fact that the empirical correlations are obtained from specific experimental data banks of pure compounds; hence, they cannot be used in a general way irrespective of the type of petroleum fraction under study. On the other hand, attempts have been made to obtain a more generalized methodology. For instance, Sinayuc and Gumrah (2004) proposed an optimization method that consists of finding the global minimum of Gibbs energy using critical properties obtained from correlations as independent variables and a genetic algorithm as an optimization tool.

Recently, Carreón-Calderón *et al.* (2012) proposed a methodology, the so-called CURR (Carreón-Uribe-Ramírez-Ramírez's) approach, to estimate the critical properties of undefined petroleum fractions using an approach based on group contribution methods. Using classical thermodynamics, a molecular pseudostructure and critical properties are assigned to a given undefined petroleum fraction by minimizing its corresponding free energy. The CURR approach was applied to light petroleum fluids, where an experimental molar distribution was available. The results obtained appeared to be acceptable, irrespective of both the equation of state and the mixing rule used. Hence, the aim of this work is to test the functionality of the CURR approach under more difficult situations; for example, when an experimental molar distribution is not provided and therefore it has to be estimated, which is usually the case of gas-condensate fluids.

1 METHODOLOGY

The CURR approach was described in detail in a previous work (Carreón-Calderón *et al.*, 2012). In that methodology, one molecular pseudostructure, a set of non-integer functional groups, is simultaneously determined for a given petroleum fraction of unknown composition. Such properties allow undefined petroleum fractions to be treated as hypothetical pure components. This means that any method developed for handling components with known molecular structure can now be applied to petroleum mixtures containing undefined petroleum fractions. Thus, petroleum fluids, which are mixtures formed of unidentified compounds mostly, turn into mixtures formed of identified and pseudocompounds.

It should be noted that this procedure has been evaluated for the characterization of undefined fractions of light petroleum, where their corresponding experimental molar distributions were available. Hence, in this work, an additional step to estimate this distribution is included because experimental molar distributions are not provided for the condensate-gas fluids under study. As will be seen in what follows, this estimation consists of splitting the heavy undefined petroleum fraction, the C_{7+} fraction, into carbon numbers by assuming a probability distribution. In Figure 1, starting with such an estimation, the stages of the CURR method described in the following sections are summarized.

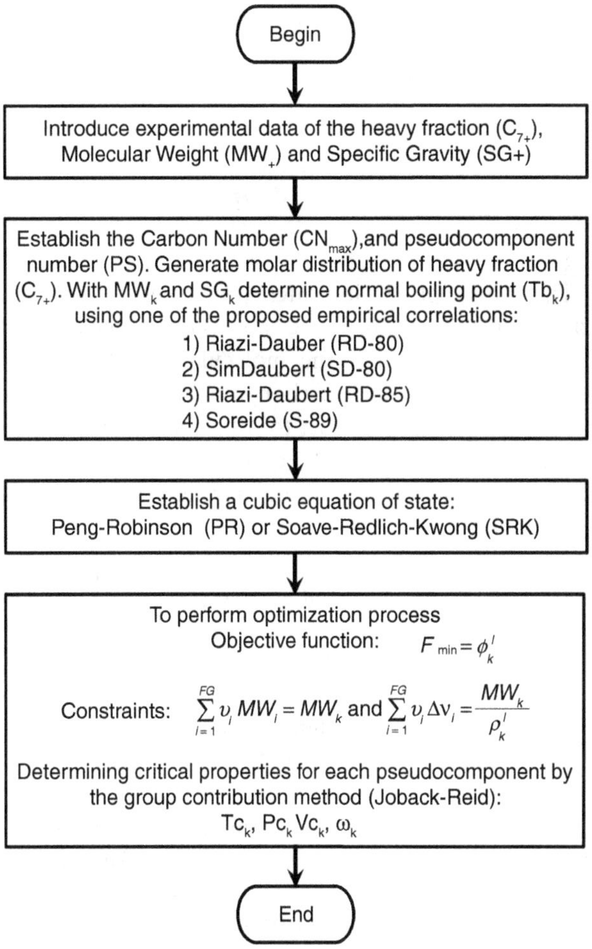

Figure 1

Flow diagram of the CURR method.

1.1 Molar Distribution of the Heavy Fraction (C_{7+})

Pedersen *et al.* (1992) proposed that, for naturally-occurring hydrocarbon mixtures, an exponential relation exists between the mole fraction of a component and the corresponding CN. They expressed this relationship using the following expressions:

$$\ln x_k = A + B(CN)_k \quad (1)$$

$$\rho_k = C + D\ln(CN)_k \quad (2)$$

$$MW_k = 14(CN)_k - 4 \quad (3)$$

where x_k symbolizes the mole fraction, ρ_k the density, MW_k the molecular weight and $(CN)_k$ the k-th carbon number. The heaviest carbon number is usually considered to be CN_{80}. The constants A, B, C and D are determined from experimental values of the mole fraction, molecular weight and density of the heavy petroleum fraction C_{7+} together with mass balances as follows:

$$x_+ = \sum_{k=+}^{CN_{max}} x_k \quad (4)$$

$$MW_+ = \frac{\sum_{k=+}^{CN_{\max}} x_k MW_k}{x_+} \quad (5)$$

$$\rho_+ = \frac{x_+ MW_+}{\sum_{k=+}^{CN_{\max}} \frac{x_k MW_k}{\rho_k}} \quad (6)$$

The subscript + indicates properties of the C_{7+} fraction to be split up. These expressions allow calculating the mole fraction, the molecular weight and the liquid density for each carbon number (Pedersen and Christensen, 2007). In order to solve the system of equations, it is considered that Equation (1) can be equally written in terms of the molecular weight instead of the carbon number, that is:

$$\ln x_k = A + BMW_k \quad (1a)$$

If this equation is inserted into Equations (4) and (5), and Equation (3) is used to determine the molecular weight for each carbon number k (MW_k), then we have a system of two equations with two unknowns (A and B). This system can be solved by any numerical method, such as Newton-Raphson. A similar procedure is performed for determining the liquid density for each carbon number k from the liquid density of the C_{7+} fraction through Equation (2) and Equation (6), but in this case, the previously determined molar distribution, MW_k against x_k, is used together with the density of the last carbon number fraction before the C_{7+} fraction, that is, the n-hexane liquid density. Once the molar distribution has been determined, the characterized petroleum fluid is formed by 80 carbon numbers and defined compounds. Hence, it is advisable to lump carbon numbers into pseudocomponents to reduce the total number of components, and consequently to make the phase equilibrium calculations easier. In Pedersen's methodology, a weight-based grouping is recommended, where each group (pseudocomponent) contains approximately the same weight amount. However, unlike Pedersen's methodology, where critical properties are found as weight mean averages, in this work, they are found from the density, molecular weight and boiling temperature of each pseudocomponent directly by the minimization procedure described below.

1.2 Determination of the Molecular Pseudostructure

In this step, one hypothetical structure, set, is non-integer functional groups are assigned to each pseudocomponent. This molecular pseudostructure may not correspond to the actual structure obtained from experimental data, but it is expected that it allows properties entering into cubic equations of state to be estimated correctly.

Critical Properties and Liquid Density Group Contribution

Estimating critical properties and the acentric factor from known chemical structure is the greatest advantage of group contribution methods. There have been various proposals for their application to characterize petroleum; for example, Ruzicka *et al.* (1983) and Avaullee *et al.* (1997a-c, 2001). In the latter works, one molecular pseudostructure is assigned to the C_{20+} fraction by solving systems of equations developed for a fixed set of non-integer functional groups, whereas, starting with the knowledge of paraffinic-naphthenic-aromatic analysis, two or three key compounds are used to represent undefined fractions from C_6 to C_{19}. For every undefined fraction, either one pseudostructure or key compounds are found by matching the experimental molecular weight and molar volume to calculations. Although the PR equation of state is also employed with C_6 to C_{19} fractions for such a purpose, it is worth pointing out that, unlike the current treatment, in those studies any minimization procedure subject to constraints was not suggested in order to obtain optimized functional groups in accordance with the thermodynamic laws. This issue allows, in principle, any group contribution method to be used in the current characterization procedure, as will be shown in future work.

Group contribution methods applied to a wider variety of chemical species have been developed to estimate the properties of a pure compound; for example, those proposed by Jöback and Reid (1987) (JR), and by Marrero and Gani (2001). In the current work, the former is suggested to be employed for estimating critical properties (T_c, P_c and V_c) of pseudocomponents; the respective expressions are:

$$T_c = \frac{T_b}{0.584 + 0.965 \sum_{i=1}^{FG} \upsilon_i t_{ci} - \left(\sum_{i=1}^{FG} \upsilon_i t_{ci} \right)^2} \quad (7)$$

$$P_c = \frac{0.1}{\left(0.113 + 0.0032 n_A - \sum_{i=1}^{FG} \upsilon_i p_{ci} \right)^2} \quad (8)$$

$$V_c = \frac{17.5 + \sum_{i=1}^{FG} \upsilon_i v_{ci}}{10^6} \quad (9)$$

TABLE 1
Correlations for estimating normal boiling point

	Normal boiling point	Source
RD80	$T_b = \frac{5}{9}\left[\left(\frac{MW \times SG^{1.0164}}{4.5673 \times 10^{-5}}\right)^{\frac{1}{2.1962}}\right]$	Riazi and Daubert (1980)
SD80	$T_b = \frac{5}{9}\left[\left(\frac{MW \times SG^{0.9371}}{1.4350476 \times 10^{-5}}\right)^{\frac{1}{2.3776}}\right]$	Sim and Daubert (1980)
RD87	$T_b = \frac{5}{9}[6.77857 \times MW^{0.401673}\, SG^{-1.58262} EXP(3.7741 \times 10^{-3} MW + 2.984036 SG - 4.25288 \times 10^{-3} SG)]$	Riazi and Daubert (1987)
S89	$T_b = \frac{5}{9}(1928.3 - 1.695 \times 10^5 MW^{-0.03522} SG^{3.266}\, e^{-4.922\times10^{-3} MW - 4.7685 SG + 3.462\times10^{-3} MW\times SG})$	Soreide (1989)

where T_c, P_c and V_c are the critical temperature, pressure and volume, respectively. Similarly, t_{ci}, p_{ci} and v_{ci} represent the corresponding contributions to T_c, P_c and V_c by the i-th functional group, being υ_i the number of the functional groups of type i. In the above equations, n_A is the number of atoms in the molecule, T_b is the normal boiling point, and FG is the total number of types of functional groups. The units for temperature, pressure and volume are K, MPa and m^3/mol, respectively. As observed from Equation (7), the normal boiling point is required for the critical temperature to be calculated. Generally, the experimental value of such a boiling point is not available; consequently, the use of some empirical correlations found in the literature is recommended, which are displayed in Table 1.

Another property required to calculate phase equilibrium calculations is the acentric factor. Unfortunately, in the JR method there is not any expression suggested; hence, in this work, we consider the following expression to estimate the acentric factor (Poling *et al.*, 2001):

$$\omega = -\frac{\ln(P_c/1.01325) + f_0(T_{br})}{f_1(T_{br})} \tag{10}$$

$$f_0(T_{br}) = \frac{-5.97616\tau + 1.29874\tau^{1.5} - 0.60394\tau^{2.5} - 1.06841\tau^5}{T_{br}} \tag{11}$$

$$f_1(T_{br}) = \frac{-5.03365\tau + 1.11505\tau^{1.5} - 5.41217\tau^{2.5} - 7.46628\tau^5}{T_{br}} \tag{12}$$

$T_{br} = T_b/T_c$ is the reduced normal boiling point and $\tau = 1 - T_{br}$.

According to Equation (2), liquid density is part of the exponential molar distribution, which implies that a methodology based on functional groups is also required to estimate liquid densities. In the CURR approach, the liquid density of a given pseudocomponent is estimated by the group contribution method GCVOL (Group Contribution method for predicting Saturated Liquid Density) (Elbro *et al.*, 1991; Ihmels and Gmehling, 2003), where the liquid density ρ is calculated from:

$$\rho = \frac{MW}{V} = \frac{MW}{\sum_{i=1}^{FG} \upsilon_i \Delta v_i} \tag{13}$$

Here, MW is the molecular weight and Δv_i is the group volume increment expressed as a function of temperature as follows:

$$\Delta v_i = \alpha_i + \beta_i T + \chi_i T^2 \tag{14}$$

where α_i, β_i, and χ_i are specific parameters of a given functional group.

Calculation of Functional Group Number

In order to estimate the number of functional groups for an undefined petroleum fraction, either the carbon number or the pseudocomponent, each fraction is considered as an independent system at a fixed pressure and temperature. It is well known that the equilibrium of the Gibbs free energy has a minimum value at a fixed value of pressure and temperature. From the PVT experiments carried out in the petroleum industry, the temperature and pressure measurements established for this purpose are 288.7 K (60°F) and 0.1013 MPa (1 atm). Thus, starting with the residual Gibbs energy, it was shown in the CURR approach that the above assumptions lead to minimizing the liquid fugacity coefficient of the respective undefined petroleum fraction (pseudocomponent), which is treated as a hypothetical pure compound. In other words, the objective function to be minimized becomes:

$$\phi_k^l(T_{ck}(\upsilon_1,\upsilon_2,\ldots,\upsilon_{FG}),P_{ck}(\upsilon_1,\upsilon_2,\ldots,\upsilon_{FG}),\omega_k(\upsilon_1,\upsilon_2,\ldots,\upsilon_{FG})) \tag{15}$$

Taking as constraints: the pseudocomponent volume expressed in Equation (13) and the corresponding molecular weight, expressed in turn by the following expression:

$$\sum_{i=1}^{FG}\upsilon_i MW_i = MW_k \tag{16}$$

Using a cubic equation of state and the fundamental expression of fugacity coefficients, Equation (15) can be rewritten as:

$$\phi_j = -\ln(Z-B)+Z-1+\frac{A}{B(\gamma+\beta)}\ln\left(\frac{Z+B\gamma}{Z-B\gamma}\right) \tag{17}$$

$$\beta=\frac{\sqrt{u^2-4w}-u}{2} \tag{18}$$

$$\gamma=\frac{\sqrt{u^2-4w}+u}{2} \tag{19}$$

$$A=\frac{Pa}{(RT)^2} \tag{20}$$

$$B=\frac{Pb}{RT} \tag{21}$$

$$Z=\frac{PV}{RT} \tag{22}$$

where R is the gas constant. In the expressions above, u and w are parameters that depend on the equation of state, as shown in Table 2. On the other hand, a and b depend on the critical properties, as will be shown in the following section.

1.3 PVT Property Calculation

The cubic equations of state were chosen here because they are relatively simple mathematical models relating the pressure-volume-temperature behavior for fluid phases, along with the fact that these equations have been used successfully in the oil industry. The Peng-Robinson (PR) (Peng and Robinson, 1976) and Soave-Redlich-Kwong (SRK) (Soave, 1972) equations of state were used in this work. The corresponding generalized expression is shown below:

TABLE 2
Equation of state parameters

Parameter	PR	SRK
u	2	1
w	−1	0
Ω_a	0.45724[a]	0.42747[b]
Ω_b	0.07780[a]	0.08664[b]
r_1	0.37464[a]	0.480[b]
r_2	1.54226[a]	1.574[b]
r_3	−0.26992[a]	−0.176[b]

[a] Peng and Robinson (1976).
[b] Soave (1972).

$$P=\frac{RT}{V-b}-\frac{a}{V^2+ubV-wb^2} \tag{23}$$

The parameters u and w define the cubic equation of state; v is the molar volume, while a and b are specific to a given component or, in the context of the current work, a given pseudocomponent through its critical properties, in accordance with the following expressions:

$$a=\frac{\Omega_a(RT_c)^2 f}{P_c} \tag{24}$$

$$f=\left[1+\left(r_1+r_2\omega+r_3\omega^2\right)\left(1-\sqrt{\frac{T}{Tc}}\right)\right]^2 \tag{25}$$

$$b=\frac{\Omega_b RT_c}{P_c} \tag{26}$$

where Ω_a, Ω_b and r_i are specific parameters for each equation of state, as can be seen in Table 2. It is important to emphasize that Equations (7) and (8) along with Equations (24) to (26) connect functional groups of pseudocomponents with the fugacity coefficient for a pure compound given by either Equation (15) or Equation (17).

TABLE 3
UNIFAC functional groups

Chemical formula	Main group	Sub-group
CH_3	1	1
CH_2	1	2
CH	1	3
C	1	4
ACH	3	9
AC	3	10
CO_2	56	117
CH_4	57	118
N_2	60	115
H_2S	61	114

Once one molecular structure and the critical properties are simultaneously determined for an undefined petroleum fraction (pseudocomponent) from the procedure described previously, the undefined petroleum fraction becomes a hypothetical pure component from the calculation standpoint. Thus, the whole mixture of hydrocarbons can be treated as a mixture of known components. This means that any approach developed for handling chemical structures can be applied to petroleum fluids regardless of the lack of experimental information about compounds forming undefined fractions and their chemical structures. To verify this assertion in the case of gas-condensate fluids, the Modified Huron-Vidal mixing rule of one parameter – MHV1 – (Horstmann *et al.*, 2005) using the group contribution method UNIFAC (Universal Functional group Activity Coefficients) as an activity coefficient model was used for PVT calculations. The set of UNIFAC functional groups used in the MHV1 mixing rule is shown in Table 3. In the same context, another mixing rule was used, the so-called Predictive Soave-Redlich-Kwong, PSRK (Horstmann *et al.*, 2005). There are other mixing rules such as MHV2 (Dahl and Michelsen 1990), Wong-Sandler (Wong and Sandler, 1992) a PPR78 (Jaubert and Mutelet, 2004; Jaubert *et al.*, 2010), that have been proved to work adequately in predicting the equilibrium of hydrocarbon mixtures, but the aim of this study is to evaluate the CURR method's performance and not mixing rules in general. In other words, we intend to characterize undefined petroleum fractions of gas-condensate fluids so that models requiring chemical structures, the UNIFAC approach inserted into mixing rules in this case, can be employed with such fluids; something that is not routinely seen in phase equilibrium calculations involving petroleum because of the unknown numbers υ_i. Otherwise, the so-called classical mixing rule or the Van Der Waals mixing rule (VDW) (Peng and Robinson, 1976; Soave, 1972), or its variations (Jaubert *et al.*, 2010), is the only available mixing rule for performing such calculations. The reader is referred to the original paper (Carreón-Calderón *et al.*, 2012) for additional information about the use of these mixing rules in the context of the CURR approach.

Although in the original CURR approach, correction in the molar volume was introduced by applying the concept of the volume translation (Peneloux *et al.*, 1982) and GCVOL approach (Eq. 13), the concept of volume translation was not introduced here because the only calculations performed were the phase equilibrium ones. However, it is important to make a distinction between the liquid density of the pseudocomponent and the liquid density of the petroleum fluid as a whole. The former is required for determining the chemical pseudostructure of the corresponding pseudocomponent through the optimization problem given by Equation (15), whereas the latter is not required for phase equilibrium calculations.

2 APPLICATIONS

The methodology described previously was used first of all to characterize thermodynamically the undefined petroleum fraction of five gas-condensate fluids found in the literature. Then, inserting the calculated chemical pseudostructures and associated critical properties into the cubic equations of state and mixing rules described above, the phase equilibrium calculations were performed. The experimental data employed in this work are shown in Table 4.

As mentioned before, first, it is necessary to determine the exponential molar distribution from the only undefined petroleum fraction provided, the C_{7+} fraction in this case (*Tab. 4*), using the split method proposed by Pedersen. Thus, carbon numbers together with their corresponding molecular weight, specific gravity and molar fraction were obtained. These carbon numbers were then lumped together into pseudocomponents (PS) of approximately the same weight fraction.

As the next step, the critical properties were estimated for every pseudocomponent from the previous step through the minimization process of Gibbs energy given by Equation (15) and related expressions. As observed in Table 4, the boiling temperatures are not available, which, according to Equation (7), are required to estimate critical temperatures; consequently, empirical

TABLE 4

Gas condensate composition, (Elsharkawy, 2002). Composition of the mole fraction. Density of the C_{7+} fraction is in g/cm^3

Name	GC1	GC2	GC3	GC4	GC5
		Mole fraction			
H_2S	0.0000	0.0000	0.0000	0.0001	0.0000
CO_2	0.0000	0.0010	0.0095	0.0045	0.0098
N_2	0.0000	0.0000	0.0077	0.0047	0.0066
C_1	0.8238	0.9522	0.5980	0.6073	0.6810
C_2	0.0428	0.0168	0.1415	0.1383	0.0700
C_3	0.0351	0.0091	0.0689	0.0633	0.0505
i-C_4	0.0161	0.0026	0.0140	0.0130	0.0179
n-C_4	0.0303	0.0033	0.0252	0.0238	0.0198
i-C_5	0.0060	0.0016	0.0102	0.0100	0.0162
n-C_5	0.0068	0.0011	0.0119	0.0109	0.0108
C_6	0.0099	0.0025	0.0175	0.0165	0.0177
C_{7+}	0.0292	0.0098	0.0956	0.1076	0.0997
MW_{C7+}	125	122.6	163	162	162
$Density_{C7+}$	0.74	0.723	0.791	0.789	0.807
Reservoir temperature (K)	278	378	416	416	398
DPP (MPa)	21.345	23.069	28.986	28.779	35.993

correlations were employed in accordance with Table 1. For illustrative purposes, the results obtained for the Gas Condensate 1 (GC1) are shown below, which correspond to the lowest Average Absolute Deviation (AAD) obtained for the Dew Point Pressure (DPP).

The molar distribution shown in Table 5 was carried out using 40 CN which were then lumped into 12 pseudocomponents. The boiling temperature correlation employed was S89 (*Tab. 1*), and the equation of state used for the minimization procedure was PR, giving rise to the critical properties also shown in Table 5. Similarly, the numbers of functional groups obtained for GC1 with CN = 40, PS = 12, Tb = S89 and EOS = PR are shown in Table 6.

As shown in Table 6, the proposed methodology allows the values of the number of functional groups to be non-integer numbers, unlike what is considered in traditional methods of group contribution. It should be noted that these values are hypothetical; that is, they do not correspond to those obtained from experimental analysis; they represent the amount of energy and mass needed from such groups, so that the free energy is minimum. Ultimately, the aim of this work is to find structures that allow bulk properties to be estimated properly for phase equilibrium purposes.

Similarly to the results obtained by Carreón-Calderón *et al.* (2012), the numbers of functional groups shown in Table 6 are not distributed randomly; the same trends can be observed. For example, although incorporated simultaneously in calculations, some functional groups vanish for some pseudocomponents. It can be seen that the lightest pseudocomponents are characterized only by paraffinic functional groups, while cyclic and aromatic functional groups are incorporated into the molecular pseudostructure for the heavier pseudocomponents. In order to accomplish a sensitivity analysis of the different variables involved in the implementation of the CURR method, different tests for the five condensed gases under study were performed. To illustrate this, the results obtained for the condensate gas GC1 are shown in Table 7, where it can be seen that for different equations of state, mixing rules and numbers of pseudocomponents, the highest average absolute deviation is 3.75 for dew point pressure.

The results are similar to those obtained previously applying the CURR approach to light petroleum fluids,

TABLE 5
Molar distribution and critical properties calculated in this work for GC1

	Molar distribution			Critical properties calculated			
	Mole fraction	MW	SG	T_c (K)	P_c (MPa)	V_c (m^3/mol)	ω
PS94	0.00908	94.0	0.6985	540.40	2.9563	4.0161E-04	0.3451
PS108	0.00626	108.0	0.7209	564.77	2.6760	4.5725E-04	0.3941
PS122	0.00431	122.0	0.7375	588.40	2.4317	5.1300E-04	0.4422
PS136	0.00297	136.0	0.7508	610.86	2.2209	5.6806E-04	0.4859
PS150	0.00205	150.0	0.7619	642.37	2.1217	6.0459E-04	0.4534
PS164	0.00141	164.0	0.7716	674.71	2.0619	6.3919E-04	0.4119
PS178	0.00097	178.0	0.7801	694.20	1.9187	6.8979E-04	0.4371
PS192	0.00067	192.0	0.7877	712.23	1.7901	7.4035E-04	0.4617
PS206	0.00046	206.0	0.7946	728.96	1.6737	7.9055E-04	0.4856
PS225	0.00054	225.7	0.8034	750.60	1.5277	8.6077E-04	0.5175
PS258	0.00033	258.6	0.8162	782.72	1.3242	9.7779E-04	0.5652
PS320	0.00016	321.0	0.8359	843.99	1.1560	1.1697E-03	0.6104

TABLE 6
Number of functional groups calculated for GC1, with 12 pseudocomponents

	Number of functional groups υ								
	$-CH_3$	$-CH_2$	>CH-	>C<	$-CH_2$- (cyclic)	>CH- (cyclic)	>C< (cyclic)	ACH	AC-C
PS94	1.7752	4.7986	0	0	0	0	0	0	0
PS108	1.7236	5.8520	0	0	0	0	0	0	0
PS122	1.6962	6.8795	0	0	0	0	0	0	0
PS136	1.7966	7.6572	0.1215	0	0	0	0	0	0
PS150	3.1988	4.5371	1.6479	0	1.1983	0	0	0	0
PS164	3.7952	2.6892	1.5863	0.3940	2.7951	0	0	0.3554	0
PS178	4.0409	2.8788	1.7202	0.4690	2.9883	0	0	0.5315	0
PS192	4.2856	3.0683	1.8546	0.5453	3.1812	0	0	0.7073	0
PS206	4.5218	3.2470	1.9758	0.6060	3.3636	0.0707	0	0.8732	0
PS225	4.8435	3.4854	2.1312	0.6736	3.6076	0.2541	0	1.0945	0
PS258	5.3759	3.8823	2.3928	0.7924	4.0134	0.5612	0	1.4629	0
PS320	6.8420	3.9583	1.0824	0	4.2046	2.3387	3.2558	1.5336	0

TABLE 7
Percent Absolute Average Deviation (AAD) of Dew Point Pressure (DPP), for gas condensate GC1

Equation of state	Mixing rule	Tb correlation	Pseudocomponent			
			1	6	12	18
			% AAD			
PR	VDW	S89	3.08	0.21	0.52	0.33
	MHV1		2.87	0.16	0.10	0.18
	PSRK		3.75	1.38	1.51	1.59
RKS	VDW		3.35	3.56	3.47	3.30
	MHV1		2.68	2.83	2.73	2.62
	PSRK		1.52	1.25	1.43	1.12

where any combination of cubic equations of state and mixing rules provided similar results provided that the properties of undefined petroleum fractions came from the minimization problem given by Equation (15) and related expressions. The possible explanation is that the CURR approach provides optimized properties, satisfying the fundamental thermodynamic relationships and the mass balance for every undefined petroleum fraction under study. Thus, the pressure-volume-temperature behavior predicted by any fluid model should improve by using such optimized properties.

The boiling point correlation used to determine the critical temperature has a higher impact on the phase envelope than both the cubic equation of state and the mixing rule, as shown in Figure 2. These results confirm that the pseudocomponent properties to be used in fluid models have a higher impact on the phase equilibrium prediction than the fluid model itself, at least in the case of cubic equations of state and their respective mixing rules. The Absolute Average Deviation (AAD) for Dew Point Pressure (DPP) for every case is shown in Table 8.

It should be noted that the results obtained with an exponential distribution employing 12 pseudocomponents had lower AAD than those obtained employing less pseudocomponents, which agrees with the results obtained by Elsharkawy (2002), who established that splitting up the plus fraction into 10 to 12 pseudocomponents leads to minimum absolute average relative deviation. Another interesting result from Table 8 is that, in contrast to Pedersen's suggestion, the best carbon number is not 80; the influence of the carbon number on phase envelope was undetectable for practical purposes. Finally, it can also be observed that the best results are

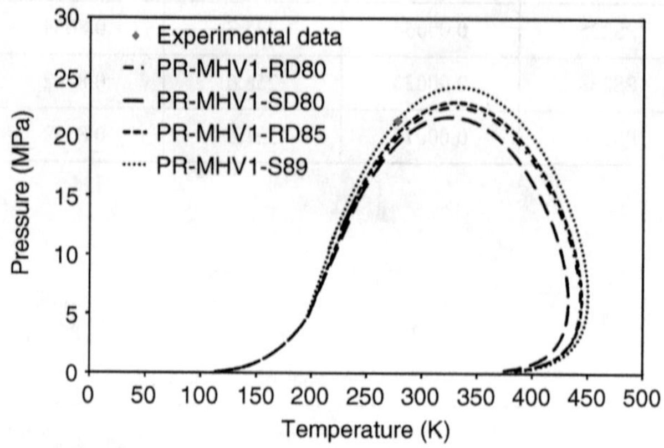

Figure 2

Phase envelope of CG1 using different correlations to estimate normal boiling temperature.

not always obtained using the same correlation for estimating boiling temperature, which shows again that properties to be used in cubic equations of state have a higher impact on the phase envelopes than the equations themselves.

Another aspect highlighted in the methodology shown in this work is that it is not necessary to adjust the parameters belonging to equations of state or their mixing rules in order to obtain a suitable estimation of the experimental data, as has been suggested elsewhere (Coats and Smart, 1986; Slot-Petersen, 1989; Merrill and Newley, 1993; Merrill *et al.*, 1994). Besides, the information needed to characterize the undefined heavy fraction is small: the molecular weight and specific

TABLE 8
Absolute Average Deviation (AAD) of equilibrium estimation of five gas condensates

Fluid	Mole fraction CH_4	Mole fraction C_{7+}	CH_4/C_{7+}	Number of pseudocomponents	CN	Tb correlation	AAD (%)
GC1	0.8238	0.0292	28.21	12	40	S89	0.10
GC2	0.9522	0.0098	97.16	12	40	S89	1.65
GC3	0.598	0.0956	6.26	12	80	SD80	2.18
GC4	0.6073	0.1076	5.64	12	80	SD80	4.79
GC5	0.681	0.0997	6.83	12	80	SD80	0.24

gravity of just one undefined petroleum fraction: the C_{7+} fraction.

CONCLUSIONS

The results showed that the gas-condensate phase envelopes depend more on the properties assigned to the undefined heavy fraction than on the pressure-volume-temperature equations modeling such fluids. The procedure applied here provides optimized values of the critical properties and chemical pseudostructures. Thus, the choice of cubic equations of state or mixing rules with adjusted parameters is of little consequence to predict phase behavior properly.

It is also concluded that the CURR method of characterizing thermodynamically undefined heavy petroleum fractions of crude oil is also applicable for thermodynamic characterization of corresponding heavy fractions belonging to gas-condensate fluids, where experimental molar distribution is not provided. The experimental data required is minimal: the fluid global composition, and the molecular weight, specific gravity and the normal boiling point of the corresponding undefined heavy fraction.

ACKNOWLEDGMENTS

The authors thank the authorities of the Mexican Institute of Petroleum in Mexico City for permission to publish this work. Funding for this research from grant D.00432 is gratefully acknowledged.

REFERENCES

Ahmed T. (1989) *Hydrocarbon phase behavior*, Gulf Publishing Houston, pp. 356-359

Avaullee L., Trassy L., Neau E., Jaubert J.N. (1997a) Thermodynamic modeling for petroleum fluids I. Equation of state and group contribution for the estimation of thermodynamic parameters of heavy hydrocarbons, *Fluid Phase Equilibria* **139**, 155-170.

Avaullee L., Neau E., Jaubert J.N. (1997b) Thermodynamic modeling for petroleum fluids II. Prediction of PVT properties of oils and gases by fitting one or two parameters to the saturation pressures of reservoir fluids, *Fluid Phase Equilibria* **139**, 171-203.

Avaullee L., Neau E., Jaubert J.N. (1997c) Thermodynamic modeling for petroleum fluids III. Reservoir fluid saturation pressures. A complete PVT property estimation. Application to swelling test, *Fluid Phase Equilibria* **141**, 87-104.

Avaullee L., Duchet-Suchaux P., Durandeau M., Jaubert J.N. (2001) A new approach in correlating the oil thermodynamic properties, *Journal of Petroleum Science and Engineering* **30**, 43-65.

Carreón-Calderón B., Uribe-Vargas V., Ramírez Jaramillo E., Ramírez de Santiago M. (2012) Thermodynamic Characterization of Undefined Petroleum Fractions using Group Contribution Methods, *Ind. Eng. Chem. Res.* **51**, 14188-14198.

Coats K.H., Smart G.T. (1986) Application of a regression-based EOS PVT program to laboratory data, *SPE Reservoir Eng.* **1**, 03, 277-299.

Dahl S., Michelsen M.L. (1990) High pressure vapour-liquid equilibrium with a UNIFAC-based equation of state, *AIChE J.* **36**, 1829-1836.

Elsharkawy A.M. (2002) Predicting the dew point pressure for gas condensate reservoirs: empirical models and equations of state, *Fluid Phase Equilib.* **193**, 147-165.

Elbro H.S., Fredenslund A., Rasmussen P. (1991) Group contribution method for the prediction of liquid densities as a function of temperature for solvents, oligomers, and polymers, *Ind. Eng. Chem. Res.* **30**, 12, 2576-2582.

Horstmann S., Jaboniec A., Krafczyk J., Fischera K., Gmehling J. (2005) PSRK group contribution equation of state: comprehensive revision and extension IV, including critical constants and function parameters for 1000 components, *Fluid Phase Equilib.* **227**, 2, 157-164.

Hosein R., McCain Jr W. (2007) Extended analysis for gas condensate systems, *SPE Annual Technical Conference and Exhibition* held in Anaheim, California USA, 11-14 Nov, SPE-110152, *SPE Reservoir Evaluation & Engineering* **12**, 01, 159-166.

Ihmels E.Ch, Gmehling J. (2003) Extension and revision of the group contribution method GCVOL for the prediction of pure compound liquid densities, *Ind. Eng. Chem. Res.* **42**, 2, 408-412.

Jaubert J.N., Mutelet F. (2004) VLE predictions with the Peng-Robinson equation of state and temperature dependent k_{ij} calculated through a group contribution method, *Fluid Phase Equilib.* **224**, 285-304.

Jaubert J.N., Privat R., Mutelet F. (2010) Predicting the Phase Equilibria of Synthetic Petroleum Fluids with the PPR78 Approach, *AIChE J.* **56**, 3225-3235.

Jöback K.G., Reid R.C. (1987) Estimation of Pure-Component Properties from Group-Contributions, *Chem. Eng. Commun.* **57**, 233-243.

Kesler M.G., Lee B.I. (1976) Improve prediction of enthalpy of fractions, *Hydrocarb Process.* **55**, 153-158.

Marrero J., Gani R. (2001) Group-contribution based estimation of pure component properties, *Fluid Phase Equilib.* **183-184**, 183-208.

Merrill R.C., Hartman K.J., Creek J.L. (1994) A comparison of equation of state tuning methods, presented at the *SPE 69th Annual Technical Conference and Exhibition*, New Orleans, LA, *SPE* 28589.

Merrill R.C., Newley T.M.J. (1993) A systematic investigation into the most suitable data for development of equation of state for petroleum reservoir fluids, *Fluid Phase Equilib.* **82**, 101-110.

Pedersen K.S., Christensen P.L. (2007) *Phase Behavior of Petroleum Reservoir Fluids*, Taylor & Francis Group, New York.

Pedersen K.S., Thomassen Q., Riazi M.R., Daubert T.E. (1980) Prediction of the composition of Petroleum Fractions, *Ind. Eng. Chem. Process Dev.* **19**, 289-294.

Pedersen K.S., Blilie A.L., Meisingset K.K. (1992) PVT Calculations on petroleum reservoir fluid using measured and estimated composition data for the plus fraction, *I&EC Research* **31**, 5, 1378-1383.

Peneloux A., Rauzy E., Fréze R. (1982) A consistent correction for Redlich-Kwong-Soave volumes, *Fluid Phase Equilib.* **8**, 1, 7-23.

Peng D.Y., Robinson D.B. (1976) A New Two-Constant Equation of State, *Ind. Chem. Fundam.* **15**, 1, 59-64.

Poling B.E., Prausnitz J.M., O'Connell J.P. (2001) *The Properties of Gases and Liquids*, 5th edn., McGraw-Hill International Edition, Singapore.

Riazi M.R., Daubert T.E. (1980) Simplify property predictions, *Hydrocarb Process.* **59**, 115-116.

Riazi M.R., Daubert T.E. (1987) Characterization parameters for petroleum fractions, *Ind. Eng. Chem. Res.* **26**, 755.

Ruzicka V., Fredenslund A., Rasmussen P. (1983) Representation of petroleum fractions by group contribution, *Ind. Eng. Chem. Process Des. Dev.* **22**, 1, 49-53.

Sim W.J., Daubert T.E. (1980) Prediction of vapor-liquid equilibria of undefined mixture, *Proces. Des. Dev.* **19**, 380.

Sinayuc C., Gumrah F. (2004) Predicting the critical properties of heptanes-plus in gas condensates: Genetic algorithms as an optimization toll, *Petroleum Society's 5th Canadian International Petroleum Conference (55th Annual Technical Meeting)*, Calgary Alberta, Canada, 8-10 June.

Slot-Petersen C. (1989) A systematic and consistent approach to determine binary interaction coefficients for Peng-Robinson equation of state, *SPE* 16941, *SPE Reservoir Engineering* **4**, 4, 488-494.

Soave G. (1972) Equilibrium constants from a modified Redlich-Kwong equation of state, *Chem. Eng. Sci.* **27**, 1197-1203.

Soreide I. (1989) Improved phase behavior predictions of petroleum reservoir fluids from a cubic equation of state, *Doktor. Ing Thesis*, Department of Petroleum Engineering Norwegian Institute of technology (NTH), Trondheim, Norway.

Whitson C.H. (1983) Characterizing hydrocarbon plus fractions, *Soc. Pet. Eng. J.* **23**, 4, 683-694.

Whitson C.H. (1984) Effect of C7+ characterization on equations of state predictions, *Soc. Pet. Eng. J.* **24**, 6, 685-694.

Whitson C.H., Anderson T.F., Soreide I. (1990) Application of gamma distribution model to molecular weight and boiling point data for petroleum fractions, *Chem. Eng. Comm.* **96**, 259.

Wong D.S.H., Sandler S.I.A. (1992) A theoretically correct mixing rule for cubic equations of state, *AIChE J.* **38**, 671-680.

14

Testing of Snorre Field Foam Assisted Water Alternating Gas (FAWAG) Performance in New Foam Screening Model

Pavel Spirov* and Svetlana Rudyk

Sultan Qaboos University - Department of Petroleum and Chemical Engineering, PO Box 34, Al-Khod 123, Muscat - Sultanate of Oman
e-mail: pavlicica3@gmail.com - svetrud77@mail.ru

* Corresponding author

Abstract — *Eclipse Functional Foam Model was used in order to provide a guideline for the history matching process (Gas-Oil Ratio (GOR), oil and gas production rates) to the Foam Assisted Water Alternating Gas method in the Snorre field, Norway, where the surfactant solution was injected in two slugs to control gas mobility and prevent gas breakthrough. The simulation showed that the first short slug was not efficient while significant GOR decrease and incremental oil production was obtained after the second longer slug in some periods. This study shows that the Eclipse foam model is applicable to the planning of water and gas injections, the testing of various surfactant properties, and the evaluation of the efficiency of the method at the field scale.*

Résumé — **Vérification des performances de la méthode FAWAG (*Foam Assisted Water Alternating Gas*) sur le champ de Snorre, en Norvège, avec un nouveau modèle de sélection des mousses** — Le modèle Éclipse "*Functional Foam*" a été utilisé pour guider le processus de calage des données de production (rapport gaz-huile, débits d'huile et de gaz) du champ Snorre, en Norvège, dans le cas d'une exploitation par une méthode d'injection alternée d'eau et de gaz avec mousse. La solution d'agents tensioactifs a été injectée en deux bouchons afin de contrôler la mobilité du gaz et d'empêcher la percée de gaz. La simulation a permis d'observer que le premier bouchon n'était pas suffisamment efficace, alors qu'une diminution significative du rapport gaz/huile et qu'une augmentation importante de la production en huile étaient constatées après le second bouchon pour certaines périodes. Cette étude démontre que le modèle Éclipse dédié aux mousses, permet de planifier les injections d'eau et de gaz, de tester les différentes propriétés des tensioactifs et d'évaluer l'efficacité de la méthode à l'échelle d'un champ pétrolier.

SYMBOLS AND NOTATIONS

t	Time
V	Block pore volume
S_w	Water saturation
C_f	Foam concentration
B_r, B_w	Rock and water formation volume
ρ_r	Rock density
C_f^a	Foam concentration adsorbed on to the rock
ϕ	Porosity
Σ	Sum over neighbouring cells
T	Transmissibility
k_{rw}	Water relative permeability
μ_w	Water viscosity
M_{rf}	Foam mobility reduction factor
P_w	Water pressure
ρ_w	Water density
D_z	Cell centre depth
Q_w	Water production rate
λ	Rate decay parameter function of oil and water saturation
S_o	Oil saturation
M_r	Reference mobility reduction factor
F_s	Mobility reduction function due to surfactant concentration
F_w	Mobility reduction function due to water saturation
F_o	Mobility reduction function due to oil saturation
F_c	Mobility reduction function due to capillary number
C_s	Surfactant concentration
C_s^r	Reference surfactant concentration
e_s	Steepness of the transition
S_o^m	Maximum oil saturation above which the foam ceases to be effective
e_o	Exponent which controls the steepness of the transition about the point where $S_o = S_o^m$
S_w^1	Limiting water saturation below which the foam cease to be effective
f_w	Weighing factor which controls the sharpness in the changes in mobility
N_c	Capillary number
N_c^r	Reference capillary number
e_c	Exponent which controls the steepness of the transition about the point where $N_c = N_c^r$
GIR	Gas Injection Rate (simulated)
GIRH	Gas Injection Rate Historical
GPR	Gas Production Rate (simulated)
GPRH	Gas Production Rate Historical
GOR	Gas Oil Ratio (simulated)
GORH	Gas Oil Ratio Historical
OPR	Oil Production Rate (simulated)
OPRH	Oil Production Rate Historical

INTRODUCTION

Foam Behaviour

The usual problem of gas injection methods is early gas breakthrough. Foam applications can solve the problems of gas mobility control and store more gas in reservoirs by building a bank behind the oil, delaying early gas breakthroughs and reducing the gas oil ratio (Skauge *et al.*, 2002). Foam is generated when gas, passing through surfactant in an aqueous phase, creates a stable dispersion of gas bubbles and lamellae trains in the liquid. With an objective of reducing the gas-oil ratio in production wells, it is essential to have foam that is immobile after placement. Stationary or trapped foam can block a large number of channels that would otherwise carry gas. When foam bubbles form in a porous medium, the bubble size typically matches the pore size of the rock. These bubbles tend not to move until they are compressed (hence, reducing their size) through the application of a higher pressure, which in turn leads to more bubbles being generated ('The Foam Model', 2012). The effect of the size of the bubbles, the longevity of the foam and the impact of the oil and water on the foam's strength and stability have been described in numerous studies. It has been observed by (Hirasaki and Lawson, 1985) that bubbles of smaller sizes are less mobile than larger bubbles. A low permeable matrix has a detrimental effect on the foam strength because of stretching and squeezing, whereas foam flowing through pores and throats causes thin films to break, foam coalescence occurs and bubble size increases, which leads to an increase in foam mobility (Ashoori *et al.*, 2010). Typically, the reduction of gas mobility depends on a range of factors, including pressure and shear rate (Prud'homme and Khan, 1996).

The physics of the foam flooding process is very complex. Low foam density causes the fractionation of the flow, while low foam viscosity leads to fingering and channelling. Foam will collapse if the water saturation is below, or the oil saturation is above, certain values. Foam also disappears through natural decay and due to the adsorption by the rock surface, which implies that the foam properties vary with space and time. At higher temperatures and more acidic media the rate of decay increases (Angstadt and Tsao, 1987). In foam flooding, the foam properties have a decisive impact on the efficiency of the method while the reduction of the interfacial tension (between oil and water) has no significant effect.

To apply foam for Enhanced Oil Recovery (EOR) one must understand both the physical interactions between foam and oil in porous media and how these interactions are represented in computer models (Farajzadeh *et al.*, 2012). Recent studies have shown that foam rheology in porous media can be characterised by foam catastrophe theory, which exhibits three foam states (weak-foam, strong-foam and intermediate states) and two strong foam regimes (high-quality and low-quality) (Afsharpoor *et al.*, 2010).

In computer models, foam is classified as strong foam (immobile foam) and weak foam (flowing foam). The model simulates the reduction of gas mobility in the presence of the foam, which is accomplished by reducing the relative permeability of the gas, increasing its viscosity, or a combination of both. The foam generation, foam movement and its effect on gas mobility are taken into consideration. The surfactant is distributed in terms of concentration in the cells. The surfactant concentration is divided into adsorbed foam, decayed foam and remaining foam, or an effective concentration that will act to change the mobility. The interaction of the foam with all available media in the reservoir, such as water, oil and rock, is formulated mathematically in order to model the project performance. The shape of the fractional flow curve changes with injection velocity; a weak foam state is obtained at low injection velocity and a strong foam state is achieved at high injection velocity (Dholkawala *et al.*, 2007). The foam can be transported by either water or gas; the two options for a tracer. Gas tracer studies measure the fraction of gas trapped within the foam at steady state (Kovscek, 1998).

Considering the beneficial effects of foam for mobility control for gas, numerous laboratory studies have been conducted (Zuta and Fjelde, 2011) but few field-scale pilot projects are available for the application of the theories and methodologies defined in the literature for foam (Shabib-Asl *et al.*, 2013).

FAWAG at Snorre Field

The Foam Assisted Water Alternating Gas (FAWAG) technique implemented in the Snorre field, Norway, has become the world's largest application of four pilot projects in different parts of the field. The pilots covered gas shut off in production wells, mobility control and the diversion of injected gas.

Most of the North Sea reservoirs are composed of massive sandstone fluvial deposits within rotated fault blocks. The properties of this reservoir type are high temperature of 90°C and high pressure of more than 383 bar at 2 475 m in Statfjord and Lunde formations. Permeability is changing in streaks in the range of 100-3 000 mD (Skauge *et al.*, 2002). The injected gas contained 70% of methane.

Rapid development of gas production after gas breakthrough might cause a severe problem for some North Sea oil fields. In these fields, the gas front will reach all the production wells almost simultaneously. Gas handling might put restrictions on further oil production and thus, any means by which to delay gas breakthrough is desirable. Side tracking of the production wells to lower parts of the formation intervals might be beneficial. Zone-isolation or blocking of the higher perforations appears difficult by traditional well treatments. For those reasons, the use of foam to block a gas cone and delay the GOR development of the wells has received much attention (Aarra *et al.*, 2002).

The project originally started in 1997 in the Central Fault Block (CFB) of the Snorre field. Owing to problems associated with the fracturing and gas leakages, the injection area was changed to the Western fault block of the Snorre field. The experience from the tests in the CFB had identified that surfactant alternating gas was the best injection method. Well P-32 was chosen as an injector to reduce the producing GOR at production well P-39, which is located at a distance of 1 550 m.

Gas, water and surfactant were injected periodically into the injector following schedule shown in Figure 1. An initial short cycle of foam injection observed in laboratory experiments (Aarra *et al.*, 2002) was more efficient than later injections. In addition, owing to the establishment of trapped gas saturation in the area after the first gas injection and mass exchange reduction

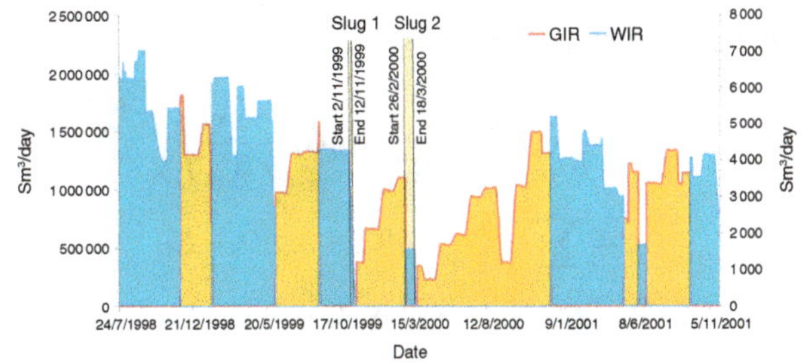

Figure 1

Schedule and injection rates: gas injection rate for well P-32G and water injection rate for well P-32W. Surfactant injections are shown in slugs.

between the oil and gas in the second gas injection period, gas breakthrough is expected to arrive earlier. Therefore, the surfactant injection of C_{14-16} commercial grade alpha olefin sulfonate or AOS-16 3-hydroxy sulfonate was applied in two slugs: an initial short foam cycle of the brine surfactant slug injection was in period from 2/11/1999 to 12/11/1999, which was followed by the gas injection from 12/11/1999 to 26/2/2000. The surfactant concentration used for the simulator was 4.9 kg/Sm^3 and the cumulative surfactant injection over nine days was 15 262 Sm^3. The second foam cycle was begun by an injection of a brine surfactant slug from 26/2/2000 to 17/3/2000 followed by gas injection from 17/3/2000 to 20/12/2000. The injection for the Water Alternating Gas (WAG) cycle started immediately after the end of FAWAG cycle, which was on 20/12/2000. The surfactant concentration for second slug was 2 kg/Sm^3 and the cumulative injection equal to 31 733 Sm^3. During the injection of the surfactant solution and gas, the injection rate was adjusted to ensure that the fracturing pressure was not exceeded.

The objectives of the field trial of the Snorre projects were as follows: to reduce the producing GOR in production well P-39 and to increase the sweep efficiency during gas injection and the storage of gas in the reservoir. The numerical simulation of the performance at the Snorre field using the Eclipse Foam Model is described in this study.

1 ECLIPSE FOAM PROPERTIES MODELING

In ECLIPSE 2009.1, the foam model has been extended to provide water as a transport phase in addition to gas. Accompanying this, a new functional model for the gas mobility reduction factor (M_{rf}) has been implemented. The foam carrier (tracer) for the functional model is bubbles generated in the water and the gas velocity is not considered.

Therefore, the foam distribution between all members of the equation C_f is described by Equation (1) as a function of water properties:

$$\frac{d}{dt}\left(\frac{VS_wC_f}{B_rB_w}\right)+\frac{d}{dt}\left(V\rho_rC_f^a\frac{1-\phi}{\phi}\right) = \sum\left[\frac{Tk_{rw}}{B_w\mu_w}(\delta P_w-\rho_w gD_z)\right]C_f + Q_wC_f-\lambda(S_w,S_o)VC_f \quad (1)$$

The basic equation for the modeling is derived from the combination of Darcy's Law and the Buckley-Leverett equation. The volume of the generated foam, when surfactant is added into water, depends on the volume of water and is calculated as:

$$\left(\frac{VS_wC_f}{B_rB_w}\right)$$

The part of the surfactant adsorbed by the rock is represented by:

$$V\left(\frac{1-\emptyset}{\emptyset}\right).\rho_r.C_f^\alpha$$

During time increment dt, the foam passing through a volume element of rock is calculated as:

$$\left[\frac{Tk_{rw}}{B_r\mu_w}(\sigma P_w-\rho_w gDz)\right]C_f$$

The foam travelling in the water phase is Q_WC_f.

The foam decay due to oil and water over time is modeled by the foam decay $\lambda(S_w,S_o)VC_f$.

The foam mobility reduction factor (M_{rf}) is a numerical multiplier taken from the functional model of Eclipse:

$$M_{rf}=\frac{1}{1+(M_r\cdot F_s\cdot F_w\cdot F_o\cdot F_c)} \quad (2)$$

For this model, the gas mobility reduction factor is modeled in terms of a set of functions that represents the individual reduction factors due to the surfactant concentration (F_s), oil saturation (F_o), water saturation (F_w) and capillary number (F_c). These are combined multiplicatively with a reference mobility reduction factor (M_r) to determine the net mobility reduction factor (M_{rf}) as shown in Figure 2.

Each of the foam controlling functions, such as F_s, F_o, F_w, and F_c is dependent on how surfactant behaves when it interacts with the porous media and regulated by two parameters: the property itself, which needs to be found by laboratory tests, and its exponent/weighting factor.

The reference mobility reduction factor M_r is typically in the range of 5 to 100 and corresponds to the normalised resistance to flow for a minimum bubble size in the absence of factors that increase bubble size. The values of (M_{rf}) can vary between 0 and 1, where "0" means no gas flow.

Variations of the five functions can produce a large number of combinations, and the effect of one property can be compensated by effect of the other property. The most important function is F_s, while the other functions tend to change it, which after multiplying with M_r results in the reduction of gas flow.

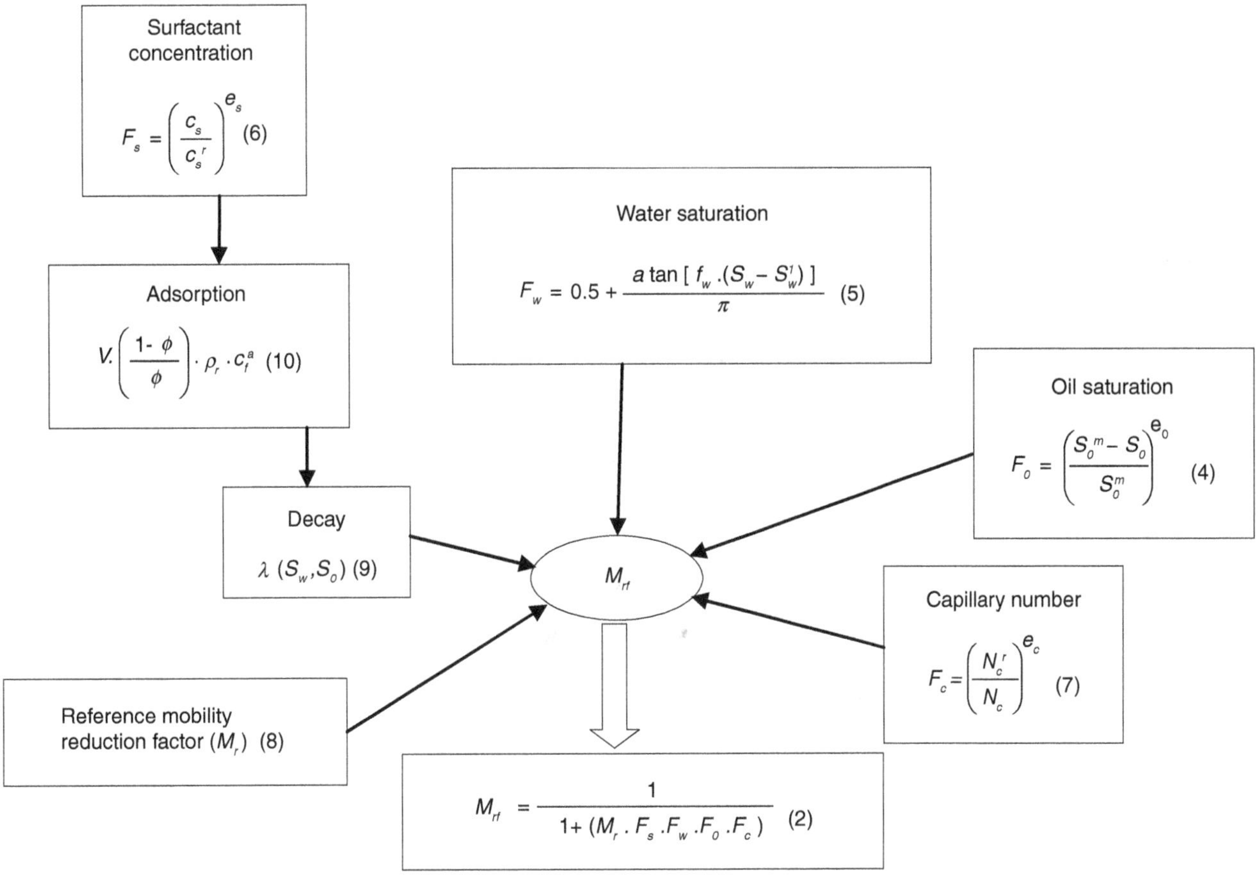

Figure 2

Concept of the foam model.

To study the contribution of each of the functions to a general performance, their alterations were initially studied as a sensitivity analysis. The details of the sensitivity of each property were described in SPE_150829 (Spirov *et al.*, 2012). The individual effect of each of the functions on foam mobility is estimated on the assumption that the other functions are equal to 1. For example, in Equation (3) foam mobility is calculated as a function of surfactant concentration F_s. To calculate the M_{rf} as a function of either capillary number (F_c) or oil saturation (F_o), or water saturation (F_w), (F_s) should be replaced with either function in this equation:

$$M_{rf} = \frac{1}{1 + (M_r \cdot F_s)} \quad (3)$$

The simulations were performed for the purpose of the sensitivity analysis to determine the variations of the property values affecting the shape and magnitude of the GOR curve with respect to historical GORH. Based on the results of the sensitivity analysis, the values providing the best match between the historical and simulated data were selected and shown in Table 1.

2 RESULTS AND DISCUSSION

2.1 Modeling of the Well Performance

In this section, we highlight a few of the simulated production responses at the producer P-39, together with a look back on historical trends compared against simulations of scenarios with and without foam.

To find the best match for the Snorre foam cycles, various combinations of variables, such as: the mobility reduction factor M_r, capillary number exponent, surfactant concentration exponent, oil (S_o), and water (S_w) saturations were tried at the given constant critical

TABLE 1
Best match of values of foam parameters used for simulation in Eclipse

Functions	Equation	Variable	Exponent
Oil saturation function (F_o)	(4)	$S_o^m = 0.1$ (best match 0.4)	$e_o = 1$
Surfactant conc. function (F_s)	(6)	$C_s^r = 0.1$ kg/Sm^3	$e_s = 1$
Capillary number function (F_c)	(7)	$N_c^r = 1.0e-7$	$e_c = 0.1$
Foam reference mobility reduction factor for foam in water phase (M_r)	(8)	$M_r = 50$	/
			Weighing factor
Water saturation function (F_w)	(5)	$S_w^1 = 0.1$	$f_w = 10$
		Foam surfactant concentration	Corresponding gas-water surface tension
Gas-water interfacial tension as a function of foam surfactant concentration	/	0 kg/Sm^3 0.01 kg/Sm^3 0.05 kg/Sm^3	10 mN/m 5 mN/m 0.1 mN/m
		Local oil or water saturation	Corresponding decay half-life
Foam oil decay function foam water decay function	(9)	0 1 Not used	3 000/day 2 500/day Not used
		Local foam concentration in the solution surrounding the rock	Saturated concentration of foam adsorbed by the rock formation
Foam adsorption	(10)	0 1 30	0 kg/rm^3 1.0e-5 kg/rm^3 1.0e-5 kg/rm^3

surfactant concentration of 0.1 and limiting capillary number of 1.0e-7. The values of $M_r = 50$, $F_o = (0.4, 1)$, $F_w = (0.1, 10)$, $F_s = (0.1, 1)$ and $F_c = (1.0e\text{-}7, 0.1)$ provided the best match.

History Gas-Oil Ratio (GORH) and the best match for foam GOR are plotted together in Figure 3. Five periods of elevated GORH values, called peaks and denoted as: 1, 2, 3, 4 and 5, are distinguishable from the beginning of the gas injection until the end of 2001. The surfactant injection periods are shown as slug 1 and slug 2 starting at the end of the 2nd peak. The present simulations aimed to obtain the best match of GORH and GOR of the peaks numbered 3, 4 and 5.

The 3rd peak of GOR was predicted two-and-a-half months later than GORH. GOR in the interval between the 3rd and 4th peaks matched in time but was lower in value. The 4th peak matched satisfactorily. GOR in the interval between the 4th and 5th peaks was higher than GORH; the start of the 5th peak was predicted 3 months earlier. Because GORH is a result of the division of oil (OPRH) and gas (GPRH) production rates, the match of the aforementioned parameters and their simulation were analysed.

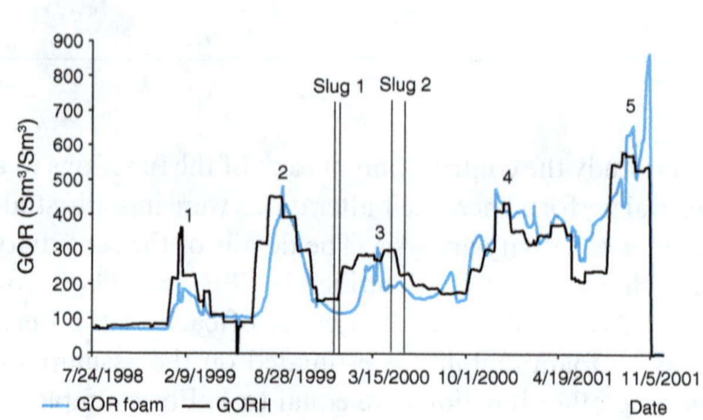

Figure 3
Best match of foam simulated GOR compared to GORH.

The foam model predicted higher oil production OPR due to the surfactant injection immediately after 1st slug, as can be seen in Figure 4. Simulated OPR closely

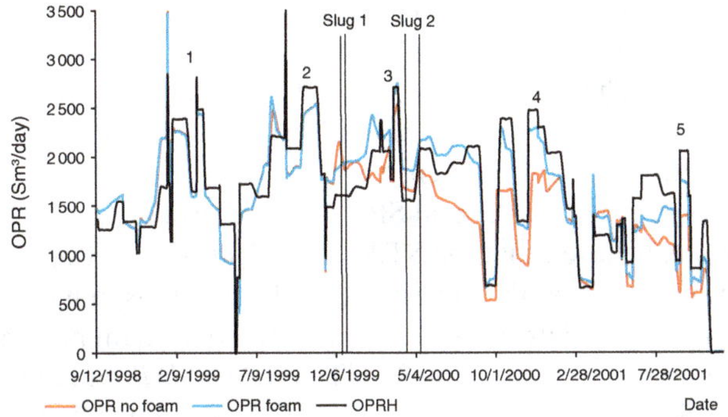

Figure 4

Best match of foam simulated OPR and no foam scenario compared to OPRH.

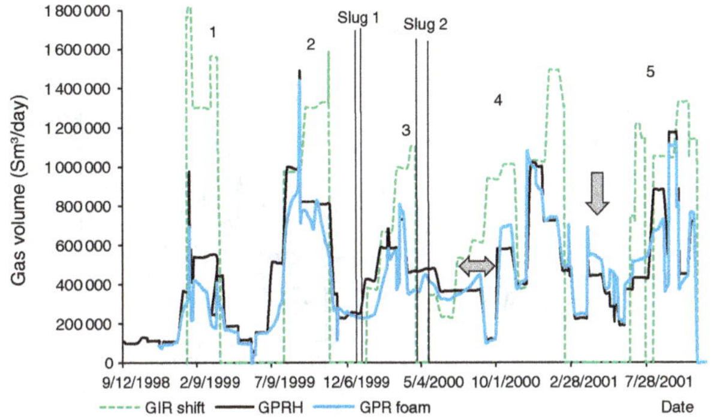

Figure 5

Comparison of gas injection GIR (plotted with 56 days of shift) and gas production GPRH. Arrows point out the delay of GPRH compared to GIR and small peak between 4th and 5th peaks occurred due to foam effect.

matched to OPRH after the 2nd slug in the periods of 3rd and 4th peaks while the oil production OPRH was much higher than the simulated OPR at the beginning of the 5th peak during about two months.

In order to analyse the response of the injection in the production well, the Gas Injection Rate (GIR) was shifted at 56 days to find the match with the GPRH and plotted additionally in Figure 5. The shift shows a period of gas travelling from the injector to the producer. To find a match with the 5th GORH peak, the GIR was shifted at 40 days indicating that the duration of the gas travel became shorter by the end of the project. Totally, 1.32×10^8 Sm^3/day was injected, whereas 1.3×10^8 Sm^3/day was produced.

The 3rd peak at GPR started with approximately two months delay predicting an immediate foam generation. The divergence between GPRH and GPR between the 3rd and the 4th peaks is not substantial. All the other peaks of GPR mimic the shape of the GPRH with a few exceptions. The general observation of Figure 4 and Figure 5 allows making conclusion that in the periods, where the foam model overestimates GPR, the OPR is underestimated and *vice versa*, causing the divergence between GOR and GORH in the corresponding periods. For example, the beginning of the 5th peak in GOR starts two months later than at OPR/OPRH and GPR/GPRH due to the underestimation of oil production and overestimation of gas production. It can be also deduced that the GPR values are simulated more accurately than OPR, and due to that the GOR values are more affected by oil production rate. The higher oil production rates at the start of the 5th peak, long after the completion of the last surfactant injection, could probably be attributed to the delayed foam effect. In addition, first three periods of GPRH coincide in time with the gas injection (GIR), while the 4th peak starts with the delay of 2.5 months, shown by the left-right arrow in Figure 5, reflecting the foam effect.

Additionally, the no-foam scenario was modeled to view how the production would have occurred if surfactants were not applied. For the no-foam model, the start of the 3rd period was delayed at one month and the simulated GOR was much higher (*Fig. 6*).

There was no delay between the 3rd peaks of GIR and GORH, indicating that the foam was not strong enough after the initial small slug, as could have been expected from laboratory investigations (Schramm, 1994). The no-foam model suggests that if the foam was not applied, the entire 4th period would have occurred five to six months earlier than it actually did and at much higher GOR values (*Fig. 6*). The model also predicted that GOR in the 5th period could have been much higher and gas would have arrived two to three months earlier.

The GPR of no-foam scenario very closely followed GPRH in the period of the 3rd peak in Figure 7, indicating that the surfactant injection did not influence gas flow after the 1st slug. The 56 days of travel time between injector and producer similar to the 1st and 2nd periods, when no surfactant was injected, confirmed that. It needs to be noted that no injections have occurred during 6 days after the 1st slug. Probably due to that the foam generation delayed, whereas the model predicted immediate foam effect. To match the history data, the simulation of the 3rd peak should be done using $M_r = 1$. The 3rd peak of GOR for the no-foam model transformed

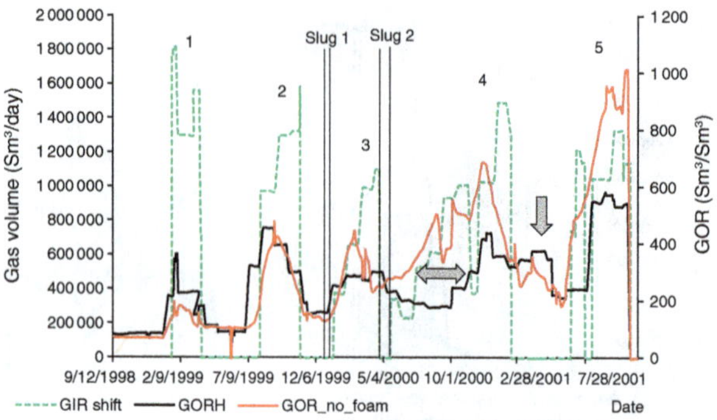

Figure 6

Comparison of GIR shifted at 56 days with GOR and GORH.

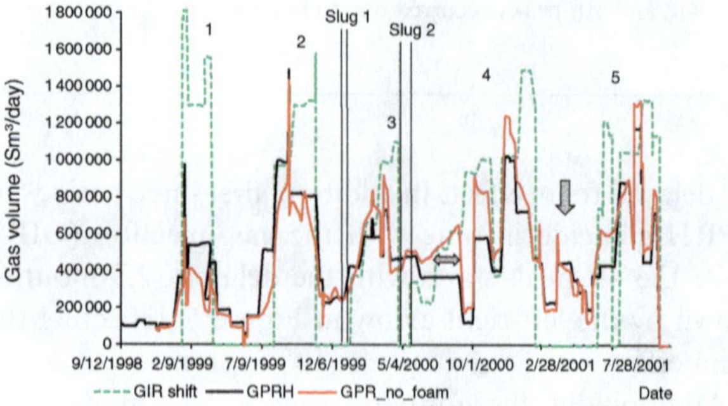

Figure 7

Comparison of gas injection GIR (plotted with 56 days of shift) and gas production GPRH. Arrows point out the delay of GPRH compared to GIR and small peak between 4th and 5th peaks occurred due to foam effect.

smoothly into the 4th; however, there is a clear boundary between the 3rd and 4th peak of GPR. Similarly to GIR-GPRH, the beginning of the 4th peak at GIR occurred approximately three-and-a-half months earlier than the 4th peak at GOR. The 2nd horn of the 4th peak of GORH, shown as the down arrow, occurred when there was no gas injection and GPRH was not high.

It could be deduced that although the view of GOR depending on both GPR and OPR distorted the real vision of the peaks distribution, the gas breakthrough at the beginning of the 4th period, shown by the left-right arrow, was significantly reduced after the 2nd slug but not postponed.

2.2 Comparison with the Previous Simulations Studies

Few studies have been devoted to foam simulation on a field scale.

Previous simulation of the foam performance for WFB of the Snorre field was performed using the STARS model. The model implemented in Eclipse by *Statoil* was converted to STARS and a sector of WFB was selected for a rapid sensitivity analysis in SPE 75157 (Skauge *et al.*, 2002) and SPE 77695 (Aarra *et al.*, 2002). It was believed that the STARS model is the best option to simulate foam under Snorre field conditions, but comparable simulators containing the essential foam parameter are LE foam model, UTCHEM, *Shell* simulator MoReS. The field-observed GORH behaviour following the second surfactant injection, performed in the period 26/2-15/3/2000, indicated a constant GORH for several months. The general trend in all the simulation runs showed an increasing GOR with time. This behaviour was very difficult to match using foam simulations. The lack of history match of the field-observed GORH development, after the second surfactant slug, made it difficult to describe the possible properties of the foam generated in the reservoir (Skauge *et al.*, 2002).

This paper describes the approbation of a new foam functional screening model in ECLIPSE 2009.1 software, using foam as the tracer in the water phase and the selection of the foam parameters that could better match the model to actual data. Previous 2002 and 2006 Eclipse versions contained tabular model. Since that time, the foam conservation equation has been fully upgraded in Equation (1) and many new properties shown in Figure 1 were not present in older versions of Eclipse.

Compared with earlier simulations performed in 2002-2005, the history match of the field data with simulated in the present study is extended for the entire period from 12 September 1998 - 1 January 2003 and includes the periods of the 4th and 5th peaks. Comparison with the simulation results from previous work in STARS (Aarra *et al.*, 2002) can be made only for the 3rd peak due to unavailability of the other data. STARS predicted the similar delay in GOR of approximately two months. We have shown that the delay is caused by the overestimation of the Oil Production Rate (OPR), because both simulators predict the immediate foam generation after the 1st slug. Whether that is due to the absence of any injections following the 1st slug is difficult to determine.

CONCLUSIONS

The simulation of the FAWAG performance using the New Eclipse Functional Foam model matches the historical data for the entire period from 12 September 1998 - 1 January 2003. The best match of GORH and GOR was obtained at $S_o^m = 0.4$.

The 1st surfactant slug did not reduce gas production rate substantially while the effect of the 2nd slug on the gas production rate decrease can be traced in the period of the 4th peak and on the oil production rate increase in the period of the 5th peak.

A comparison of runs with and without foam showed that although the presence of foam reduced the GORH and GPRH considerably, it did not stop or delay gas flow.

To improve the match of the GOR with history data, some discrepancies between over- and underestimations of GPR and OPR should be eliminated. Multiple M_r's should be used for different foam cycles rather than just one single value.

ACKNOWLEDGMENTS

The authors would like to thank *Statoil ASA* for authorizing data presented in this paper, *Schlumberger* Technology Centre and Arif Khan for advisory and software support.

REFERENCES

Aarra M.G., Skauge A., Martinsen H.A. (2002) FAWAG: A Breakthrough for EOR in the North Sea, *SPE Annual Technical Conference and Exhibition*, San Antonio, Texas, 29 Sept.-2 Oct.

Afsharpoor A., Lee G., Kam S. (2010) Mechanistic simulation of continuous gas injection period during Surfactant-Alternating-Gas (SAG) processes using foam catastrophe theory, *Chemical Engineering Science* **65**, 11, 3615-3631.

Angstadt H., Tsao H. (1987) Kinetic Study of the Decomposition of Surfactants for EOR, *SPE Reservoir Engineering* **2**, 4, 613-618.

Ashoori E., van der Heijden T., Rossen W. (2010) Fractional-flow theory of foam displacements with oil, *SPE Journal* **15**, 2, 260-273.

Dholkawala Z.F., Sarma H.K., Kam S.I. (2007) Application of fractional flow theory to foams in porous media, *Journal of Petroleum Science and Engineering* **57**, 1, 152-165.

Farajzadeh R., Andrianov A., Krastev R., Hirasaki G., Rossen W.R. (2012) Foam-Oil Interaction in Porous Media: Implications for Foam Assisted Enhanced Oil Recovery, *SPE EOR Conference at Oil and Gas West Asia*, Muscat, Oman, 16-18 April.

The Foam Model (2012) *ECLIPSE Reference Manual 2012.2*, Schlumberger, 341-376.

Hirasaki G., Lawson J. (1985) Mechanisms of foam flow in porous media: apparent viscosity in smooth capillaries, *Old SPE Journal* **25**, 2, 176-190.

Kovscek A.R. (1998) Reservoir Simulation of Foam Displacement Processes, *7th UNITAR International Conference on Heavy Crude and Tar Sand*, Beijing, China, 27-30 Oct.

Prud'homme R.K., Khan S.A. (1996) *Foams: Theory, Measurements, and Applications*, New York, Marcel Dekker Inc.

Schramm L.L. (1994) *Foams: Fundamentals and applications in the petroleum industry*, American Chemical Society.

Shabib-Asl A.M., Ayoub A.M., Alta'ee A.F. (2013) Application of Foam during Foam Assisted Water Alternating Gas (FAWAG) Process: A Review, *International Oil and Gas Symposium and Exhibition*, Sabah, Malaysia, 9-11 Oct.

Skauge A., Aarra M.G., Surguchev L., Martinsen H.A., Rasmussen L. (2002) Foam-Assisted WAG: Experience from the Snorre Field, *SPE/DOE Improved Oil Recovery Symposium*, Tulsa, Oklahoma, 13-17 April.

Spirov P., Rudyk S., Khan A. (2012) Foam Assisted WAG, Snorre Revisit with New Foam Screening Model, *SPE North Africa Technical Conference and Exhibition*, Cairo, Egypt, 20-22 Feb.

Zuta J., Fjelde I. (2011) Mechanistic Modeling of CO_2-Foam Processes in Fractured Chalk Rock: Effect of Foam Strength and Gravity Forces on Oil Recovery, *SPE Enhanced Oil Recovery Conference*, Kuala Lumpur, Malaysia, 19-21 July.

15

About the Use of Quality Indicators to Reduce Information Loss When Performing Upscaling

Christophe Preux

IFP Energies nouvelles, 1-4 avenue de Bois-Préau, 92852 Rueil-Malmaison Cedex - France
e-mail: christophe.preux@ifpen.fr

Abstract — *Hydrocarbon reservoirs are characterized by the spatial distributions of petrophysical properties. These spatial characteristics are usually derived from well data and seismic information. To study a reservoir, the engineers build a fine geological model, also called a geostatistical model, to represent the field. The purpose is to capture as well as possible the peculiarities and heterogeneity of the true reservoir. At this stage, performing a flow simulation with such detailed geological models is just too time-demanding. Therefore, a possibility consists of upscaling the geological model to an upscaled mesh, thus resulting in a coarse reservoir model for which fluid flow can be numerically simulated in a reasonable amount of time. The coarse grid blocks of this reservoir model are attributed equivalent petrophysical properties related to the properties populating the fine grid blocks. These properties are upscaled, and so they do not capture all of the details of the fine model. In this paper, we investigate the potential of various numerical and easy to compute criteria, which help evaluate the information loss due to the upscaling process. Our final purpose is to provide and access the reliability of quality indicators, which make it possible to evaluate the quality of the upscaled reservoir model. The potential of this systematic and integrated study is illustrated with two types of numerical experiments based upon the SPE10 case. First, we apply different upscaling methods to determine coarse reservoir models. Quality indicators are computed for each of them so that we identify the most suitable upscaling methods. Then, the upscaled models are input to flow simulators to check the accuracy of our quality estimations. Second, we also investigate the influence of coarsening and try to determine from the computed quality indicators the coarse cell size above which too much information is lost.*

Résumé — À propos de l'utilisation d'indicateurs de qualité afin de réduire la perte d'information lors d'un *upscaling* — Les réservoirs pétroliers sont caractérisés par une distribution spatiale de propriétés pétrophysiques. Ces caractéristiques spatiales sont généralement dérivées de données de puits et sismiques. Les ingénieurs construisent alors un modèle géologique fin, aussi appelé modèle géostatistique, pour représenter le champ. Le but est de capturer aussi bien que possible les particularités et l'hétérogénéité du réservoir. À ce stade, réaliser une simulation d'écoulement avec ces modèles géologiques aussi détaillés est trop coûteux en temps de calcul. Par conséquent, une possibilité consiste à « *upscaler* » le modèle géologique sur une grille grossière, afin d'obtenir un modèle de réservoir grossier pour lequel l'écoulement peut être simulé numériquement dans un laps de temps raisonnable. On attribue alors des propriétés pétrophysiques équivalentes aux cellules de la grille grossière. Même si ces données équivalentes sont obtenues à partir des propriétés de la grille fine, elles ne tiennent pas compte de tous les détails du modèle géologique fin. Dans cet article, nous étudions le potentiel de divers critères quantitatifs et facile à calculer, qui aident à évaluer la perte d'information due au processus d'*upscaling*. Notre but final est de fournir et d'évaluer des indicateurs de qualité, qui permettent d'évaluer la qualité du modèle de réservoir. L'intérêt de cette étude est illustrée avec deux types d'expériences numériques basées sur le cas SPE10. Tout d'abord, nous appliquons différentes méthodes d'*upscaling* afin d'obtenir différents modèles de réservoir. Les indicateurs de qualité sont calculés pour chacun d'eux afin d'identifier les méthodes d'*upscaling* les plus appropriées. Ensuite, les modèles réservoir sont fournis à un simulateur d'écoulement afin de vérifier l'exactitude de nos estimations de qualité. Deuxièmement, nous étudions également l'influence du nombre de mailles afin de déterminer à partir des indicateurs de qualité le nombre de maille limite à partir duquel trop d'information est perdue.

INTRODUCTION

The Upscaling Process

In the petroleum industry, the study of a field begins with the construction of a fine geological model. This model must capture the heterogeneity of the actual reservoir. As it is used to properly describe the spatial variations in petrophysical properties within the reservoir, reservoir engineers usually build high-resolution models with up to hundreds of millions of grid blocks. Inputting such models at once to flow simulators is not feasible in practice as it is extremely CPU time-demanding. Therefore, engineers are used to upscaling these geological models to upscaled meshes. The resulting coarse reservoir models represent the same field, but with fewer cells. The upscaling process consists in assigning equivalent petrophysical properties to the coarse grid blocks based upon the knowledge of the petrophysical properties populating the fine grid blocks. This has to be performed ensuring that fluid flow simulations for the fine or the coarse models yield very similar results. The usual properties to be upscaled are:

- discrete properties such as rock types or regions;
- continuous properties such as porosity, permeability or saturation.

In this paper, we focus on the upscaling of continuous properties, particularly permeability. The key assumption of the upscaling process is that fluid flow equations are the same on the fine (geological) and coarse (reservoir) scales. The upscaling process is split into two steps as shown in Figure 1:

- homogenization of the petrophysical properties;
- coarse discretization.

Clearly, the counterpart of upscaling is the information loss: upscaling tends to smooth fine heterogeneity features. This induces errors, which have to be quantified. Many upscaling methods exist to determine coarse reservoir models and many comprehensive reviews are available in the scientific literature [1-5]. In this paper, "upscaling" refers to an average in the physical space; the coarse permeability is also an equivalent (related to an average in the physical space) permeability and not an effective one (related to an average in the probability space). The notion of equivalent and effective permeability is discussed in [5].

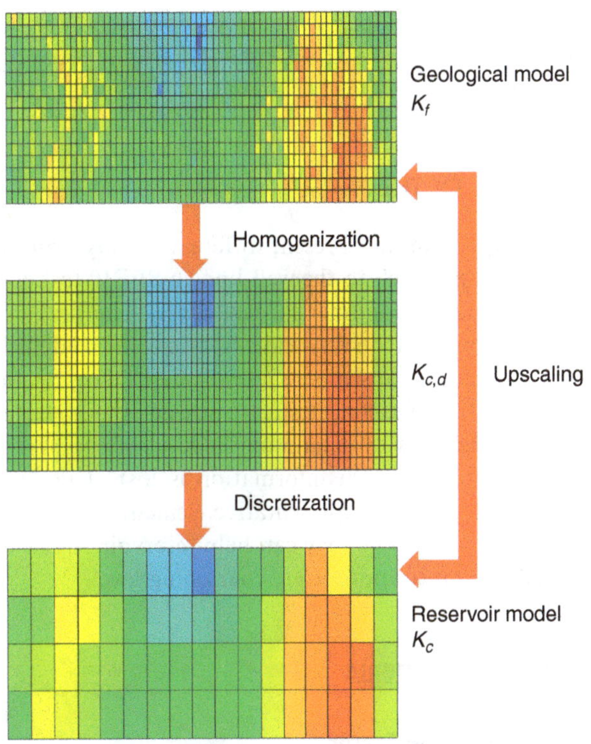

Figure 1

The two upscaling steps: the homogenization of the properties and the coarse discretization.

The Different Errors

The two steps mentioned above, that is to say homogenization and discretization, imply different consequences [6]: homogenization smooths the permeability field and grid coarsening increases truncation errors. To study these different effects, Sablok and Aziz [6] define three kinds of errors:

- the total upscaling error;
- the discretization error due to grid coarsening;
- the smoothing error due to the smoothing of fine heterogeneity.

The total (or resulting) error is a non-intuitive combination of the two others: it is in no way a simple addition. This can be explained as follows. The discretization error and the smoothing error have different natures and can balance each other. This makes the analysis of information loss more complicated. In addition, error analysis is often based upon numerical fluid flow responses such as breakthrough time or production, which are complex functions of reservoir and simulation parameters: the multiphasic parameters such as relative permeability or capillary pressure are non-linear functions of saturation. Moreover, PVT parameters are non-linear functions of pressure. All these parameters make the error analysis very complex by fluid flow response. This is the reason why we look for quality criteria capable of handling these errors separately. Various authors [7-11] proposed different criteria to quantify these errors as recapped hereafter.

OBJECTIVES OF THE PAPER

In this paper, we first present different quality criteria. Then, we come back to the well-known SPE10 test case and introduce two numerical test cases. The first one aims to show that quality indicators can help identify the most suitable upscaling methods depending on the case studied. The second test case stresses the potential of these indicators to determine the level of coarsening above which too much information is lost. This level can be related to the representative elementary volume. Finally, we show how they can help reservoir engineers to select the most appropriate coarsening level.

1 UPSCALING CRITERIA

1.1 The Upscaling Factor

The upscaling factor [12, 13], also known as the upscaling ratio [14] or aggregation rate, is often used in reservoir simulation for upscaling purposes. The dependence of upscaled permeability distribution as a function of the upscaling factor has been treated by several previous works [15, 16]. It corresponds to the number of fine grid blocks, which are gathered all together to create a coarse grid block. It thus defines the coarsening level between the geological and the reservoir models. The computation of this factor is straightforward. With N_f and N_c the numbers of cells in the fine and coarse reservoir grids, respectively, the upscaling factor (Fu) is:

$$Fu = \frac{N_c}{N_f} \tag{1}$$

The smaller the Fu, the coarser the reservoir model. However, this factor usually depends on direction. Therefore, we denote $N_{f,x}, N_{f,y}, N_{f,z}$ and $N_{c,x}, N_{c,y}, N_{c,z}$ the numbers of cells in the geological and reservoir grids along directions X, Y and Z. This leads to the following three upscaling factors as proposed by [8]:

$$Fu_i = \frac{N_{c,i}}{N_{f,i}} \quad \text{for } i = x, y, z \tag{2}$$

As mentioned above, the upscaling factor following [12, 13] is the number of fine grid blocks in a coarse grid block. This definition is suitable provided the fine and coarse grids are Cartesian and the coarse grid is built by aggregating entire fine cells. However, as soon as the coarse grid is a Corner Point Geometry (CPG) grid, it no longer holds. In this case, the number of fine cells included in a coarse cell can vary depending on the target coarse cell. Thus, we prefer to refer to formulae 1 and 2 because they apply to any grids, whatever the coarse grid. This definition is very general and is independent of the upgridding method used to create the coarse grid [17, 18]. Upgridding methods make it possible to adapt the gridding to geological heterogeneity. Therefore, the number of fine cells in a coarse reservoir cell is not constant per coarse cell. In addition, for a given upscaling factor, the impact of upscaling, in terms of information loss, will be less significant for a model with many fine cells than for a model with less fine cells. This motivated us to propose a new criterion, which allows us to aggregate more cells when there are many fine cells. This was already envisioned by [19], who investigated the impact of upscaling on a 3D model describing a field with meander belts and operated with steam-assisted gravity drainage.

1.2 The Upscaling Extent Parameter

It is quite intuitive that the higher the degree of coarsening, the greater the errors. To quantify the link between the upscaling factor and the information loss, [11] introduces the Upscaling Extent parameter, denoted *UE*. This was modified as follows by [9, 10] for practical purposes:

$$UE = 1 - \frac{\ln(N_c)}{\ln(N_f)} \tag{3}$$

Following this notation, when $UE = 1$, the upscaled model is as coarse as possible and the resulting errors are potentially significant. On the other hand, when $UE = 0$, the upscaled model is the same as the fine one and detailed information is fully preserved. We can propose an explicit relation between the upscaling factor and the upscaling extent parameter:

$$\ln Fu = -UE \ln\left(N_f\right) \tag{4}$$

For a given upscaling method, the information loss is roughly proportional to the upscaling extent parameter [11]. Therefore, it is essential to evaluate the *UE* parameter when studying the potential of upscaling methods. Comparisons between different upscaling methods are meaningful provided the *UE* levels are the same. Finally, the *UE* concept can be propagated along axes *X*, *Y* and *Z*, just like the upscaling factor [8]. In this case, we obtain:

$$UE_i = 1 - \frac{\ln\left(N_{c,i}\right)}{\ln\left(N_{f,i}\right)} \quad \text{for } i = x, y, z \tag{5}$$

1.3 The Data Range Indicator

Different studies stressed that upscaling removes extreme permeability values. It actually contributes to reducing large values and increasing small ones, which results in a narrower range, the range being the difference between the lowest and highest values. Based upon this analysis, [10] defined the *ILIa* Information Loss Indicator as:

$$ILIa = \frac{R_c}{R_f} = \frac{\max_c(\mathrm{K}) - \min_c(\mathrm{K})}{\max_f(\mathrm{K}) - \min_f(\mathrm{K})} \tag{6}$$

R_f and R_c are the ranges computed for the property of interest, that is permeability *K*, for the fine and the coarse models, respectively. Thus, according to this definition, no information is lost when $ILIa = 1$, while all of it is lost when $ILIa = 0$. Let us note that the upscaling extent parameter and the *ILIa* parameter vary inversely. In order to handle parameters all varying in the same way, we suggest using the *ILIr* Indicator (*r* for range):

$$ILIr = 1 - \frac{R_c}{R_f} \tag{7}$$

Thus, no information is lost when the *ILIr* parameter is 0, whereas all of it is lost when it is 1. This criterion makes it possible to control the range of values of the studied petrophysical property, but does not provide anything about their statistical properties. The following two sections investigate the potential of other criteria to address this issue.

1.4 The variance Indicator

[8] proposed another indicator, named the variance indicator, which is related to residuals. Briefly, residuals correspond to the difference between the coarse property values $K_{c,d}$ transferred to the fine grid and the fine property values K_f (*Fig. 1*). This quantifies the error related to the smoothing of the heterogeneities observed on the fine scale. Thus, residuals are defined as:

$$K_r = K_f - K_{c,d} \tag{8}$$

The variance indicator corresponds to the variance of the residuals. In the case of a Cartesian grid, it is defined as:

$$\sigma^2 = E\left[(K - E(K))^2\right] \approx \frac{1}{N}\sum_{j=1}^{N}\left(K_j - \bar{K}\right)^2 \tag{9}$$

σ^2 and $E(x)$ are the variance and the mean of *K*, respectively. The sample mean or arithmetic mean of *K* is denoted $\bar{K}$ and *N* is the number of sampled values. The variance measures the dispersion of a distribution. The purpose of an indicator relative to variance is to check whether the variance of the property on the coarse scale is the same as its variance on the fine scale. Keeping this goal in mind, we focus on the variance of the residuals, which must be close to zero. We introduce the *ILIv* variance indicator:

$$ILIv = \frac{E\left[(K_r - E(K_r))^2\right]}{E\left[(K_f - E(K_f))^2\right]} \tag{10}$$

Thus, $ILIv = 0$ when the variance of the residuals is zero (white noise). In this case, there is no information loss. Expanding residual variance yields:

$$E\left[(K_r - E(K_r))^2\right] \approx \frac{1}{N}\sum_{j=1}^{N}(K_{r,j} - \bar{K}_r)^2$$

$$= \frac{1}{N}\sum_{j=1}^{N}(K_{f,j} - \bar{K}_f)^2$$

$$+ \frac{1}{N}\sum_{j=1}^{N}(K_{c,d,j} - \bar{K}_{c,d})$$

$$((K_{c,d,j} - \bar{K}_{c,d}) - 2(K_{f,j} - \bar{K}_f)) \quad (11)$$

Clearly, the upscaling process tends to narrow the range of values. Thus:

$$K_{c,d,j} - E(K_{c,d}) \leq 2(K_{f,j} - \bar{K}_f) \quad (12)$$

As a result, the second term on the left-hand side of the equation is negative, meaning that:

$$E\left[(K_r - E(K_r))^2\right] \leq E\left[(K_f - E(K_f))^2\right]$$

This means that the $ILIv$ indicator is less than 1 in most of the cases studied. When this indicator tends to 1, the variance of residuals and the variance of the fine distribution are almost the same. This means that the upscaling process is too strong and that too much information is lost.

1.5 The Cardwell and Parsons Bounds

This section comes back to the bounds proposed by Cardwell and Parsons [20]. These bounds were derived from a variational principle [21]. They are written as (*Fig. 2*):

$$K_{c,x}^{\min} = \mu_a^z(\mu_a^y(\mu_h^x(K_{f,x}))) \quad (13)$$

$$K_{c,x}^{\max} = \mu_h^x(\mu_a^z(\mu_a^y(K_{f,x}))) \quad (14)$$

where:
- μ_a^i is the arithmetic mean along direction i;
- μ_h^i is the harmonic mean along direction i.

The interested reader can refer to [20] and [22] for detailed formulations in the case of Cartesian grids. Let us consider CPG grids and denote $v_{i,j,k}$ the volume

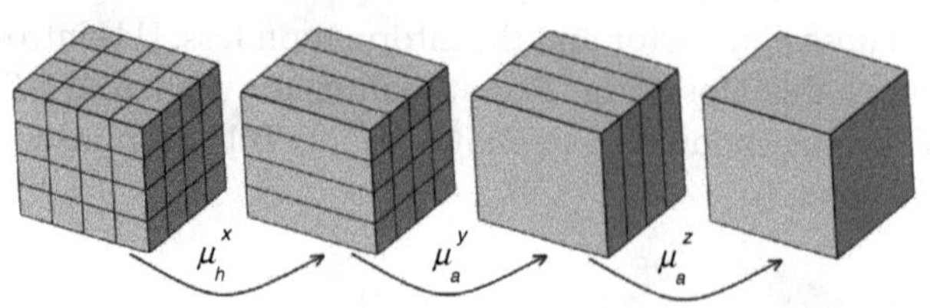

Figure 2

Lower bound computation.

of a fine cell. We assume that a coarse cell consists of $N_xN_yN_z$ cells and define:

$$V_{i,j} = \sum_{i=1}^{N_x} v_{i,j,k} \quad (15)$$

$$V_k = \sum_{j=1}^{N_y} V_{j,k} \quad (16)$$

$$V = \sum_{k=1} N_z V_k \quad (17)$$

This leads to the following expression for $K_{c,x}^{\min}$:

$$K_{c,x}^{\min} = \frac{1}{V}\sum_{k=1}^{N_z}\sum_{j=1}^{N_y} V_{j,k}\frac{V_{i,k}}{\sum_{i=1}^{N_x}\frac{v_{i,j,k}}{k_{i,j,k}}} \quad (18)$$

Therefore, referring to Cardwell and Parsons bounds, we define the Cardwell and Parsons indicator:

$$Error_i = \log_{10}\frac{K_{c,i}^{\max}}{K_{c,i}^{\min}} \quad (19)$$

This provides a grid property, that can be used to anticipate where the approximation of the upscaled permeability is poor. When $Error = 0$, $K_{c,i}^{\min} = K_{c,i}^{\max}$ and the upscaling is trivial. When it is 1, $K_{c,i}^{\max}$ is ten times larger than $K_{c,i}^{\min}$.

1.6 The QQplot

Sablok and Aziz [6] stressed the usefulness of QQplots [7] to analyze information loss. This is especially true when the stochastic realizations used to represent petrophysical property variations are not derived from two-point statistics methods. QQplots make it possible to compare at once the distribution on the coarse scale with the distribution on the fine scale. To build the QQplot between K_f and $K_{c,d}$, we proceed as follows. First, we sort all K_f and $K_{c,d}$ values in ascending order. These distributions having the same number of values, we note K_f^{sort} and $K_{c,d}^{sort}$ the sorted distributions of K_f and $K_{c,d}$. The QQplot is the discrete function:

$$K_f^{sort} \rightarrow K_{c,d}^{sort} \tag{20}$$

If we compare two identical distributions, we obtain the bisector. Thus, the more accurate the upscaling process, the closer the QQplot to the bisector.

2 UPSCALING METHODS

In this section, we recap the basics of the upscaling methods applied in this paper. The interested reader can refer to [3] for a detailed review.

2.1 Basic Algebraic Methods

We note V and v the volume of the coarse and fine cells, K and k the permeability of the coarse and fine cells and J the set of fine cells contained in the coarse cell:

In such conditions, the arithmetic mean is written as:

$$K_i^{arith} = \frac{1}{V_i} \sum_{j \in J} v_j k_j \tag{21}$$

While the harmonic mean is:

$$\frac{1}{K_i^{harmo}} = \frac{1}{V_i} \sum_{j \in J} \frac{v_j}{k_j} \tag{22}$$

Another useful simple algebraic average is the geometric mean:

$$K_i^{geom} = \frac{1}{V_i} \prod_{j \in J} k_j^{\frac{v_j}{V_j}} \tag{23}$$

2.2 More Sophisticated Algebraic Methods

The following three upscaling methods are more complex:

The Power-Law Method

Journel *et al.* [23] introduced the power-law method, which can be written as:

$$K_c = \frac{1}{V_c} \left(\int_{V_c} K(x)^p dV \right)^{\frac{1}{p}} \tag{24}$$

with $p \in [-1, 1]$. When p is -1, 0 and 1, the power-law method is equivalent to the harmonic mean, the geometric mean and the arithmetic mean, respectively. A study of this method and a determination of the power exponent are available in [24].

The Algebraic Isotropic Method

The algebraic isotropic method [25] is rooted in the Cardwell and Parsons approach [20]. In [25], Guérillot *et al.* proposed computing the geometric mean of the two Cardwell and Parsons bounds:

$$K_x^{Alg.Iso.} = \sqrt{K_{c,i}^{\min} K_{c,i}^{\max}} \tag{25}$$

$K_{c,x}^{\min}$ and $K_{c,x}^{\max}$ are given by Equations (13) and (14).

The Bound Combination Method

Duquerroix *et al.* [26] and Kruel-Romeu [27] suggested to generalizing the algebraic isotropic method to account for field anisotropy. The upscaled properties are then derived from the following relationship, known as the bound combination method:

$$K_x^{BC} = \sqrt[6]{\left(K_{c,x}^{\min}\right)^2 \left(K_{c,x}^{\max}\right)^2 K_{c,x}^3 K_{c,x}^4} \tag{26}$$

The two additional terms, compared with Equation (25), are estimated from the following expressions:

$$K_{c,x}^3 = \mu_a^y\left(\mu_h^x\left(\mu_a^z\left(K_{f,x}\right)\right)\right) \tag{27}$$

$$K_{c,x}^4 = \mu_a^z\left(\mu_h^x\left(\mu_a^y\left(K_{f,x}\right)\right)\right) \tag{28}$$

Notation μ stands for mean. The a and h subscripts specify whether this mean is arithmetic or harmonic.

The Numerical Local Method

The last upscaling method considered hereafter is the numerical local method. This one is based upon the simulation of Darcy flow with given boundary conditions for each group of fine grid blocks within coarse grid blocks. In this paper, we use no-flow boundary conditions. Flow is assumed to be linear and steady-state. This method is described in several papers [28-31]. It was shown to provide satisfactory results provided the coarse cell size is very large compared with the correlation length of permeability [32].

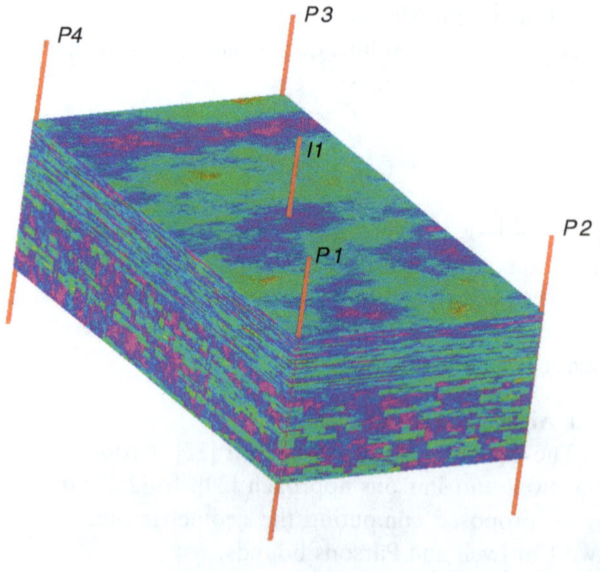

Figure 3

Porosity distribution within the whole SPE10 model.

3 SENSITIVITY STUDY AND NUMERICAL EXPERIMENT

3.1 The SPE10 Test Case [33]

In this section, we consider the second case of the Tenth SPE Comparative Solution [12]. The geological model consists of part of a Brent sequence. It includes two units. The top part of the model (21.336 m (70 ft), 35 layers) belongs to the Tarbert formation and corresponds to a prograding near-shore environment. Its lower part (Upper Ness, 30.48 m (100 ft), 50 layers) is fluvial. The geometry of this model is very simple: there is no top structure or fault. The fine scale model is discretized over a regular Cartesian grid with $60 \times 220 \times 85$ grid blocks (*i.e.* 1.122×10^6 grid blocks). The total dimensions are $365.76 \times 670.56 \times 51.816\,\text{m}^3$ ($1\,200 \times 2\,200 \times 170\,\text{ft}^3$) with grid blocks of $6.096 \times 3.048 \times 0.6096\,\text{m}^3$ ($20 \times 10 \times 2\,\text{ft}^3$). The production scheme is based upon a classical inverted "5-spot" pattern with four producers P located in the corners and an injector I in the center (Fig. 3). The coordinates of the wells are (i: X-coordinate, j: Y-coordinate):

- P_1: $i = 0, j = 0$;
- P_2: $i = 60, j = 0$;
- P_3: $i = 60, j = 220$;
- P_4: $i = 0, j = 220$;
- I: $i = 30, j = 110$.

All wells are vertical and completed throughout the whole formation (Fig. 3). The productivity index of each well is computed using the Peaceman method. The injection rate is 795 m^3/day (5 000 bbl/day) in reservoir conditions with a maximum injection bottom-hole pressure of 689.18 bar (10 000 psi). The four producers produce at 275.67 bar (4 000 psi) bottom-hole pressure. All wells have an internal diameter of 0.3048 m (1.0 ft).

The water properties are:

- formation volume factor $B_w = 1.01$;
- compressibility $c_w = 4.5 \times 10^{-5}$ bar^{-1};
- viscosity $\mu_w = 0.3$ cp.

On the other hand, we consider a dead oil with the PVT properties reported in Table 1.

TABLE 1

Oil PVT table

Pressure (bar)	B_o	μ_o
20.68	1.05	2.85
55.13	1.02	2.99
551.34	1.01	3.0

The relative permeabilities are derived from the Brooks-Corey models [34]:

$$k_{rw} = \left(\frac{S - S_{wi}}{1 - S_{wi} - S_{or}}\right)^2 \tag{29}$$

$$k_{r0} = \left(\frac{1 - S - S_{wi}}{1 - S_{wi} - S_{or}}\right)^2 \tag{30}$$

where $S_{wi} = S_{or} = 0.2$.

The upscaling and flow simulation results presented hereafter were obtained using the CobraFlow software and the PumaFlow simulator [35, 36], respectively. In this work, we did not study the numerical dispersion due to the mathematical model of the simulator. Upscaling and coarsening cause numerical dispersion. In this work, the simulator adapts the time step itself to obtain correct results. To study the numerical dispersion precisely, we should study the mathematical method used in the simulator and using a constant CFL criterion [37].

3.2 How Important is the Choice Of Upscaling Method?

The spatial distribution of petrophysical properties within the two SPE10 units is very different (*Fig. 3*). On the other hand, it is essential to consider a stationary distribution for investigating the influence of upscaling methods on information loss. We first attempted to apply the various criteria defined above without decoupling the two units, which made the interpretation and

TABLE 2
Upscaling factor [8]

Global	X	Y	Z
0.0125	0.25	0.25	0.2

TABLE 3
Upscaling extent parameter [10]

Global	X	Y	Z
0.3360	0.3386	0.2570	0.4527

understanding of the results very difficult. A few tests pointed out that it is required to estimate the upscaling criteria on each unit separately when there are significant differences in petrophysical properties. As a result, we focus on the top part of the SPE10 model. The following subsections are dedicated to the comparison of different upscaling methods referring to the criteria defined above. This analysis makes it possible to select the best method to use. In all cases studied, the coarse scale reservoir grid is fixed.

3.2.1 Upscaling Methods Used

The reservoir grid encompasses 15 × 55 × 17 cells. We consider the 7 upscaling methods introduced in Section 2:

- arithmetic mean;
- harmonic mean;
- geometric mean;
- power-law method;
- algebraic isotropic method;
- bound combination method;
- numerical local method.

The upscaling factor and upscaling extent parameter computed for the case studied are reported in Tables 2 and 3. These parameters have the same values whatever the upscaling method.

The upscaling extent parameter (Tab. 3) is higher along direction Z. This suggests that upscaling along this direction is likely to be more problematic. The analysis of the upscaling factor leads to a similar conclusion. Its value is smaller along direction Z (Tab. 2), meaning that coarsening is stronger in this direction.

3.2.2 The Cardwell and Parsons Bounds

The results derived from the Cardwell and Parsons bounds (*Eq. 19*) are recapped hereafter. Figure 4 shows that the X- and Y-permeability errors are very similar as

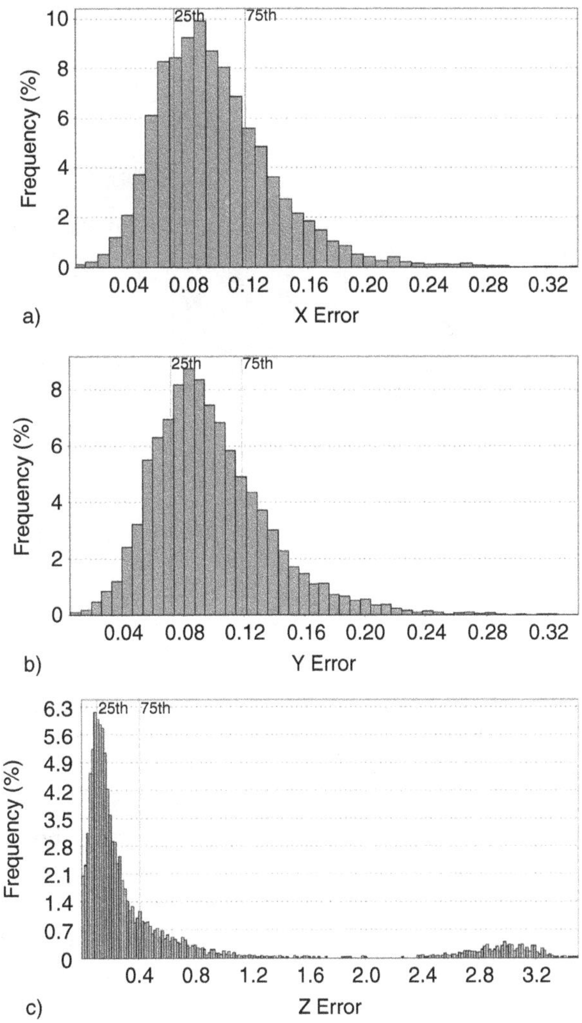

Figure 4

Histograms of the Cardwell and Parsons errors computed for a,b) the horizontal permeabilities and c) the vertical permeabilities.

$K_x^f = K_y^f$ for the SPE10 test case. In addition, the upscaling factors are the same for both X and Y axes (*Tab. 2*). The slight differences evidenced from Figure 4 result from the dimensions of the coarse cells which vary with direction. Generally speaking, the error is small. In terms of percentiles, 75% of the grid blocks show Cardwell and Parsons errors of less than 0.12 for horizontal permeabilities. This means that the coarsening effect induced by upscaling is not too strong, or in other words, that heterogeneity is moderate in every grid block. The results are clearly different for vertical permeabilities Kz. 75% of the cells show an error of less than 0.41 and 25% of them less than 0.11. Basically, 25% of the grid blocks are characterized by an error greater than 0.41.

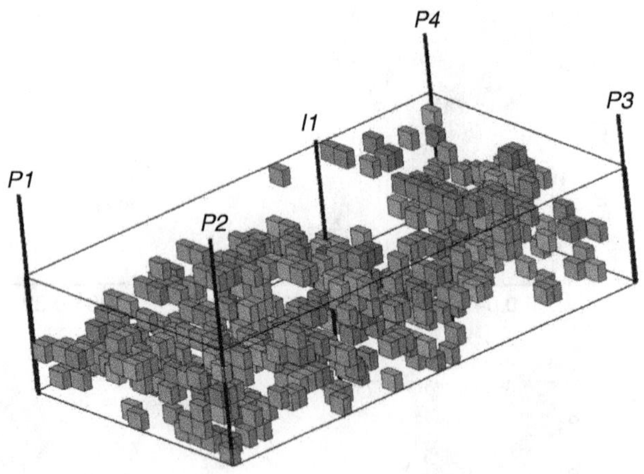

Figure 5

Localization of the grid blocks with Cardwell and Parsons errors for vertical permeabilities larger than 2.

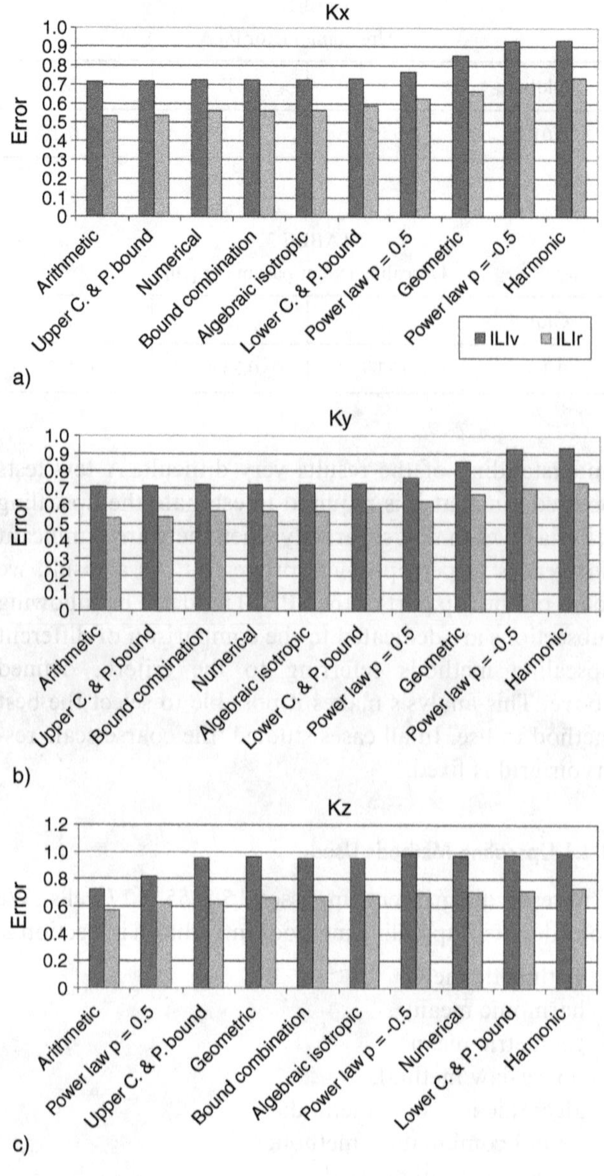

Figure 6

Data range and variance indicators, respectively *ILIr* and *ILIv*, for a, b) the horizontal permeabilities and c) the vertical permeabilities derived from different upscaling methods.

This means that the upper Cardwell and Parsons bound is 2.57 times larger than the lower bound for 25% of cells. In addition, a second peak is evidenced with errors in the range between 2.4 and 3.3: all these cells correspond to an upper bound 251 times larger than the lower bound. These values are very high and stress that the 0.2 upscaling factor applied along the *Z* axis is too significant as it is (without any specifically designed upgridding technique). Moreover, in the *Z* direction, the permeability is more heterogeneous: the permeability is organized on fine layers with very different permeability. Therefore, any upscaling technique can be applied when dealing with the horizontal directions for the case studied. However, the vertical upscaling has to be performed more carefully. Figure 5 shows the locations of the grid blocks with *Z* error values higher than 2. At first glance, we observe that they can be anywhere, which once again highlights the difficulty in determining an appropriate technique for upscaling the vertical permeabilities.

3.2.3 Variance, Data Range Indicators and QQplots

The data range and variance indicators, denoted *ILIr* and *ILIv* as explained in the previous sections, are displayed in Figure 6. A general comment is that both *ILIr* and *ILIv* indicators evolve the same way whatever the upscaling technique applied.

Whatever the direction of interest, the data range and variance indicators are minimal for the arithmetic method and maximal for the harmonic method. This feature is also evidenced for the QQplots, especially for vertical permeabilities. All errors are bounded by those estimated for the arithmetic and harmonic methods. The indicators also respect these bounds, also called fundamental inequality or Wiener bounds [38], because they are always valid [3, 20, 39, 40]. Thus, in the case studied, the data range and variance indicators can be considered as simply-computed criteria, which allow us to rank the various upscaling methods. Another point is that the bound combination and the algebraic methods yield similar results. For the example studied, the horizontal

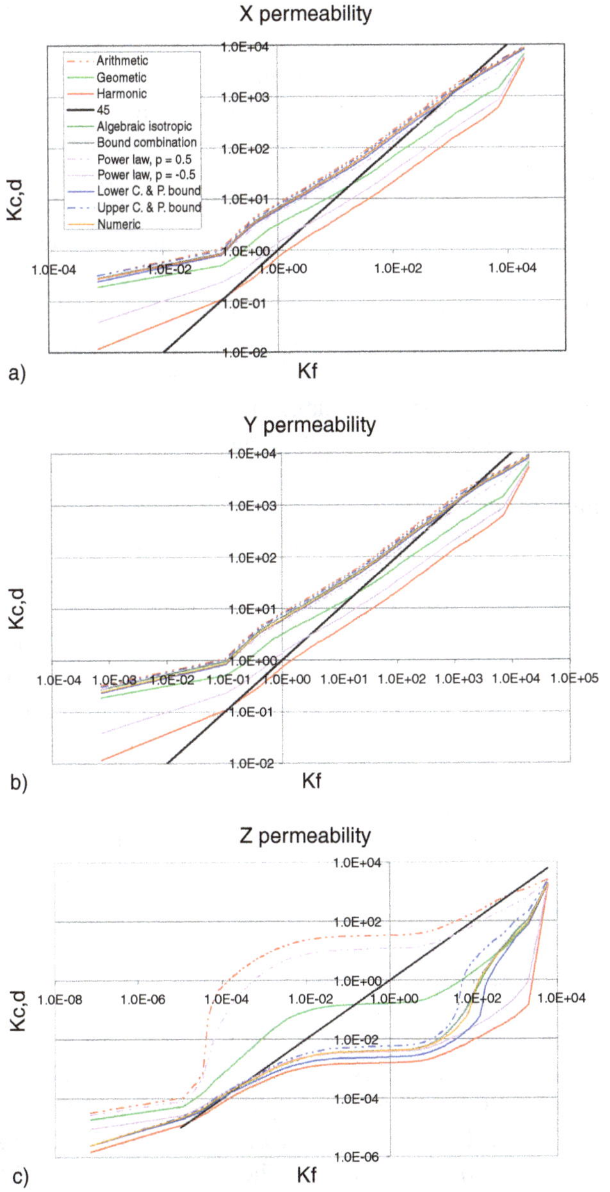

Figure 7

QQplots for a, b) the horizontal permeabilities and c) the vertical permeabilities derived from different upscaling methods.

permeability is isotropic: K_x is the same as K_y. In this special case, the two methods are similar. Now, in Figure 7, we can analyze the results given by the Cardwell and Parsons bounds on the QQplot. These bounds define an envelope, narrower than the one derived from the harmonic and arithmetic methods. According to Cardwell and Parsons [20] and Le Loch [41], an accurate estimate of the upscaled permeability should range between the Cardwell and Parsons bounds. This also holds for the data range and the variance indicators as well as the QQplots. Also, as explained above, a good method is expected to be associated with zero data range and variance indicators. Moreover, if we study the result given by the power-law method for different values of p [$p = -1$ (harmonic), $p = -0.5$, $p = 0.5$, $p = 1$ (arithmetic)], we note that the results are included in the envelope given by the Wiener bounds. So, an appropriate method consists of fitting the exponent p to reduce the difference between the QQplot curve and the bisector. The analysis of these various criteria must help us select an appropriate upscaling method. In addition, it is worth noting that the numerical local upscaling method yields for K_x and K_y the same results as the bound combination and the algebraic isotropic methods. Such behavior was actually expected. The two latter techniques are known to be very close to the numerical local method. However, if considering especially K_z (*Fig. 6*), it is shown that the numerical local method departs from the results derived from the bound combination and the algebraic isotropic methods. Considering the QQplots on top of the data range, variance indicators can contribute to refining our choice of the upscaling method. In Figure 7, we observe that the harmonic average tends to reduce the upscaled values, while the arithmetic method leads to the opposite behavior. In this case, the differences between the numerical local method and the bound combination and algebraic isotropic methods are very small. The three curves turn out to be almost the same for both K_x and K_y. On the other hand, the numerical local method lightly departs from the two algebraic methods for K_z. In Sections 3.2.1 and 3.2.2, we checked that the upscaling of vertical permeabilities is more challenging than the upscaling of horizontal permeabilities. The analysis of the indicators that we suggested here lead us to the same conclusion. Whatever the upscaling method, the data range and variance indicators are larger for vertical permeabilities than for horizontal permeabilities. In addition, the area between the QQplots for the two Cardwell and Parsons bounds is larger for vertical permeabilities than for horizontal permeabilities. To consolidate these conclusions, we decided to move one step ahead and confront the simulation results.

3.2.4 Convergence of the Quality Indicators and the Simulation Results

Figure 8 depicts the bottom-hole pressure at the injector. For three upscaling methods (geometric, harmonic and power-law with an exponent of -0.5), the injector reaches the maximal bottom-hole pressure. The harmonic method provides a lower bound for permeability,

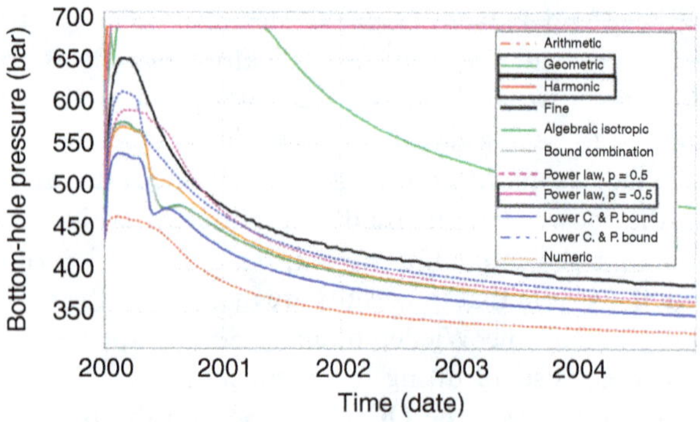

Figure 8

Comparison of the bottom-hole pressures at the injector for reservoir models derived from different upscaling methods (black = fine grid results).

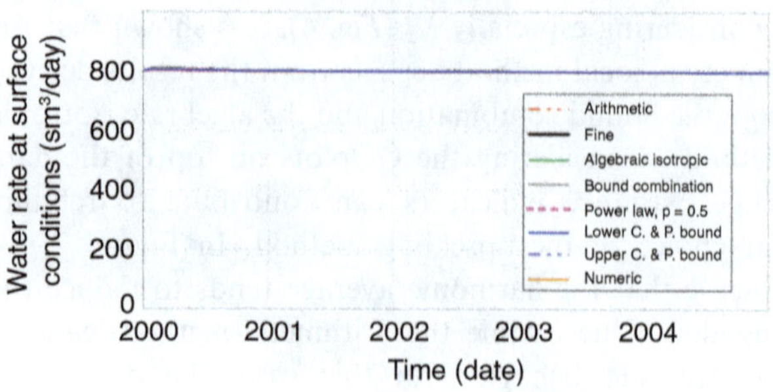

Figure 9

Comparison of the water injection rates at surface conditions for reservoir models derived from different upscaling methods (black = fine grid results).

the upscaled productivity index for the injector is also too low and the pressure drops. The same conclusion can be established for the power-law method with an exponent of −0.5. Coming back to the QQplots in Figure 7, we note that these methods should not be recommended in this case: these three methods give very different QQplots compared with the others (for *Kx* and *Ky*). In such conditions, there is no way to make valuable comparisons with these three methods.

The water injection rates at surface conditions obtained for different upscaling methods are plotted in Figure 9. Excluding these three methods, we have a constant injection rate. We can also make valuable comparisons with these seven other methods. In conclusion, we can note that if the upscaling method is not chosen intelligently, the upscaling process can drastically change the well behavior and mislead the reservoir engineer. The water-oil ratios simulated for different reservoir models obtained from different upscaling methods are plotted in Figure 10.

The increase in the water-oil ratio by the arithmetic method is not always noticeable (see, for instance, wells P1, P3 and P4). It is known that for isotropic fields, the bound combination method and the algebraic isotropic method give the same results for both indicators and QQplots. Now, focusing on the simulated water-oil ratios, an anomalous behavior is evidenced for wells P3 and P4. The results at well P3 obtained for the upscaled models are all wrong compared with the fine model. The overall behavior of the ratio is not respected. On the other hand, the curves obtained for well P4 evolve the same way, but the water-oil ratios are underestimated. This may be related to the too large value of the upscaling factor. In this special case, it seems that the most relevant methods are the bound combination and the algebraic methods, particularly when considering wells P1 and P4. However, none of these methods makes it possible to perfectly reproduce the results simulated for the geological model. A surprising result is pointed out for the numerical local method. Clearly, this method does not yield the best approximation. For wells P3 and P4, the results obtained are very close to those derived from the bound combination and the algebraic methods. However, for the other two wells, they are very different. The water-oil ratio is overestimated for well P1, while it is the opposite for well P4. This stresses the difficulties in selecting a suitable upscaling method.

The results in terms of oil rates are reported in Figure 11. In Figure 7, we observed that the power-law method with an exponent of 0.5 gives results very similar to those derived from the arithmetic average. The permeabilities derived from these two methods (specially K_z) are slightly higher compared with the bisector. The same phenomenon is shown for oil rates. Overall, for all wells, the methods identified as best in the previous sections (bound combination, algebraic and numerical local) overestimate oil rates. More precisely, the two algebraic methods give similar results. The numerical local method gives a correct result for oil rates, that is a result very close to those provided by the two algebraic methods. If referring to well P1, the arithmetic average and the power-law method with an exponent of 0.5 drastically overestimate the oil rates. Coming back to the QQplots in Figure 7, we note that these methods should not be recommended in this case. Therefore, the QQplot results are in good agreement with the flow results simulated for the coarse reservoir models. In conclusion, it seems that the bound combination, algebraic isotropic and numerical methods are the best choice.

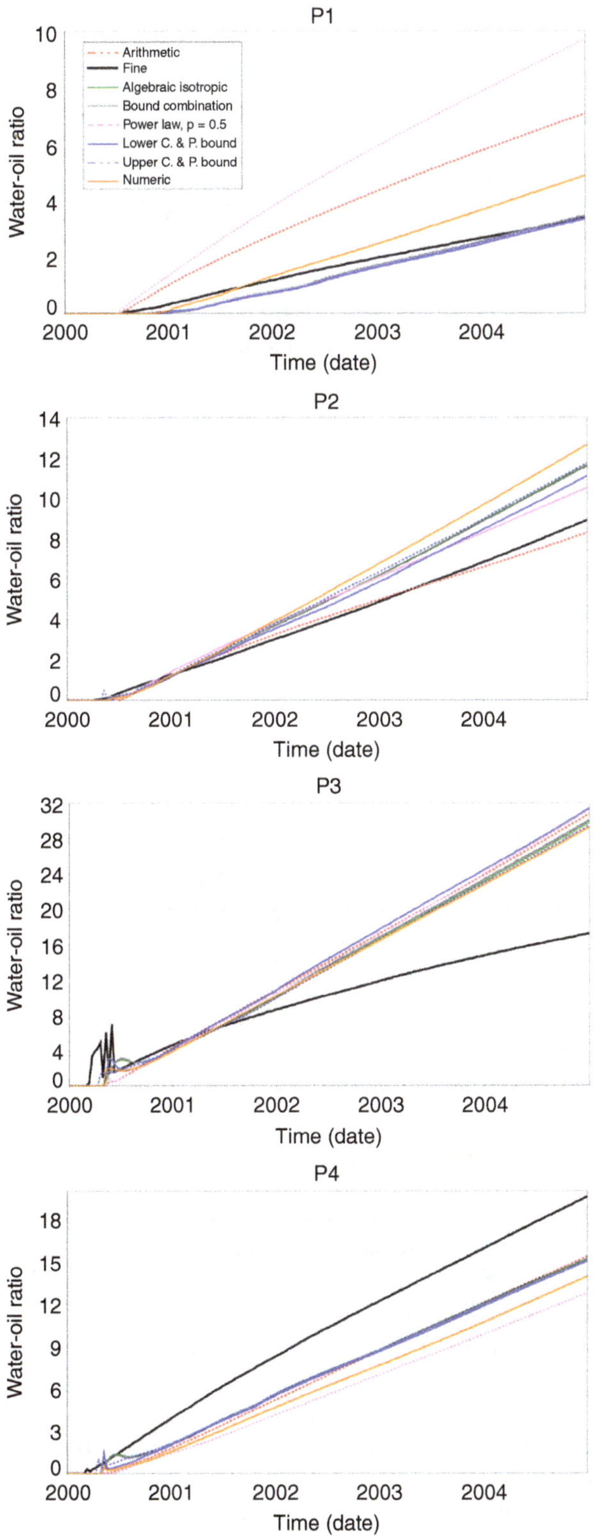

Figure 10

Comparison of the water-oil ratios for reservoir models derived from different upscaling methods (black = fine grid results).

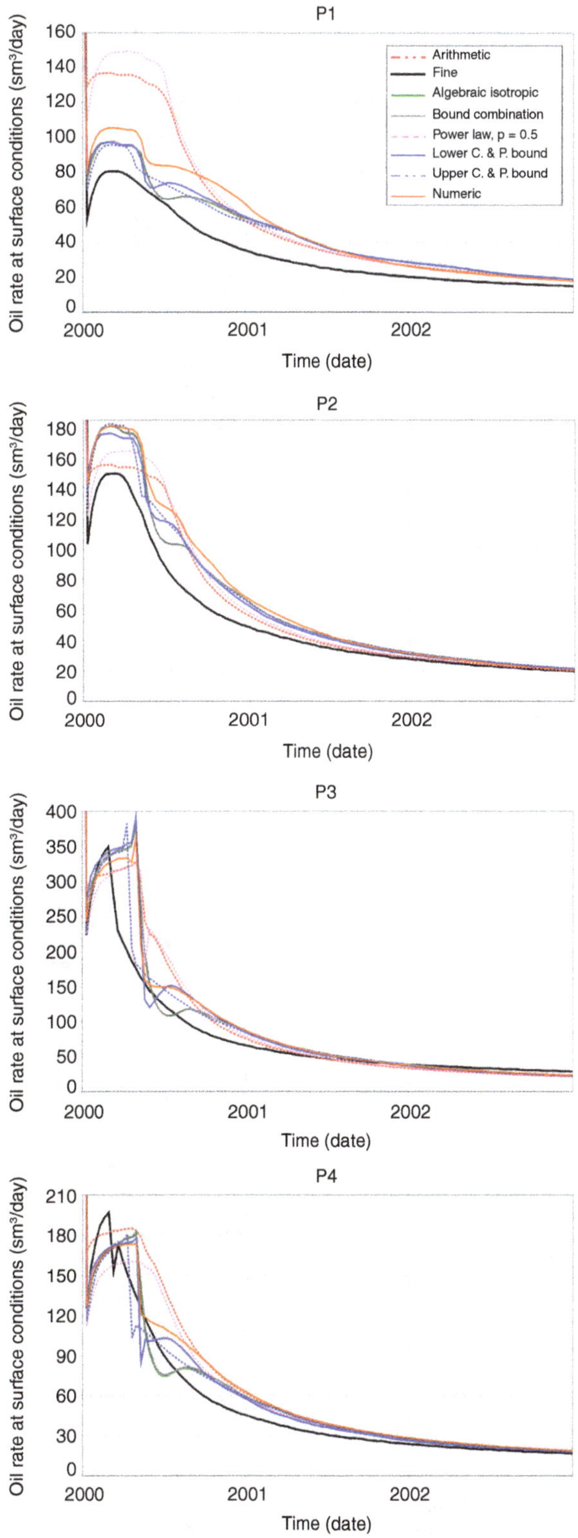

Figure 11

Comparison of the oil rates at surface conditions for reservoir models derived from different upscaling methods.

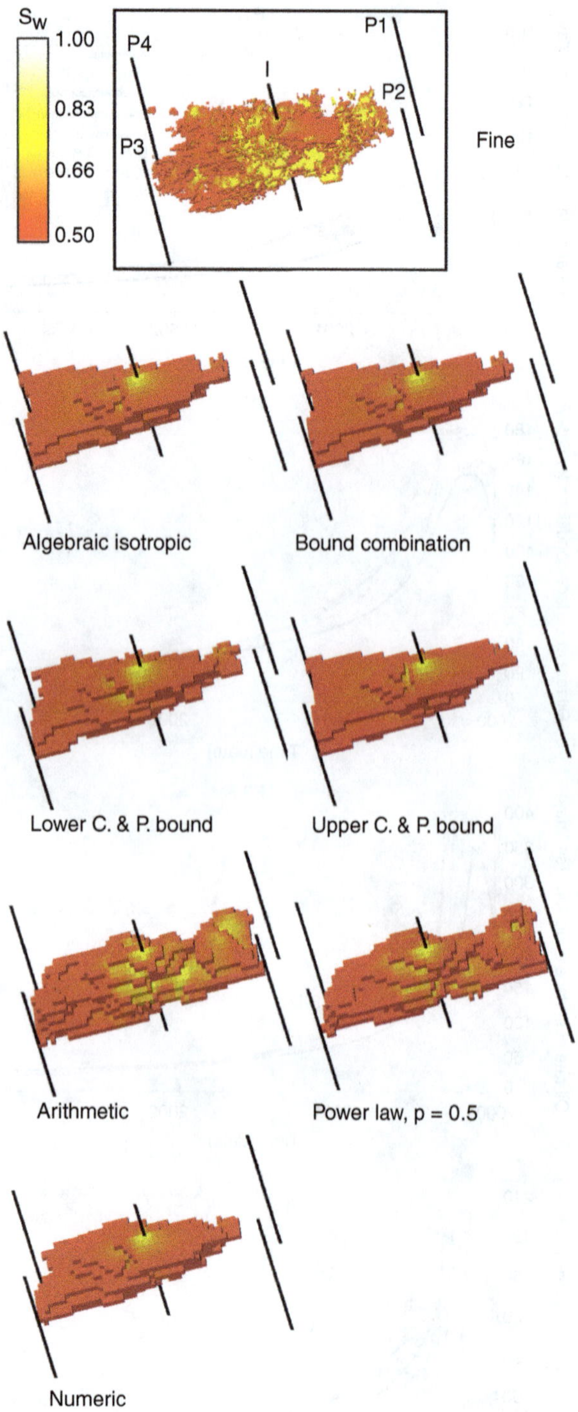

Figure 12

Comparison of water saturation for reservoir models derived from different upscaling methods after 530 days of injection.

We focus on saturation plots to analyze the impact of upscaling. Figure 12 shows water saturation after 530 days of injection.

The upscaling process smooths the displacement front, but its general form is more or less preserved. As expected, the bound combination method and the algebraic isotropic method give the same results in terms of water saturation. As mentioned earlier, anomalous behavior was evidenced for the water-oil ratios simulated for wells P3 and P4. For these two wells, Figure 9 shows that the front is very different from the one obtained on the fine scale. In the case of the oil rate, we observed that the power-law method with an exponent of 0.5 and the arithmetic average lead to similar water displacements. Whatever the upscaling method applied to this test case, the displacement front tends to extend, especially for wells P3 and P4. This may be due to coarsening, or more precisely, to numerical diffusion. For wells P1 and P2, the front also extends with the arithmetic mean and the power-law method (with an exponent of 0.5). These two methods are known to be optimistic. Therefore, permeability overestimation adds to numerical diffusion, which accentuates front distortion.

3.3 How Important is the Choice of the Upscaling Factor?

Section 3.2 investigated the influence of the choice of the upscaling method on the simulation results. Now, given an upscaling method, we propose to focus on the upgridding step, that is, on the analysis of the upscaling factor impact. The application case considered hereafter is the bottom part of the SPE10 case. This unit includes very thin and long channels. As such, the flow simulation results are expected to be very sensitive to the upscaling factor. The upscaling method used is the bound combination one [25] (*Eq. 26*).

3.3.1 The Upscaling Factor and the Upscaling Extent Parameter

In the first step, we study the relative effect of coarsening in different directions. We envision two different configurations: first along the X direction, and second along the Y direction. The dimensions of the grids used are reported in Tables 4 and 5 along with the upscaling factor values.

The *UE* Upscaling Extent parameter is computed for the two coarsening configurations of interest. Both global (*Eq. 3*) and directional (*Eq. 5*) parameters are represented in Figure 10. The *UE* decreasing at small upscaling factors is stronger for the upscaling along direction Y than for the upscaling along direction X. According to [10], the information loss is proportional to the *UE*. Therefore, the *UE* curves depicted in Figure 13 can be used to evaluate the information loss. This idea drives our understanding of the results presented in the following subsections.

TABLE 4

Number of grid blocks along axis X (N_x) *versus* the corresponding upscaling factor. Upscaling is applied along the X direction only ($N_y = 220, N_z = 50$)

N_x	Upscaling factor
30	0.5
20	0.333
15	0.25
12	0.2
10	0.166
6	0.1
5	0.0833
3	0.05
2	0.033

TABLE 5

Number of grid blocks along axis Y (N_y) *versus* the corresponding upscaling factor. Upscaling is applied along the Y direction only ($N_x = 60, N_z = 50$)

N_y	Upscaling factor
110	0.5
55	0.25
44	0.2
22	0.1
11	0.05
10	0.04545
5	0.022727
3	0.01818
2	0.00909

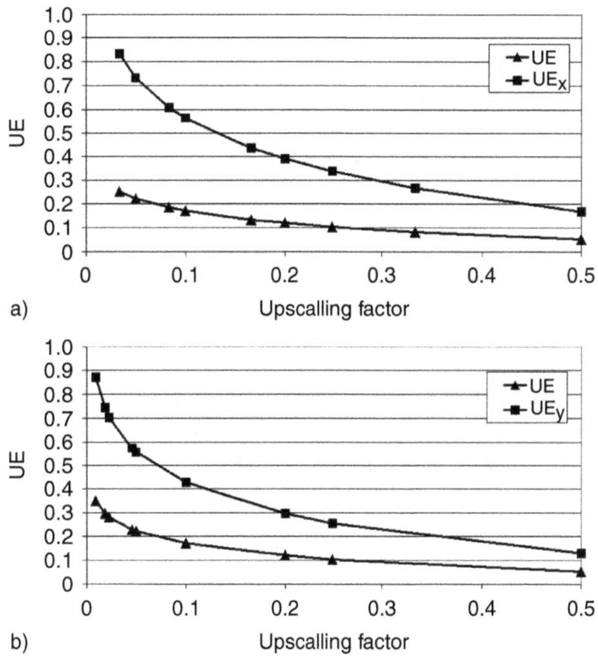

Figure 13

The Upscaling Extent (UE) parameter *versus* the upscaling factor for the two coarsening directions considered (coarsening along a) axis X only, b) axis Y only). The global UE parameter is plotted together with the directional UE (UE_x and UE_y).

In the Figure 13, we remark that for the coarsening along direction X, the UE and UE_x variations are more linear like the UE and UE_y variations. Qi and Hesketh [10] stated that the information loss is proportional to the UE criterion. Therefore, these curves can also help explain the next results.

3.3.2 The Data Range and Variance Indicators

Before going further, we detail the key features of the selected upscaling method when applied to this special case. As explained above, the bound combination method is a generalization of the algebraic isotropic method. It depends on the Cardwell and Parsons bounds. The effective permeability, when computed along direction X, is given by Equation (26). In the case studied, we upscale the model either along X or Y. Then, with the upscaling along X, a coarse cell encompasses a single cell along directions Y and Z and several along direction X (*Fig. 14*). So, if only upscaling in one direction is carried out then it is trivial that the harmonic average might be chosen along the coarse cell and arithmetic average for flow transverse to the cell. In this case, the bound combination method boils down to this configuration.

In fact, based upon this configuration, expressions (13, 14, 27) and (28) can be simplified, resulting in:

$$K_{c,x}^{\min} = K_{c,x}^{\max} = K_{c,x}^{3} = K_{c,x}^{4} = \mu_h^x(K_{f,x}) \quad (31)$$

$$K_{c,y}^{\min} = K_{c,y}^{\max} = K_{c,y}^{3} = K_{c,y}^{4} = \mu_a^x(K_{f,y}) \quad (32)$$

$$K_{c,z}^{\min} = K_{c,z}^{\max} = K_{c,z}^{3} = K_{c,z}^{4} = \mu_a^x(K_{f,z}) \quad (33)$$

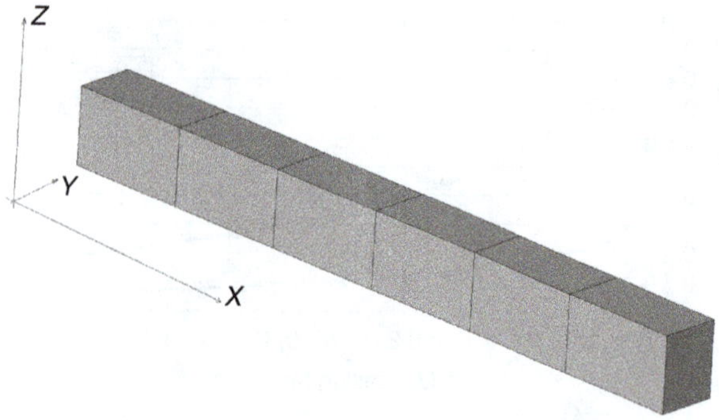

Figure 14

Fine cells included in a coarse cell (upscaling along direction X only; upscaling factor of 0.166).

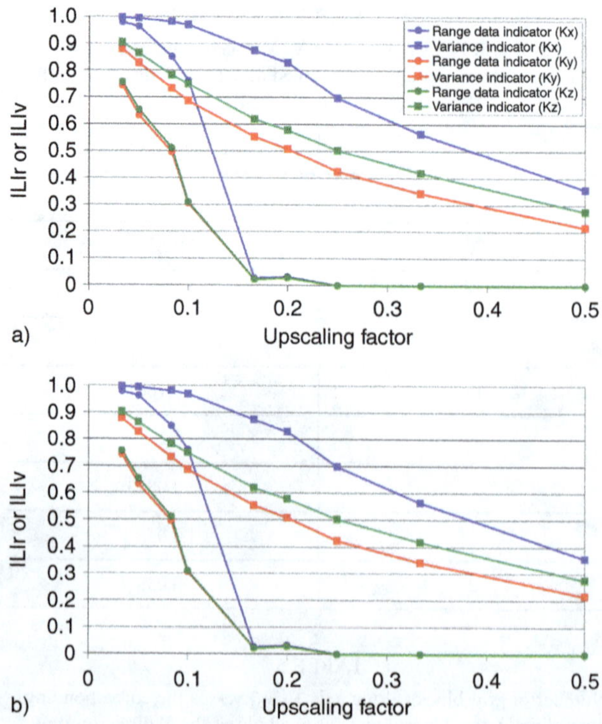

Figure 15

Data range ($ILIr$) and variance ($ILIv$) indicators obtained when performing upscaling along a) X and b) Y.

In this case, the bound combination method yields:

$$K_x^{BoundComb.} = \mu_h^x(K_{f,x}) \quad (34)$$

$$K_y^{BoundComb.} = \mu_a^x(K_{f,y}) \quad (35)$$

$$K_z^{BoundComb.} = \mu_a^x(K_{f,z}) \quad (36)$$

As a conclusion, when aggregating cells along direction X only, the bound combination method boils down to the harmonic mean for Kx and the arithmetic mean for Ky and Kz. The same result is obtained when considering the algebraic isotropic method instead of the bound combination method. This is a known result in upscaling ([42] p. 30).

Figure 15 shows that the indicator values obtained for Ky and Kz are pretty close for the upscaling along X. Similarly, the results for Kx and Ky are very close for the upscaling along Y. These similarities and the simplification of the bound combination method arise because there is only one direction of upscaling. The variations in the data range indicators provide useful insights about the maximal value that can be used for the upscaling factor. When along X, the upscaling process is erroneous for upscaling factors less than 0.25. Even with an upscaling factor of 0.166, the smoothing effect due to upscaling can be considered as too strong. When along direction Y, the upscaling process is characterized by a critical upscaling factor of 0.1. The overall variations in the variance indicator being linear, it is difficult to draw conclusion. The only remarkable point is that the values obtained with the harmonic mean are higher than with the arithmetic mean. This is also true for the data range indicator.

3.3.3 Effects of Coarsening on the QQplot

Whatever the coarsening direction, Figures 16 and 17 show that the smaller the upscaling factor, the more the QQplot departs from the bisector. In other words, the smaller the upscaling factor, the less accurate the upscaling. In addition, these figures point out that the upscaling process is wrong for the intermediate permeability values in between 1 mD and 10 mD, while the extreme values are properly captured. Moreover, as explained previously in the paper the bound combination method boils down to the harmonic mean in the direction of upscaling, and to the arithmetic mean in the other directions. And the harmonic mean will be pessimistic in the direction of upscaling (QQplot below the bisector) and the arithmetic average will be optimistic (QQplot above the bisector). The results obtained for vertical permeabilities are more intricate, especially for high upscaling factors. Two bumps are observed: one when fine permeabilities are in the range $[10^{-4}\,\text{mD}; 10^{-2}\,\text{mD}]$ and the other one when they are in the range $[10^{-1}\,\text{mD}, 10\,\text{mD}]$. However, these two bumps merge when the upscaling factor decreases.

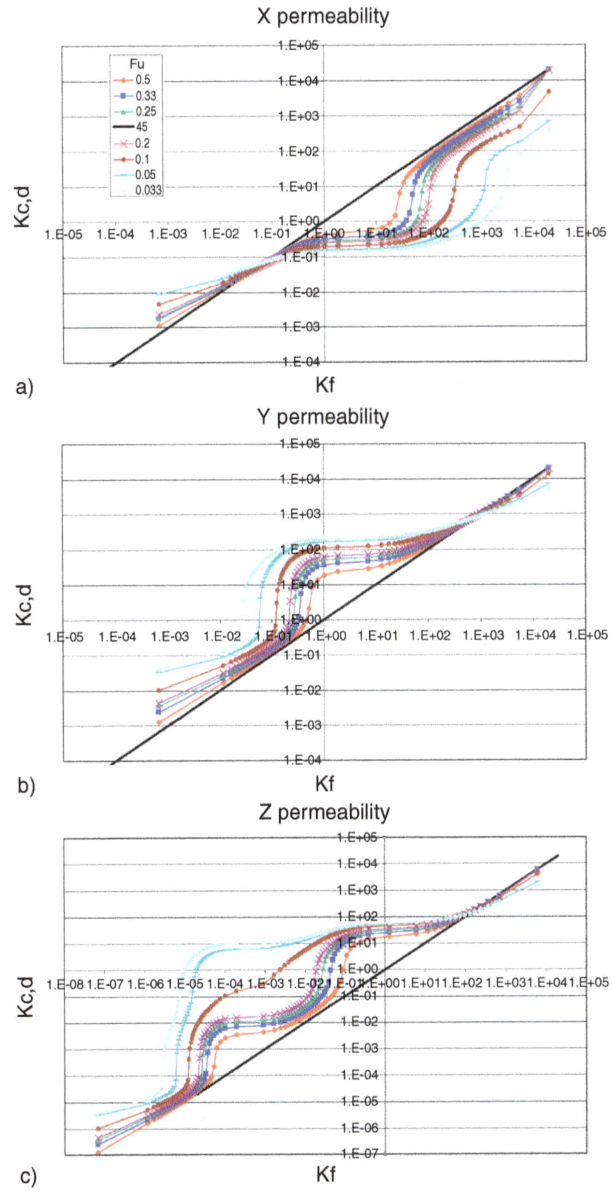

Figure 16

QQplots for a) horizontal X, b) horizontal Y and c) vertical Z permeabilities when performing upscaling along X.

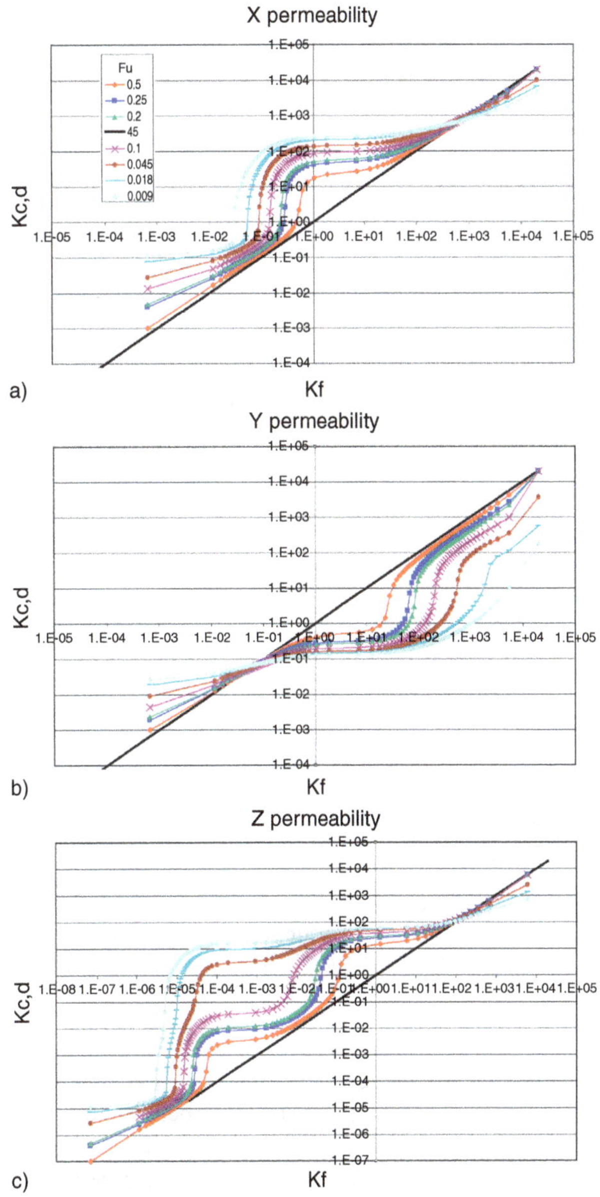

Figure 17

QQplots for a) horizontal X, b) horizontal Y and c) vertical Z permeabilities when performing upscaling along Y.

This phenomenon related to systems close to their percolation threshold is detailed in [16].

3.3.4 Convergence of the Quality Indicators and the Simulation Results

This section presents the production responses simulated for the various reservoir models obtained, considering the two upscaling configurations described above: coarsening along X and coarsening along Y.

Coarsening along X

Let us first focus on the responses simulated for the injector, which is located in the middle of the field. Figure 18 depicts the bottom-hole pressure and the water injection rate at surface conditions for the different upscaling factors. As expected, the injection rate is almost constant: the simulator has difficulties maintaining a constant injection rate, but the difference is negligible (a maximal difference of 10 m^3/day for a constant rate of 795 m^3/day). The difference in bottom-hole pressure

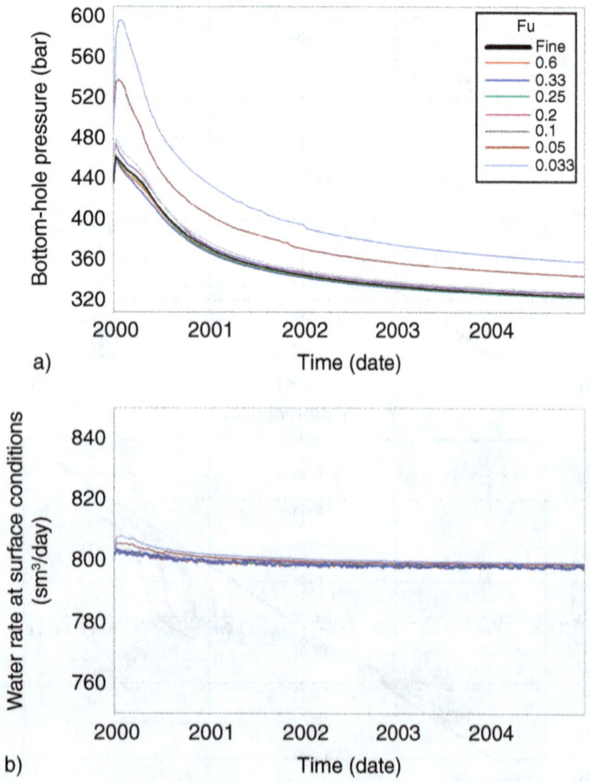

Figure 18

a) Bottom-hole pressures and b) water injection rates at surface conditions at the injector for various upscaling factors – upscaling is performed along direction X (black = fine grid results).

tells us about the average permeability between wells and about the upscaled productivity index. The upscaling modifies this average permeability and the productivity index: we also see a difference in BHP. Small Fu upscaling factors lead to very inaccurate results. When $Fu < 0.1$ (that is $UE > 0.17$ or $UE_x > 0.56$), bottom-hole pressures and water injection rates are both overestimated. When $Fu > 0.16$ (that is $UE < 0.134$ or $UE_x < 0.438$), the fine and coarse simulated production responses are pretty close. This is consistent with the results for the *ILIr* data range indicator shown in Figure 15. When the upscaling factor is more than 0.16, the *ILIr* values obtained indicate very bad results. When Fu decreases, that is when the UE increases; the responses simulated for the injector deteriorate. For $Fu > 0.33$ (for $UE < 0.082$ or $UE_x < 0.27$), the results for the coarse models are in very good agreement with the responses simulated for the fine model. Moreover, the water injection rates computed are smoother when $Fu \leq 0.25$ (Fig. 18a). This may be induced by the homogenization of the permeability field. Thus, for the case studied and considering only the responses at the injector, the upgridding can be regarded as reasonable as long as the upscaling factor Fu is more than 0.33. In the case of the producers, we focused on the water-oil ratios (*Fig. 22*) as these responses are very sensitive to the upscaling factor. The breakthrough is actually a key factor when managing reservoir production. It is therefore crucial to build upscaled reservoir models able to properly reproduce the breakthrough simulated for the fine model. For a small upscaling factor, we remark that the results are very inaccurate. For $Fu < 0.1$ ($UE > 0.17$ or $UE_x > 0.56$), the bottom-hole pressure and the water injection rate are overestimated. For $Fu > 0.16$ ($UE < 0.134$ or $UE_x < 0.438$) the results are quite similar to the fine results. This conclusion can also be drawn from the *ILIr* result (Fig. 15). For an upscaling factor higher than 0.16, the *ILIr* results are very bad. We remark that when Fu lowers and so when the UE increases, the results deteriorate at the injector. The results for $Fu > 0.33$ (for $UE < 0.082$ or $UE_x < 0.27$) are in very good agreement with the fine results. Moreover, the water injection rate results are smoother for $Fu \leq 0.25$: this may be due to the homogenization of the permeability field. As a conclusion, the upgridding is reasonable for the injector results when $Fu > 0.33$. This value gives an idea of the Representative Elementary Volume of this case. This also has to be investigated (*Fig. 19*) for the responses simulated at the producer. For the four producers, the results are very poor when $Fu < 0.16$ ($UE > 0.134$ or $UE_x > 0.438$). This is the same as for the injector. Second, the results obtained for wells P2, P3 and P4 are in very good agreement with the fine results when $Fu > 0.33$ (for $UE < 0.082$ or $UEx < 0.27$). However, for well P1, all results are relatively bad whatever the upscaling factor: too much water is produced at this well. The upscaled permeabilities around this well are probably overestimated. Despite these poor results for well P1, we concluded that the breakthrough was correctly captured provided Fu was more than 0.33.

Coarsening along Y

The same study was performed when performing the upscaling along the Y direction. The bottom-hole pressures and the water rates simulated for the injector are displayed in Figure 20. It is worth noting that the maximal bottom-hole pressure is 689.18 bar (10 000 psi).

Figure 20 shows that the injector reaches the maximal bottom-hole pressure when $Fu < 0.05$ ($UE > 0.224$ or $UE_y > 0.556$). At this point, as soon as the pressure constraint is reached, the well production is based upon a

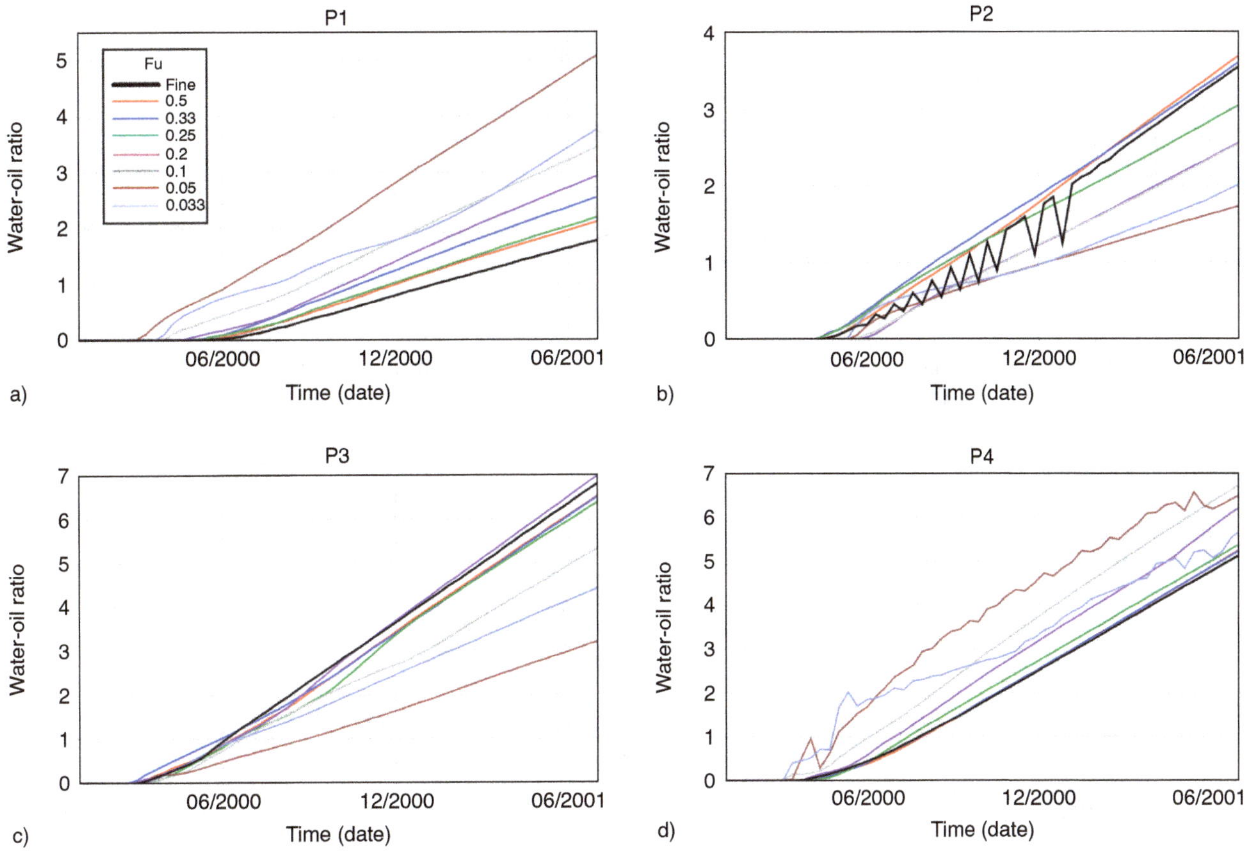

Figure 19

Water-oil ratios at producers a) P1, b) P2, c) P3 and d) P4 for various upscaling factors – upscaling is performed along direction *X* (black = fine grid results).

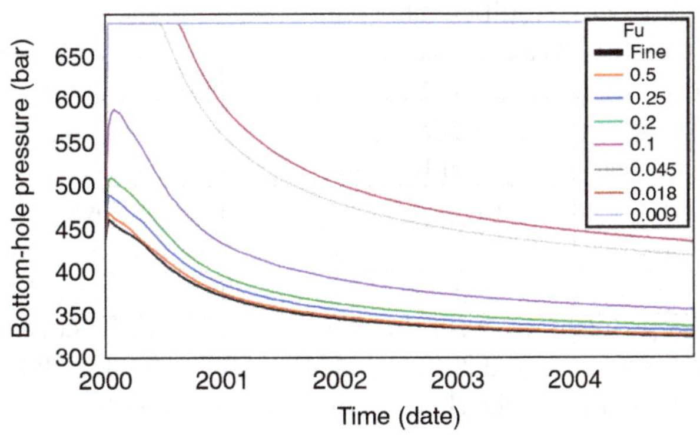

Figure 20

Bottom-hole pressures at the injector for various upscaling factors – upscaling is performed along direction *Y* (black = fine grid results).

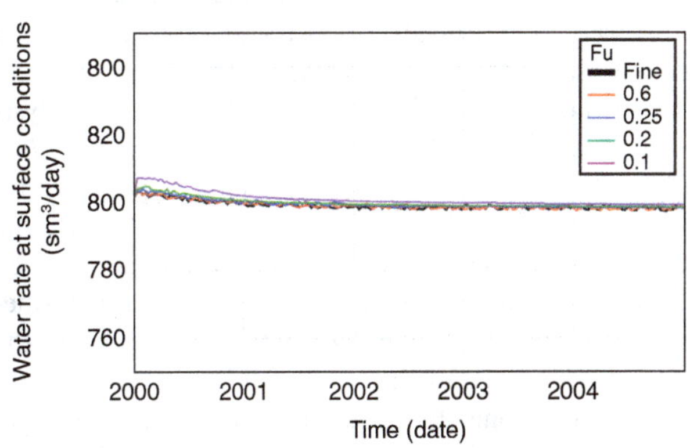

Figure 21

Constant water injection rate for various upscaling factors – upscaling is performed along direction *Y* (black = fine grid results).

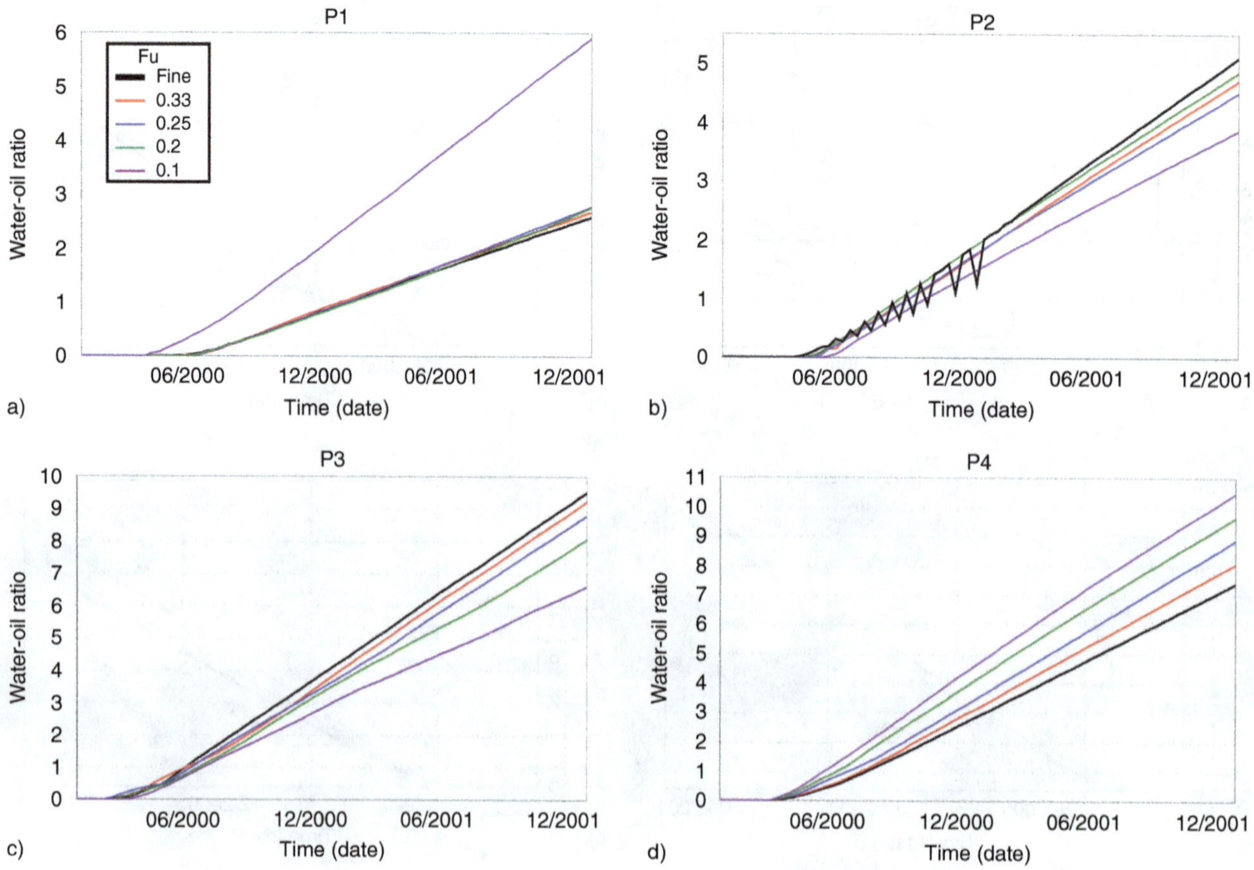

Figure 22

Water-oil ratios for various upscaling factors – upscaling is performed along direction Y (black = fine grid results).

constant pressure and the water injection rate falls. In such conditions, there is no way to make valuable comparisons. A major result is that too strong a coarsening can drastically change the dynamic behavior of the field. This phenomenon could be predicted from the analysis of the data range indicators. Figure 15 shows that *ILIr* values strongly decrease for upscaling factors within the range $[0.05;0.1]$. When *Fu* is more than 0.01, the *ILIr* values are very small. Another insight can also be detected from the analysis of QQplots for vertical permeabilities (*Fig. 17*). When $Fu < 0.05$, the shape of the QQplots changes and the two small bumps combine to form a single one. Based upon this result, we restrict our attention to the water injection rates simulated for $Fu > 0.05$.

Figure 21 shows that water injection rates are lightly overestimated when *Fu* is too small. The simulator has difficulties maintaining a constant injection rate (oscillations), but the difference is negligible (a maximal difference of 10 m^3/day for a constant rate of 795 m^3/day). In addition, the upscaling process contributes to homogenizing permeabilities. Thus, the smaller *Fu*, the smoother the results. We can also observe that the results are very good when $Fu > 0.2$. When $Fu = 0.1$, the rates simulated for the upscaled models are not very accurate, but the main features of well behavior are preserved.

In the case of the producers, we focus on the water-oil ratios. At first glance, the results obtained for $Fu = 0.1$ are wrong. This is particularly obvious for well P1. For this problematic well, the results turn out to be correct as soon as $Fu \geq 0.2$. For wells P3 and P4, the simulated water-oil rates clearly depend on *Fu*: the smaller the *Fu* (the higher the *UE*, *ILIr* and *ILIem*), the worse the results. Depending on the criterion considered, the water-oil ratio is underestimated for well P3 while it is overestimated for well P4. As a conclusion, the upgridding is reasonable when $Fu \geq 0.2$: to obtain correct results we must choose an upscaling factor greater than 0.2.

CONCLUSION

In this paper, we studied and detailed different criteria, which make it possible to evaluate the validity of upscaling methods and to quantify the information loss resulting from their application. We studied two test cases, the first one to see whether indicators or criteria help to identify the best upscaling method to use, and the second one to identify the most relevant upscaling factor. We showed that the QQplot and the variance indicator are valuable preprocessing tools to investigate the impact of uspcaling before performing any fluid flow simulation. Then, estimating the Cardwell and Parsons bounds helps identify the cells or directions for which upscaling may lead to erroneous results if not carefully and properly done. The analysis of Cardwell and Parsons bounds combined together with the search for the zero data range and variance indicators provide useful insights into the validity of the upscaled results obtained. In addition, we observed that for the first test case the bound combination, algebraic and numerical local methods give the most satisfactory results. Combining these methods to upscale the absolute permeabilities with a method to upscale well productivity indices [43] would contribute to improving the results further. The second test case pointed out that these different criteria are also very useful for selecting an appropriate upscaling factor. In this case, an upscaling factor greater than 0.2 gives correct results in terms of the oil rate or water-oil ratio. This study treats one example of heterogeneous medium: the well-known SPE10 test case. A perspective of this work could be to test how these different upscaling indicators suit different types of heterogeneity.

ACKNOWLEDGMENTS

The author would like to thank Dr. Mickaele Le Ravalec-Dupin for her valuable help and advice in writing this paper.

REFERENCES

1 Durlofsky L.J. (2003) Upscaling of geocellular models for reservoir flow simulation: a review of recent progress, *Proceedings of the 7th International Forum on Reservoir Simulation*, Buhl/Baden-Baden, Germany, 23-27 June.

2 Farmer C.L. (2002) Upscaling: a review, *International Journal for Numerical Methods in Fluids* **40**, 63-78

3 Renard Ph., de Marsily G. (1997) Calculating equivalent permeability: a review, *Advances in Water Resources* **20**, 5-6, 253-278.

4 Wen X.H., Gómez-Hernández J.J. (1996) Upscaling hydraulic conductivities in heterogeneous media: an overview, *Journal of Hydrology* **183,** ix-xxxii.

5 Sánchez-Vila X., Guadagnini A., Carrera J. (2006) Representative hydraulic conductivities in saturated groundwater flow, *Reviews of Geophysics* **44**, 3.

6 Sablok C., Aziz K. (2005) Upscaling and discretization errors in reservoir simulation, *SPE simulation symposium*, Houston, Texas, USA, 31 Jan.-2 Feb., SPE 93372.

7 Lodder R., Hieftje G. (1988) Quantile Analysis: a method for characterizing data distributions, *Applied Spectroscopie* **42**, section 8.

8 Preux C., (2011) Study of evaluation criteria for reservoir upscaling, *73rd EAGE Conference & Exhibition incorporating SPE EUROPEC 2011*, Vienna, Austria, 23-26 May.

9 Qi D. (2010) Upscaling Extent *vs.* Information Loss in Reservoir Upscaling, *Petroleum Science and Technology* **28**, 12, 1197-1202.

10 Qi D., Hesketh T. (2004) Quantitative Evaluation of Information Loss in Reservoir Upscaling, *Petroleum Science and Technology* **22,** 11, 1625-1640.

11 Qi D., Wong P.M., Liu K. (2001) An Improved Global Upscaling Approach for Reservoir Simulation, *Petroleum Science and Technology* **19**, 7, 8, 779-795.

12 Christie M.A., Blunt M.J. (2001) Tenth SPE Comparative Solution Project: A comparison of Upscaling Techniques. SPE66599, *SPE Reservoir Simulation Symposium*, Houston, Texas, 11-14 Feb.

13 Gautier Y., Blunt M.J., Christie M. A. (1999) Nested gridding and streamline-based simulation for fast reservoir performance prediction, *Computational Geosciences* **3**, 3-4, 295-320.

14 Salazar M.O., Villa J.R. (2007) Permeability upscaling techniques for reservoir simulation. *Latin American & Caribbean Petroleum Engineering Conference*, Buenos Aires, Argentina, 15-18 April, SPE106679.

15 Fleckenstein J.H., Fogg G.E. (2008) Efficient upscaling of hydraulic conductivity in heterogeneous alluvial aquifers, *Hydrogeology Journal* **16**, 7, 1239-1250.

16 Boschan A., Noetinger B., (2012) Scale dependence of effective hydraulic conductivity distributions in 3D heterogeneous media: a numerical study, *Transport in Porous Media* **94**, 1, 101-121.

17 Hosseini S.A., Kelkar M. (2008) Analytical Upgridding method to preserve dynamic flow behavior, *SPE Annual Technical Conference and Exhibition*, Denver, 21-24 Sept., SPE 116113.

18 King M.J., Burn K.S., Wang P., Muralidharan V., Alvardo F., Ma X., Datta-Gupta A. (2006) Optimal Coarsening of 3D Reservoir Models for Flow Simulation, *SPE Reservoir Evaluation & Engineering* **9**, 4, 317-334.

19 Deschamps R., Guy N., Preux C., Lerat O. (2011) Impact of Upscaling on 3D Modelling of SAGD in a Meander Belt, *SPE Annual Technical Conference and Exhibition* **4**, 3328-3343, Society of Petroleum Engineers - Denver, Colorado, USA, 30 Oct.-2 Nov., SPE 147035.

20 Cardwell W.T., Parsons R.L. (1944) Average permeabilities of heterogeneous oil sands, *Transactions of the American Institute of Mining, Metallurgical and Petroleum Engineers* **160**, 34-42.

21 Renard Ph., Le Loc'h G., Ledoux E., de Marsily G., Mackay R. (2000) A fast algorithm for the estimation of the equivalent hydraulic conductivity of heterogeneous porous media, *Water Resources Research* **36**, 12, 3567-3580.

22 Li D., Kumar A. (2001) A New Efficient Averaging Technique for Scaleup of Multimillion-Cell Geologic Models, *SPE Reservoir Evaluation & Engineering* **4**, 4, August.

23 Journel A.G., Deutsch C.V., Desbarats A.J. (1986) Power averaging for Block effective permeability. *56th California Regional meeting of SPE*, Oakland, SPE 15128.

24 de Dreuzy J.-R., de Boiry P., Pichot G., Davy P. (2010) Use of power averaging for quantifying the influence of structure organization on permeability upscaling in on-lattice networks under mean parallel flow, *Water Resour. Res.* **46** W08519.

25 Guérillot D., Rudkiewicz J.L., Ravenne C., Renard G., Galli A. (1990) An integrated model for computer aided reservoir description: from outcrop study to fluid flow simulations, *Revue de l'IFP* **45**, 1.

26 Duquerroix J-P.L., Lemouzy P., Noetinger B., Romeu R.K. (1993) Influence of the Permeability Anisotropy Ratio on Large-Scale Properties of Heterogeneous Reservoirs, *68th Annual Technical Conference and Exhibition of the Society of Petroleum Engineers*, Houston, Texas, 3-6 Oct., SPE 26612.

27 Kruel-Romeu R. (1994) Écoulement en milieu poreux hétérogènes: prise de moyenne de perméabilités en régime permanent et transitoire (fluid flow through heterogeneous media: averaging permeabilities for the steady-state and transient cases), *PhD Dissertation*, University of Paris VI.

28 Christie M.A. (1996) Upscaling for reservoir simulation, *Journal of Petroleum Technology* **48**, 11, 1004-1010.

29 Noetinger B., Zargar G. (2004) Multiscale Description and Upscaling of Fluid Flow in Subsurface Reservoirs, *Oil & Gas Science and Technology, Revue de l'IFP* **59**, 2, 119-139.

30 Pickup G.E., Hern C.Y. (2002) The Development of Appropriate Upscaling Procedures, *Transport in Porous Media* **46**, 119-138.

31 Pickup G.E., Ringrose P.S., Jensen J.L., Sorbie K.S. (1994) Permeability tensors for sedimentary structures, *Mathematical Geology* **26,** 2, 227-250.

32 Jikov V.V., Kozlov S.M., Oleinik O.A. (1994) *Homogenization of Differential Operator and Integral Functionals*, Springer Verlag

33 Web site: www.spe.org/csp/datasets/set02.htm

34 Brooks R.H., Corey A.T. (1964) Hydraulic properties of porous media, *Hydrological Papers* (3), Colorado State University.

35 Gratien J.M., Guignon T., Magras J.F., Quandalle P., Ricois O. (2007) Scalability and Load Balancing Problems in Parallel Reservoir Simulation, *SPE Reservoir Simulation Symposium*, Houston, Texas, 26-28 Feb., SPE 106023.

36 Web site: http://www.beicip.com/index.php/eng/software/openflowplatform

37 Preux C., McKee F. (2011) Study and Approximation of IMPES Stability: the CFL Criteria, Proceeding of the *6th International Symposium on Finite Volumes for Complex Applications*, 6-10 Juin, Pragues, Czech Republic, Fort J., Fürst J., Halama J., Herbin R., Hubert F. (eds), Springer, Vol. 1, pp. 713-721.

38 Wiener O. (1912) Abhandlungen des mathematisch, *Physischen Klasse des Königlichen Sächsischen Gesellschaft des Wissenschaften* **32**, 509.

39 Dagan G. (1979) Models of groundwater flow in statistically homogeneous porous formations, *Water Resources Research* **15,** 1, 47-63.

40 Matheron G. (1967) Éléments pour une Théorie des Milieux Poreux, Masson, Paris.

41 Le Loc'h G. (1987) Étude de la composition des perméabilités par des méthodes variationnelles (Study of the composition of permeabilities using variational methods), *PhD Dissertation*, École Nationale Supérieure des Mines de Paris.

42 Hunt G.H. (2005) Percolation theory for flow in porous media, *Lecture Notes in Physics* **674**, 67-87, Springer, Berlin Heidelberg.

43 Preux C., Ding D., Duval T. (2010) A new method of near-well upscaling for reservoir simulations based on optimization, *Third International Conference on Porous Media and its applications in Science, Engineering and Industry*, Montecatini, Italia, 20-25 June.

16

Fast Charging and Smart Charging Tests for Electric Vehicles Batteries Using Renewable Energy

Oscar Mauricio Forero Camacho[1] and Lucian Mihet-Popa[1,2*]

[1] Electrical Engineering Department-DTU, Frederiksborgvej 399, 4000 Roskilde - Denmark
[2] Politehnica University of Timisoara, V. Parvan 2, 300223 Timisoara - Romania
e-mail: oscarmforero@gmail.com - lmih@elektro.dtu.dk - lucian.mihet@upt.ro

* Corresponding author

Abstract — *Electric Vehicles (EV) technologies are still relatively new and under strong development. Although some standardized solutions are being promoted and becoming a new trend, there is an outstanding need for common platforms and sharing of knowledge and core technologies. This paper presents the development of a test platform, including three Li-ion batteries designed for EV applications, and three associated bi-directional power converters, for testing impacts on different advanced loadings of EV batteries. Different charging algorithms/profiles have been tested, including constant current and power, and forced and pulsed power. The aim of the tests has been to study the impact of smart charging and fast charging on the power system, on the battery state of health and degradation, and to find out the limitations of the batteries for a Smart Grid. The paper outlines the advantages and disadvantages of both tests in terms of regulation of the aggregated local power, power capacity and the power exchange with the grid. The smart charging tests performed have demonstrated that even with a simple control algorithm, without any forecasting, it is possible to provide the required charging and at the same time the power system services, reducing the peak power and the energy losses in the power connection line of the power exchange with the national grid.*

Résumé — **Tests de recharge rapide et intelligente de batteries pour voitures électriques utilisant des énergies renouvelables** — Les développements technologiques liés aux voitures électriques (EV, *Electric Vehicles*) sont encore relativement nouveaux et s'intensifient. Même si certaines solutions standardisées sont encouragées et deviennent une nouvelle tendance, il y a un besoin exceptionnel pour des plateformes communes, mais aussi pour le partage de connaissances et de technologies de base. Cet article présente le développement d'une plateforme de tests, comprenant trois batteries Li-ion rechargeables conçues pour des voitures électriques, et trois convertisseurs de puissance bidirectionnels associés, permettant de tester l'impact de différentes stratégies de recharge sur les batteries. Différents algorithmes / profils de charge ont été testés, y compris à courant et puissance constants, et à puissance pulsée et forcée. L'objectif de ces tests est d'étudier l'impact de stratégies de recharge intelligente et rapide sur le système électrique, sur l'état électrique et la dégradation de la batterie, et aussi pour découvrir les limites des batteries dans un réseau électrique intelligent (*Smart Grid*). Cet article présente les avantages et les inconvénients de ces deux tests en termes de régulation de la puissance agrégée locale, de la puissance électrique installée et de l'échange de puissance électrique avec le réseau.

Les tests de recharge intelligente effectués ont aussi démontré que même avec un algorithme de contrôle simple, sans aucune prévision, il est possible de fournir la charge requise et dans le même temps les services du système d'alimentation, ce qui réduit la puissance de crête et les pertes d'énergie avec la connexion d'alimentation au réseau national d'électricité.

INTRODUCTION

Advanced energy storage devices and converters are introduced and tested recently for the next generation of Electric Vehicles (EV) for the smart grids. It is quite clear that the available technologies and its drivers are voltage-level independent and network structure independent [1-5]. The interest in these applications has been long-standing but the Battery Energy Storage System (BESS) technology had not matured sufficiently to warrant much interest. This is changing rapidly with the new generation of BESS using Li-ion batteries and state of the art power electronics converter topologies and control techniques [6-8].

Battery Management System (BMS) functionalities are very important for the optimal use and handling of batteries. The functionalities of the BMS could play a critical role to make an EV battery robust towards charging tests [9-13].

The different types of EV batteries have different characteristics regarding cost, energy density, safety, energy efficiency, degradation etc. making them suitable for different EV applications, such as Plug-in Hybrid Electric Vehicle (PHEV), many small cycles, full EV (many full cycles), fast charging or Vehicle-to-grid V_2G (cycles at high SOC (State of Charge)) [4-5].

Today, batteries are still a critical component in the EV industrialization due to their high price, technical complexity, limited records of long term operational data, etc. One of the most commonly selected Li-ion battery types for commercial electric vehicles is Li-ion manganese oxide based due to its overall balanced performance in energy/power densities, lifetime, cost and safety. At the same time, new Li-ion battery types started to show competitive key feature. Lithium-iron phosphate type was becoming a promising choice for PHEV's due to its high power/energy ratio favorable for Hybrid Electric Vehicle (HEV) and PHEV as well as improved safety and cycle life. On the other hand, mixed oxide based Li-ion types such as NMC (Ni, Mn and Co based) showed high energy density, which is crucial for the driving range started to be an interesting choice for EV's [4, 14-19].

Smart charging is one of the main challenges in EV mobility development. It is seen as indispensable to facilitate the optimal integration of EV to power system in order to avoid technical bottlenecks and the corresponding unnecessary investments in the electricity network [4, 20-22].

V_2G allows using EV to provide ancillary services to the electrical power grid. Additional grid services contribute to the optimization of energy use, for example, by the management of energy and power balance. Interaction with power grid needs bi-directional communication between EV and charging spots and bi-directional power flow. Therefore, the focus is on issues such as: reverse energy flow, control, communication and safety, which are the key issues in the future smart city concepts [23-25].

Fast charging should be able to charge a significant amount of energy (25÷75% of the battery's capacity) into the battery within 5÷15 minutes, depending on the actual application (typically for 150-200 km) [4-5].

Objectives

This paper is focus on testing of electric vehicle batteries to study the impact of smart charging and fast charging on the power system and on the battery degradation. Two different types of EV battery packs have been tested: a 16 kWh Lithium-Ferro type, optimized for hybrid EV applications, and a 26 kWh Lithium polymer type, optimized for full EV applications. The objectives of the smart charging are to address not only the charging needs specified by the user, but also to address the needs for power regulation in the power system. The objectives of the fast charging are to test different charging profiles and to study the impact on the charging efficiency and on the batteries.

Outline of the Paper

In the next section, we present some details about experimental facilities of SYSLAB. Section 2 provides the description of the topology, operation and battery management system with details about monitoring, protection and control of the battery pack systems. The experimental set-up, algorithm tests profile and results are presented in Section 3 for smart charging and fast charging, as well. Finally, concluding remarks are also presented.

1 SYSLAB-EXPERIMENTAL FACILITY OF A DISTRIBUTED ENERGY SYSTEM ARCHITECTURE

SYSLAB is a laboratory for research in distributed control and smart grids with a high share of renewable energy production. Its experimental facility is a Wind/PV/Diesel Hybrid Mini-Grid with local storage and a novel control infrastructure. The facility is spread across three sites located several hundred meters apart. At each of the three sites, there is a switchboard that allows the components installed at the site to be connected to either of two bus bars. The bus bars can be either connected to the national grid or can be part of an isolated system, as it is shown in Figure 1.

It includes two wind turbines (10 kW and 11 kW), three PV-plants (7.2 kW and 2 × 10 kW), a Diesel gen-set (48 kW/60 kVA), an intelligent office building with controllable loads (10 kW), a number of loads (75 kW, 3 × 36 kW), a Vanadium Redox Battery of 15 kW/120 kWh and three containers, each with an EV battery pack and a bidirectional power converter (*Fig. 2*).

Under the tests reported in this paper the SYSLAB power system has been operated as a micro-grid connected to the national grid, emulating a local part of a larger power system. In addition to the EV batteries, the micro-grid includes wind power, solar power and flexible loads representing future consumers. The batteries can switch between being connected either to the micro-grid or direct to the national grid through dedicated grid connections, emulating grid connected state or driving state, respectively, as can also be seen in Figure 2.

2 TOPOLOGY, CONTROL AND OPERATION OF THE BATTERY SYSTEM

This section covers the description of the battery management system with details about monitoring and

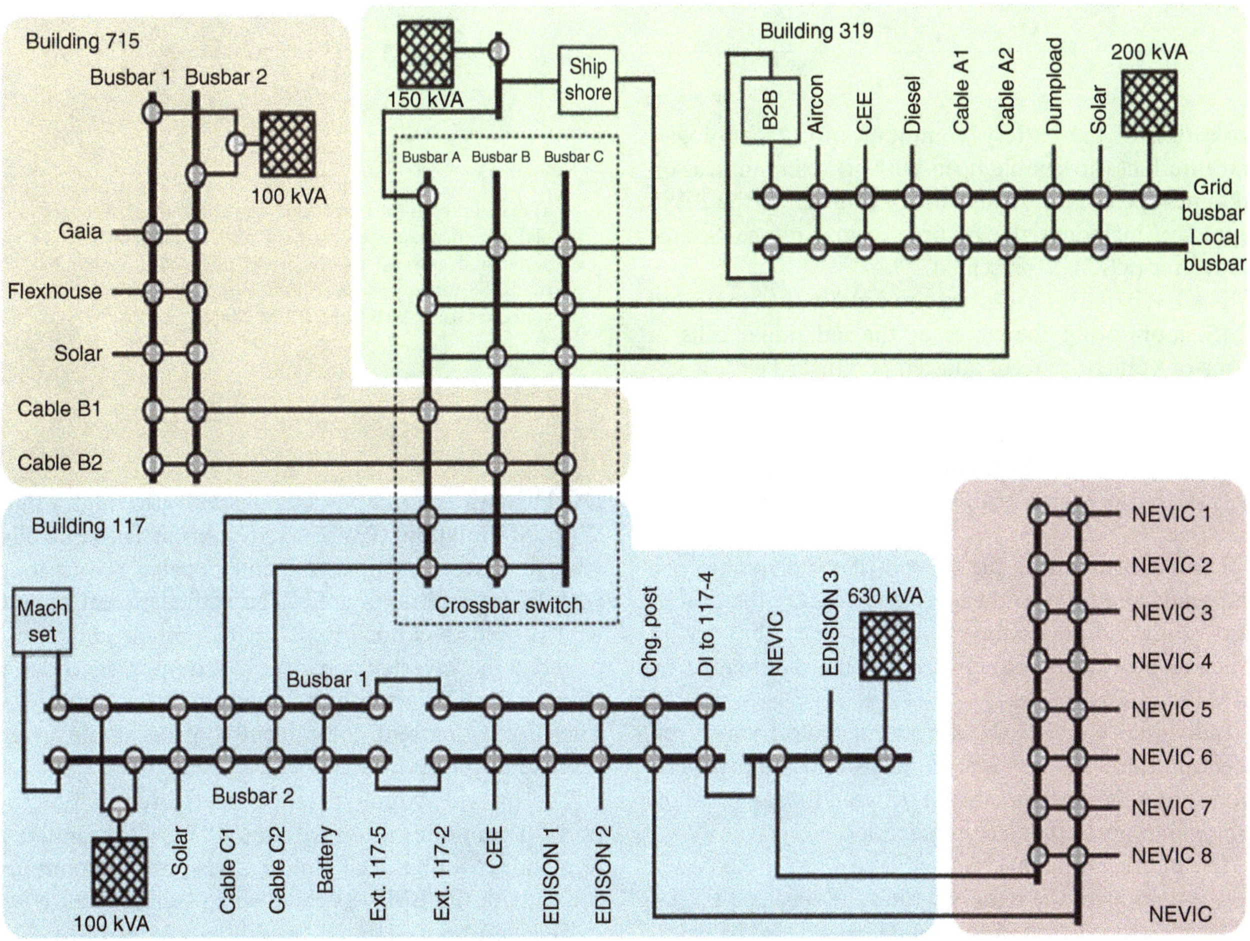

Figure 1

SYSLAB - DTU Risø's new laboratory for intelligent, active and distributed power systems.

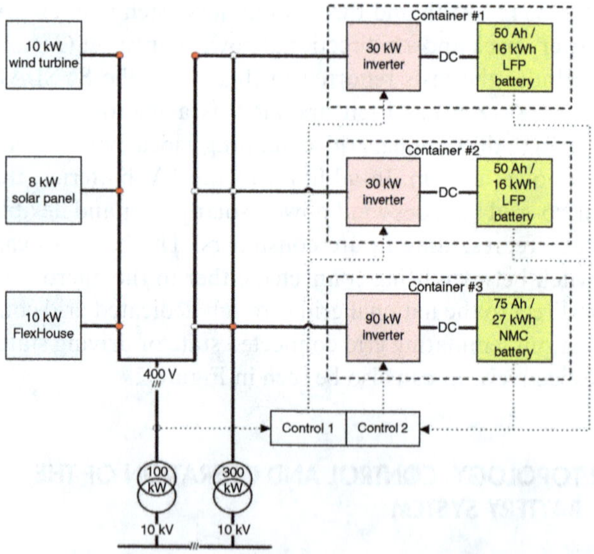

Figure 2

The power grid connection options of the EV batteries and their inverters to the SYSLAB power systems.

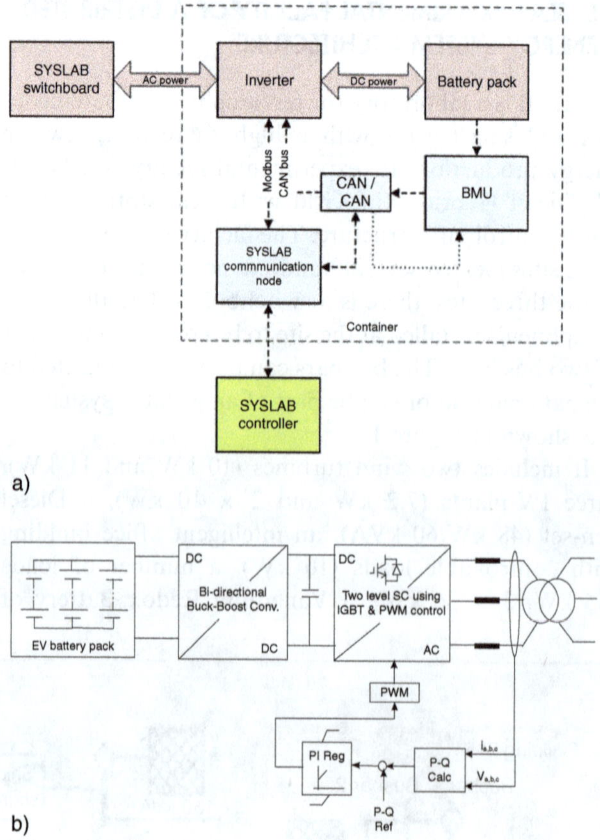

Figure 3

a) Details about the communication and control of the test platform including EV battery pack, bidirectional converter and external controller; b) the block diagram of the BESS topology with its two-level bi-directional power converter using IGBT and PWM control.

protection of the battery, communication technologies and interfaces to enable open network communication and interoperability. Also, a block diagram of the BESS topology, including the bi-directional Voltage Source Converter (VSC), is presented.

The EV batteries packs are equipped with a dedicated BMS, monitoring the states of the individual cells in terms of voltage, current and temperature. The test set-up (as it is shown in *Fig. 3*) covers an individual BMS, a Controller Area Network (CAN-CAN) converter and a bidirectional power converter; the set-up also protects the battery from being overloaded by external load requests.

The BMS monitors the state of the individual cells, and sends a request to the inverter to reduce the power, controlling the power flow to and from the battery, in order to protect the battery from being overloaded by an external load request.

Individual CAN-CAN converters have been developed and inserted between the BMS and the inverters, monitoring the communication between them and establishing an appropriate communication.

The bidirectional inverter is able to control the power flow to and from the battery within the specified limitations using its integrated control panel and also by external communications. The block diagram of the BESS topology, including the bidirectional power converter is shown in Figure 3b. The power converter contains a DC-DC bidirectional buck-boost converter and a Pulse With Modulation (PWM) VSC. A current-controlled voltage source, using active and reactive power loops with PI-regulators, is used. The active current component is used to control the DC link voltage and, consequently, the inverter active power output, in order to balance the EV battery and inverter active output power. The reactive current component controls the inverter's reactive output power.

The inverter communication ports are: an Ethernet port for communication with the SYSLAB communication node (Modbus TCP) and a CAN port for communication with the BMS, as can also be seen in Figure 3a.

The various test load patterns are controlled from dedicated controllers implemented on SYSLAB. From the SYSLAB controller the power flow to/from the batteries can be controlled in either current or power control mode independent of the sign of the control parameter. The power flows to and from the batteries are fully controlled by the SYSLAB controller.

3 EXPERIMENTAL TESTS AND RESULTS

The test platform includes three test benches in individual containers (for safety), each with an EV battery pack and a bidirectional converter to control the power flow to and from the battery, as it was shown in Figure 3.

Two of the inverters have capacities of 80 A, and the 3rd one has a capacity of 210 A, used for fast charging testing.

Two types of batteries were tested: a *BYD* 50 Ah/ 16 kWh Lithium-Ferro-Phosphate [25], suitable for HEV applications, and a *Kokam* 75 Ah/26 kWh Lithium-Polymer (NMC) [26], suitable for EV applications. More details can be find out in Appendix. Each battery pack is equipped with a dedicated BMS to monitor the voltages, currents and temperatures of the individual cells, and, if necessary, sending requests to the inverters to reduce the current in order to protect the battery against overloading.

3.1 Smart Charging

Smart charging covers all types of regulated charging with an additional objective, to the fully charging of the battery with maximum power. The objective of this test is to address not only the charging needs specified by the user but also the needs for power regulation in the power system. The "smart" control of the charging ensures full charging of the battery within a given period, but schedules the charging within the period relative to the local power balance.

This test is made with real components of the SYSLAB, including a 10 kW wind turbine, an 8 kW PV system and the consumption from a 10 kW office building, as can be seen in Figure 4.

The test emulates that the EV (battery) alternates between driving and being connected to the grid for charging. While "driving" (discharging the battery), the inverter is connected directly to the national grid and while "charging" the inverter is connected to the local power system.

The tests have been performed with a series of 6 hours driving-charging tests cycles with the driving loads emulating real driving patterns (based on vehicle driving data recorded in the Copenhagen area by GPS loggers, as part of the AKTA project [27]), as can be seen in Figure 5.

In Figure 6 is presented an example of the smart charging test cycling of the BYD battery with a driving load based on AKTA driving pattern.

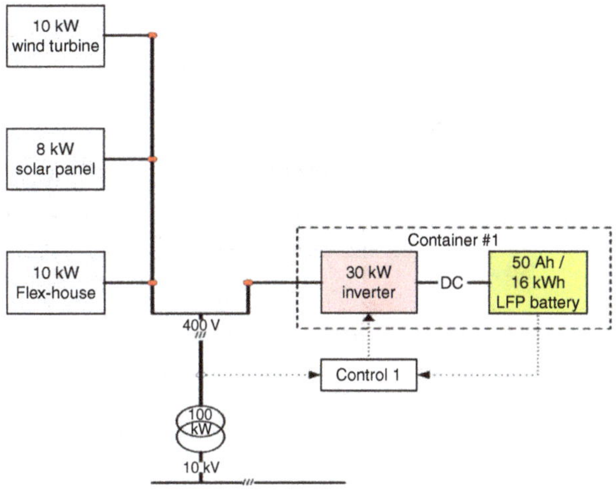

Figure 4

The set-up for smart charging tests.

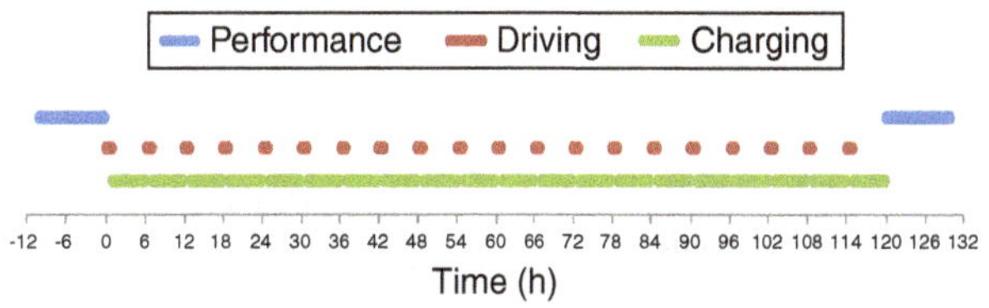

Figure 5

The "smart" charging tests alternates between one hour driving (time in which a single driving day is perform) and six hours charge modes and performance test at regular intervals.

As can be seen in Figure 6 are shown the BYD battery parameters which have been monitored during the smart charging test, such as the battery pack DC power, current and voltage (from the top), the maximum and minimum cell voltage and a comparison between the battery temperature, the maximum and minimum cell temperature during the test.

The developed smart charging strategy focuses on fulfilling the charging needs and constrains of EV owner (in this case, the battery must be fully charged in maximum 6 hours) and also focus on supplying power system services such as power regulation and load shifting. Additionally, the battery constrains are taken into account, so that a maximum charging current is set to a limit of C/3.

Hence, the charge process is accomplished by fully charging the battery within the specified time and scheduling the charge according to the power balance of the local grid:

$$P_{generation} = P_{consumption} + P_{losses}$$

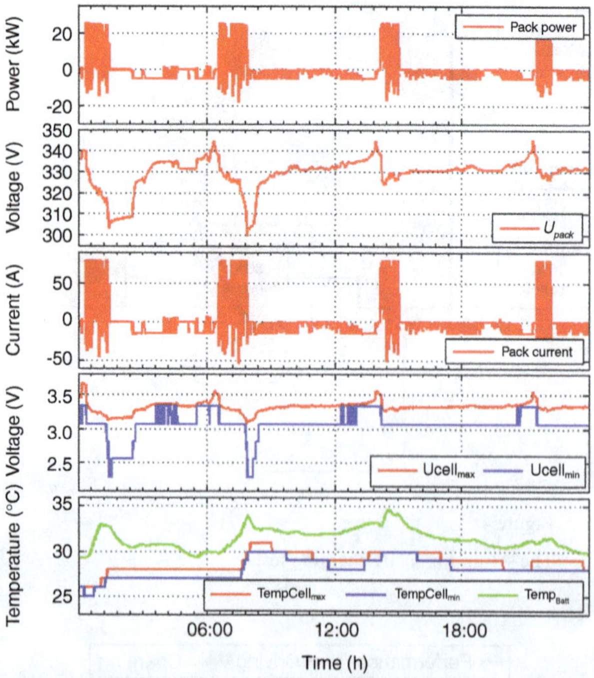

Figure 6

Driving load based on AKTA driving pattern and the EV battery load model.

in which:

$$P_{generation} = P_{wind} + P_{PV}$$

and

$$P_{consumption} = P_{Flex\text{-}house}$$

The designed control algorithm for smart charging, implemented as SYSLAB controller block (*Fig. 3*), is based on: the higher local aggregated excess power, P_x, the less charging flexibility, C_x, and the higher charging power, P_c. The local aggregated excess power, P_x, is defined as:

$$P_x = P_w + P_s - P_h \quad (1)$$

where P_w is the wind turbine power, P_s is the PV solar power and P_h is the Flex-house power.

The charging flexibility, C_x, which represents a number between $[0 \div 1]$, where 0 means no flexibility and 1 means full flexibility, could be defined as:

$$C_x = 1 - T_{c\_min}/T_c \quad (2)$$

where T_c is the remaining charging time and its minimum value could be express as:

$$T_{c\_min} = E_c/P_{c\_max} \quad (3)$$

in which E_c represents the remaining required charging energy and P_{c_max} is the maximum charging power.

In charging mode the charging power is giving by:

$$P_c = k_p \cdot (P_x - P_{x,e}) - k_c \cdot \log(T_c \cdot P_{c\_max}/E_c - 1) \quad (4)$$

where $P_{x,e}$ is the "expected" aggregated excess power at the actual time of the day and k_p and k_c are weighting constants. Since the first terms from (4) represent the power system needs k_p expresses the deviation of the actual aggregated excess power (positive when the actual aggregated excess power is higher than expected and negative otherwise). Also, because the second part from (4) represents the charging needs, negative when the charging flexibility is high and positive when flexibility is low, k_c can be set up accordingly.

A flow diagram of the active power control algorithm for smart charging, based in Equations (1-4) and the assumptions presented in this section, is show in Figure 7.

An example of the smart charging is presented in Figure 8, monitoring the current and the voltage of the battery pack, which are the inputs for the inverter. It can also be seen how the charging current is controlled according to the defined algorithm, based on (1) ÷ (4).

The impact of the smart charging on the power exchange to the public grid is illustrated in Figure 9. In Figure 9a is presented the output power during a day (for around 6 hours) for each component-power production (wind power and solar power), power consumption (representing here by Flex-house) and the charging power of the battery (P_c). In Figure 9b is presented a comparison between total power (P_x) exchange to the public grid with and without considering charging power of the battery (P_c). The influence of the charging power of the EV on the total power of DER components, in order to balance the grid power, is also pointed out in Figure 9b.

3.2 Fast Charging

The objectives of the fast charging tests are to test different charging profiles impact on the charging efficiency and on the battery. This method has been tested with many cycles of fast charging and slow discharging at different power levels, SOC levels and charging energy.

In Figure 10 is presented a block diagram of the set-up with the high power inverter and EV battery used for the fast charging tests.

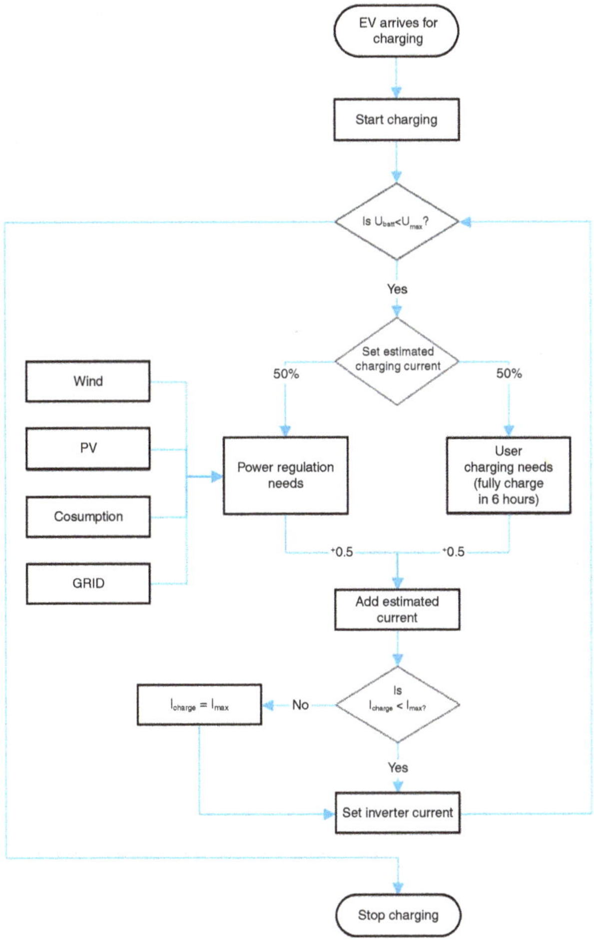

Figure 7

A flow chart diagram of the active power control algorithm for the smart charging strategy.

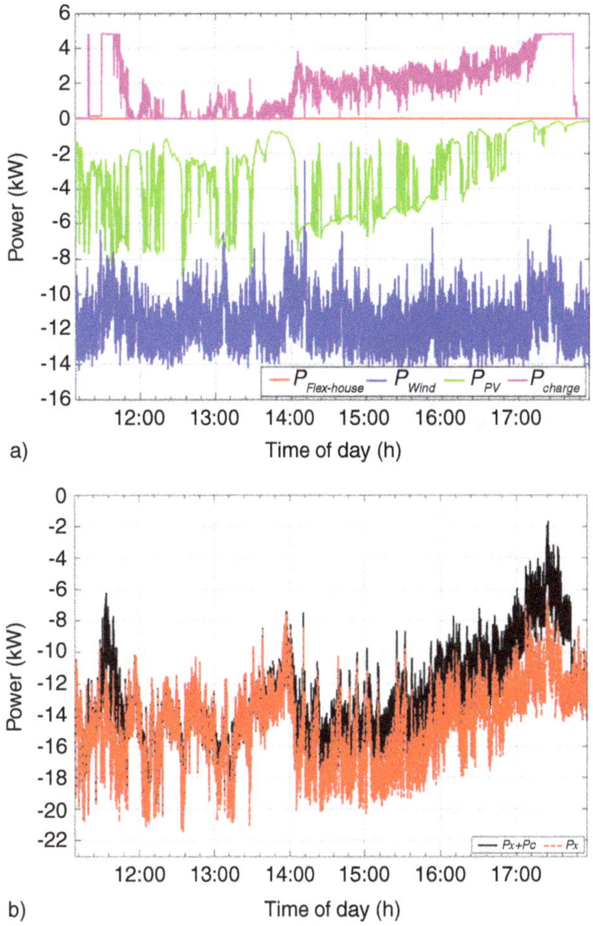

Figure 9

a) Power exchange to the public grid of the individual components, and b) a comparison of the power exchange to the power grid with and without the smart charging power.

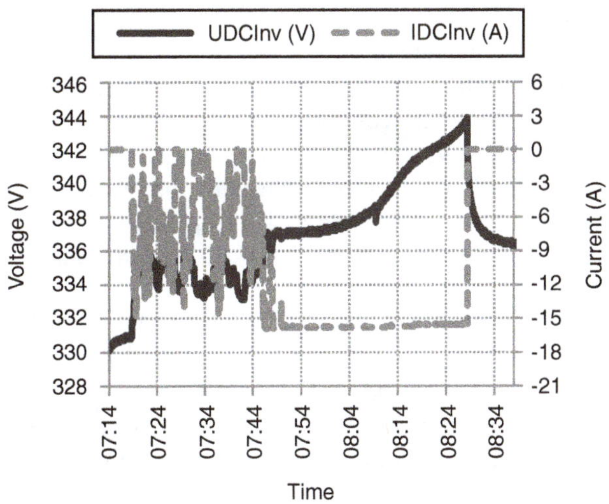

Figure 8

The battery packs current and voltage during a smart charging cycle.

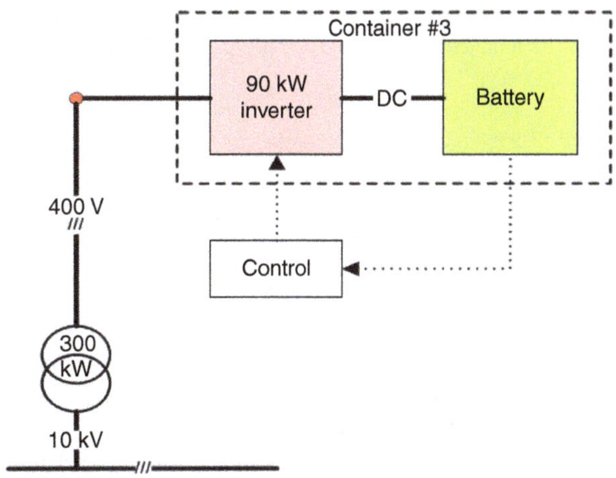

Figure 10

The power connection of the EV battery and its inverter for fast charging tests.

TABLE 1
An overview of the different charging profiles tested

Profile	Charging characteristics
Constant Current (CC)	The battery is charged with CC until the required energy has been charged or the charging power is limited by battery constrains
Constant Power (CP)	The battery is charged with constant power until the required energy has been charged or the charging power is limited by the battery constrains
Forced Power (FP)	The charging power is gradually reduced until the required energy has been charged or the charging power is limited by the battery constrains

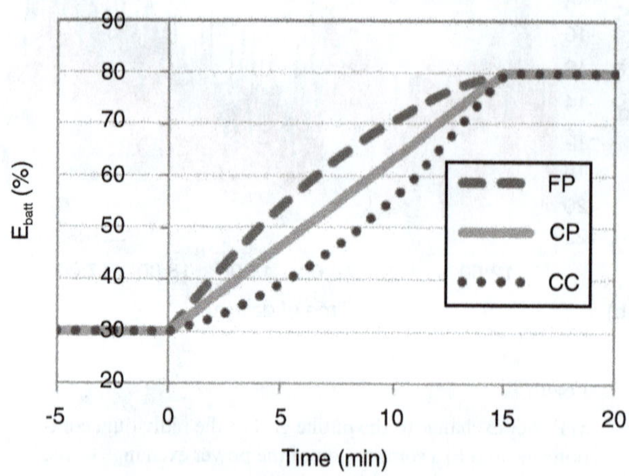

Figure 11

The estimated stored energy in the battery during the planned fast charging strategies for 50% energy charged in 15 minutes for three charging profiles: CC, CP and FP.

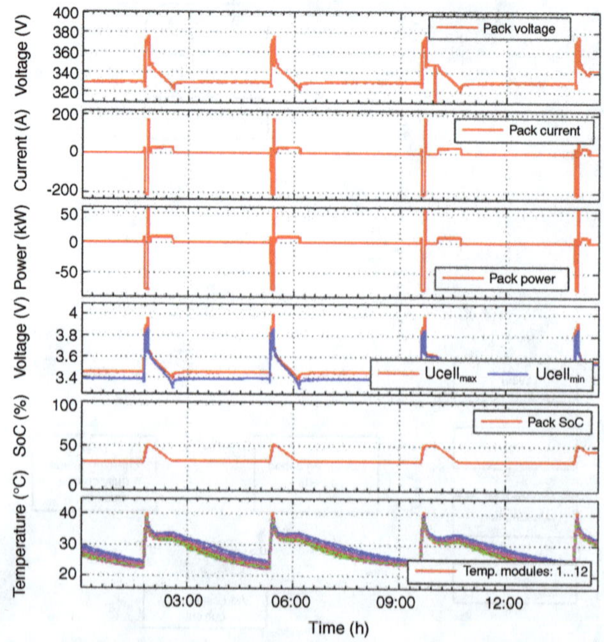

Figure 12

Example of a repeated fast charging test cycling of the *Kokam* battery with a charge energy of 13 kWh (40% of SOC) at constant current of 3C (210 A) and discharging with C/3 (25 A).

Different charging profiles have been tested. An overview of these profiles is given in Table 1, including Constant Current (CC), Constant Power (CP), and Forced Power (FP).

To point out the difference between charging profiles a comparison is shown in Figure 11. The different charging profiles reach a given amount of charged energy within a given time, 15 minutes in Figure 11, and independently of the profile.

With the limited cooling of the batteries in the tests, there were found very little differences between the different charging profiles.

An example of the repeated fast charging tests cycles with a constant power is presented in Figure 12.

The temperature increases very fast from around 25°C to around 40°C and as the voltage increases with the SOC level the charging current decreases during the fast charging.

Since the voltage of the *Kokam* battery increases constantly with the SOC over the full range this makes possible to use it in combination with temperature to estimate the SOC, and allows identifying the fully charged/discharged states by checking the specified maximum/minimum cells voltage. It means that the current must be limited when it is closed to fully charged/discharged.

In Figure 13 is presented a detail with a comparison between CC, CP and FP profiles during fast charging tests of the *Kokam* battery. The SOC is difficult to be estimated under normal operation and because of that cannot be used as an accurate measure, but only as an indicator.

The SOC will decrease/increase linearly when the battery is discharged/charged with constant power, but not when the battery is discharged/charged with constant current (and variable voltage).

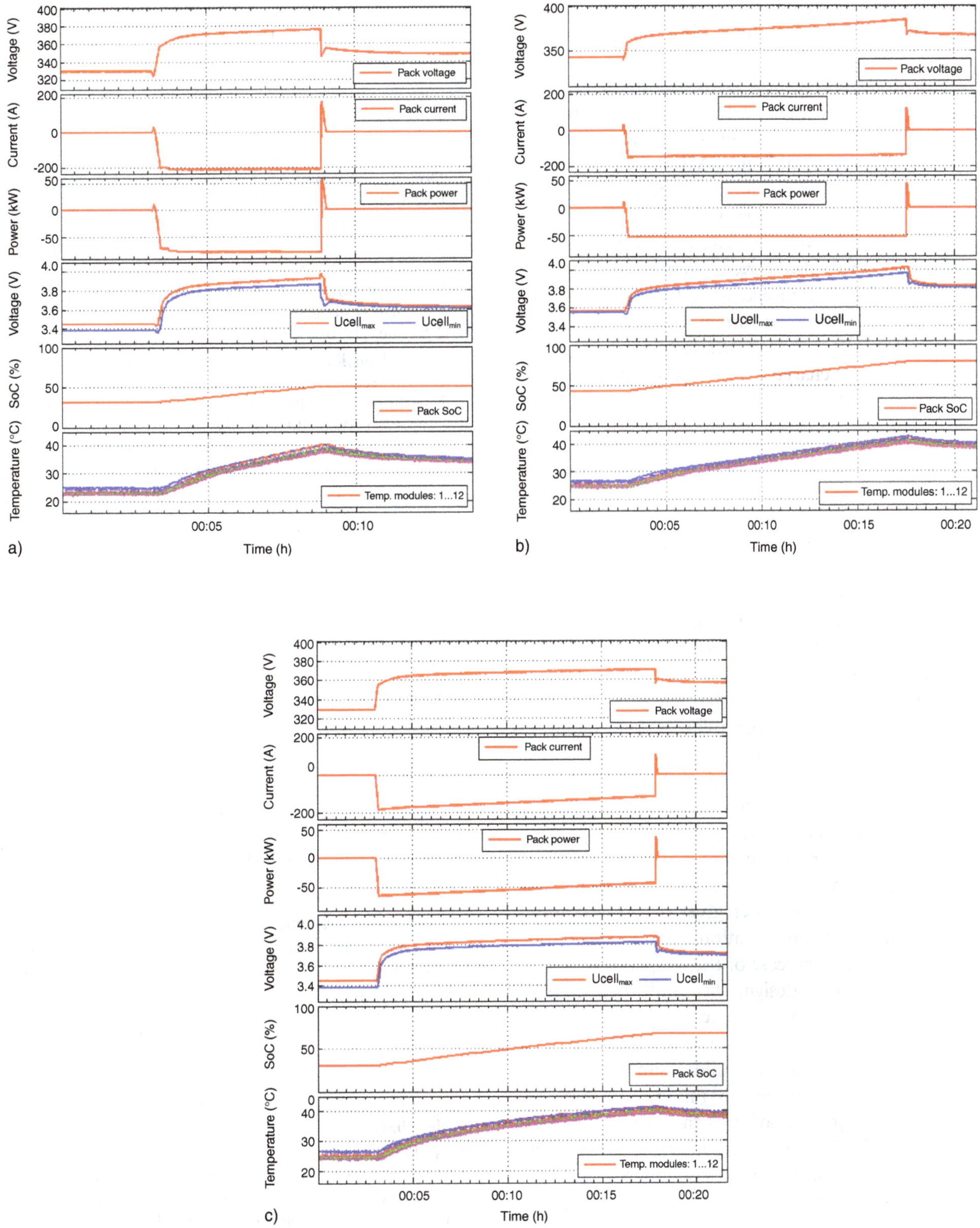

Figure 13

Details of the fast charging with a charge energy of 6.5 kWh (50% of SOC) for a) constant current at 2 C (210 A), b) constant power at 52 kW, and c) forced power from 60 kW to 44 kW.

As can also be seen in Figure 13, the charging power to the EV battery is not only limited by the capacity of the charging unit, but also by the maximum charging current, the temperature and by the SOC. As the internal cell temperature cannot be measured directly, and therefore not accurately, the charging current is gradually reduced by the Battery Management Unit (BMU) when the temperature reaches the given maximum temperature.

As a comparison between CC (*Fig. 13a*) and CP (*Fig. 13b*), when the voltage increases with the SOC, the charging power will also increase, during the CC profile, whereas during the CP profile, when the voltage increases with the SOC, the charging current decreases. During the FP profile (*Fig. 13c*) as the voltage is lowest at the beginning of the charging, the charging current is very high, as well.

CONCLUSIONS

In this paper, we have presented the development of a test platform, including three Li-ion batteries, designed for EV applications, and three associated bi-directional power converters, for testing impacts on different advanced loadings of EV batteries.

The smart charging tests performed have demonstrated that even with a simple control algorithm it is possible to provide a required charging energy within a given period of time and to provide power system services in terms of regulation of the aggregated local power, reducing the peak power and the total energy exchange of the power exchange with the national grid. Fast charging tests have been performed with constant power, constant current and forced power at charging rates corresponding to a double (2C) and triple (3C) capacity. The constant power gives the lowest impact on the grid and the main limitation of the *Kokam* battery seems to be the fast increase of the temperature.

For future battery design, this could be addressed through build-in cooling systems and/or through increased thermal capacity of the cells.

As a future work we will build a simulation model based on the tests reported in this paper which will describe the dynamic behavior of the batteries packs as a function of SOC and cell temperature to be used in a power system analysis tool to design different scenarios for a future smart grid.

ACKNOWLEDGMENTS

This work was supported in part by the Project-EDISON, No. 10580/2011-2012.

The authors would like to thanks to Per Nørgard for interesting discussion and his valuable suggestions during the tests.

REFERENCES

1 Dow L., Marshall M., Xu L., Aguero J.R., Willis H.L. (2010) A novel approach for evaluating the impact of electric vehicles on the power distribution system, *IEEE PES General Meeting 2010*, 25-29 July, Minneapolis, USA.

2 Wu X., Toshihiro M., Hisashi F. (2012) Three-phase high frequency transformer isolated AC to DC converter for EV battery quick charging, *Proc. of 7th IEEE-ECCE Conf.*, 2-5 June, Harbin, China.

3 Richardson K., Dahl-Jensen D., Elmeskov J., Hagem C., Henningsen J., Korstgård J.A., Kristensen N.B., Morthorst P.E., Olesen J.E., Wier M. (2010) Green energy - the road to a Danish energy system without fossil fuels, *Danish Commission on Climate Change Policy*.

4 Norgard P., Forero Camacho O.M., Rao N. (2012) Electric vehicles in a distributed and integrated market using sustainable energy and open network, *Riso-R-1753 Report*, Roskilde, Denmark.

5 Website: http://www.edison-net.dk/, A. Foosnæs, Electric vehicle technology, *EDISON WP1.1 report*, 2010

6 Ekanayake J., Jenkins L., Wu J., Yokoyama A. (2012) *Smart Grid: Technology and Applications*, John Willey & Sons, Chichester, England.

7 Simoes M.G., Roche R., Kyriakides E., Suryanarayanan S., Blunier B., McBee K.D., Nguyen P.H., Ribeiro P.F., Miraoui A. (2012) AQ comparison of smart grid technologies and progresses in Europe and the U.S., *IEEE Transactions on Industry Applications* **48**, 1154-1162.

8 Cha T.C., Zhao H., Wu Q., Saleem A., Østergard J. (2012) Coordinated control scheme of Battery Energy Storage System (BESS) and distributed generations (DGs) for electric distribution grid operation, *Proc. of the IEEE IECON 2012 – the 38th Annual Conference on Industrial Electronics Society*, 25-28 Oct., Montreal-Canada.

9 Sista S., Sista A. (2013) Intelligent BMS solution using AI prognostic SPA, *Lecture notes in Electrical Engineering Journal* **192**, 4, 755-764.

10 Wu C., Wan J., Zhao G. (2012) Addressing human factors in electric vehicle system design: building an integrated computational human-electric vehicle framework, *ELSEVIER Journal of Power Sources* **214**, 319-329.

11 Yan X., Li W., Gu J., Xiao X., Li W. (2012) A simulated system of battery management system to test electric vehicles charger, *Proc. 2012 IEEE IEVC Conf.*, pp.183-197.

12 He H., Xiong R., Guo H., Li S. (2012) Comparison study on the battery models used for the energy management of batteries in electric vehicles, *ELSEVIER Energy Conversion and Management Journal* **64**, 113-121.

13 Davide A. (2010) *Battery Management Systems for Large Lithium-Ion Battery Packs*, Artech House book, London. ISBN 13 978-1-60807-104-3.

14 Lee S.S., Kim H.T., Hu S.J., Cai W.W., Abell J.A. (2010) Joining technologies for automotive lithium-ion battery manufacturing - A review, *International Manufacturing Science and Engineering Conference*, 20-22 May, Michigan, USA.

15 Xin S., Guo Y.G., Wan L.J. (2012) Nanocarbon networks for advanced rechargeable lithium batteries, *Accounts for Chemical Research Journal* 45, 10, 1759-1769.

16 Sundararagavan S., Baker E. (2012) Evaluating energy storage technologies for wind power integration, *Solar Energy Journal* **86**, 9, 2707-2717.

17 Marinescu D.G., Tabacu I., Serban F., Viorel N., Tabacu S., Vieru I. (2013) Plug-in hybrid vehicle with a lithium iron phosphate battery traction type, *Lecture Notes in Electrical Engineering Journal* **191**, 3, 449-461.

18 Rao N. (2012) Interplay between EV battery and Power Grid, *EV battery forum*, Barcelona, Spain.

19 Website: http://www.trafikstyrelsen.dk/DA/Groen-Transport/Elbiler/~/media/7700C7669AB44627BCDC7CDE075ADBE3.ashx

20 *International Electrotechnical Commission IEC 61851-1 Standard* (2010) Electric vehicle conductive charging system.

21 Kenichi M. (2012) Information and communication technology and electric vehicles-Paving the way towards a smart community, *IEICE Transaction on Communication System for Electric Vehicle Charging Journal* **E95B**, 6, 1902-1910.

22 Fei X., Xianzhang L., Yebiao Z., Hongchao L., Ciurei G. (2012) A complex network-based method to evaluate smart charging and swapping network for electrical vehicles, *Power System Technology Journal* **36**, 9, 20-24.

23 Yujin L., Hak-Man K., Sanggil K., Tai-Hoon K. (2012) Vehicle-to-grid communication system for EV charging, *Integrated Computer-Aided Engineering Journal* **19**, 1, 57-65.

24 Sortomme E., El-Sharkawi M.A. (2012) Optimal scheduling of vehicle-to-grid energy and ancillary services, *IEEE Transactions on Smart Grid* **3**, No. 1.

25 Website: http://www.raesystems.com/sites/default/files/downloads/MSDS_Lithium_Ion_Battery.pdf, BYD Battery parameters, BYD Company Limited FV 50 cells

26 Website: http://www.kokam.com/product/product_pdf, Technical specifications of Kokam battery

27 AKTA (2003) The road pricing experiment in Copenhagen, *DTU Technical Report*, Roskilde, Denmark.

APPENDIX

TABLE A1
Main specification of the battery packs tested in this paper

Manufacturer	*BYD*	*Kokam*
Model	FV 50 F3DM Fe	SLPB 75
Type	Lithium-Ferro-Phosphate	Lithium Polymer
Typical application	PHEV	EV
Nr. of modules	10 (each with 10)	12 (each with 8)
Nr. of cells	100	96
Nominal voltage	320 V	350 V
Capacity	50 Ah/16 kWh	75 Ah/26 kWh
Max cell voltage	3.6 V	4.15 V
Min cell voltage	2 V	3 V
Max charging current	100(300) A	225 A
Max discharging current	300(600) A	450 A
Max temperature	55°C	45/55°C
Nr. of battery packs	2	1

17

Comparative Study on the Photocatalytic Hydrogen Production from Methanol over Cu-, Pd-, Co- and Au-Loaded TiO_2

P.P.C. Udani and M. Rønning*

Department of Chemical Engineering, Norwegian University of Science and Technology, NO-7491 Trondheim - Norway
e-mail: charitha.udani@ntnu.no - magnus.ronning@ntnu.no

* Corresponding author

Abstract — *Photocatalytic hydrogen production from a methanol-water solution was investigated in a semi-continuous reactor over different metal-loaded TiO_2 catalysts under UltraViolet (UV) light irradiation. The catalysts were mainly prepared by the incipient wetness impregnation method by varying the metal weight ratio in the range of 1-10 wt%. The effects of metal loading and H_2 pre-treatment on the photocatalytic activity were investigated. In addition, the activity of the catalysts was also compared with a reference Au-TiO_2 catalyst from the World Gold Council (WGC). The photocatalysts were characterized by using X-Ray Diffraction (XRD) and N_2 physisorption before and after the activity measurements. The photocatalytic activity decreased in the order of Pd > Au > Cu > Co in the comparative study of Cu-TiO_2, Co-TiO_2, Au-TiO_2 and Pd-TiO_2. Optimum hydrogen evolution was achieved with 5 wt% Pd-TiO_2 and 5 wt% Cu-TiO_2.*

Résumé — Étude comparative de production d'hydrogène par photocatalyse à partir de méthanol et à l'aide de différentes phases actives (Cu, Pd, Co et Au) supportées sur TiO_2 — La production d'hydrogène assistée par photocatalyse à partir d'une solution diluée de méthanol a été réalisée dans un réacteur semi-continu avec l'aide de différents catalyseurs supportés par TiO_2 et sous irradiation UV. Les catalyseurs ont été préparés par imprégnation par capillarité en faisant varier le pourcentage massique du métal choisi entre 1 et 10 %. L'influence de ce pourcentage, le prétraitement sous H_2 et le matériau semi-conducteur utilisé pour la photocatalyse ont été étudiés. D'autres expériences ont été menées concernant l'activité catalytique des matériaux sélectionnés et les résultats ont été comparés avec ceux obtenus avec un catalyseur de référence provenant du *World Gold Council* (WGC), or supporté sur TiO_2. Les photocatalyseurs ont également été caractérisés par Diffraction des Rayons X (DRX) et physisorption de l'azote, avant et après réaction. Il a été observé que l'activité photocatalytique diminuait pour les phases actives dans l'ordre décroissant suivant: Pd > Au > Cu > Co ; résultats obtenus lors de l'étude comparative entre les catalyseurs Cu-TiO_2, Co-TiO_2, Au-TiO_2 et Pd-TiO_2. La production optimale d'hydrogène a été obtenue avec les catalyseurs 5 % massique Pd-TiO_2 et 5 % massique Cu-TiO_2.

INTRODUCTION

Hydrogen is often regarded as a potential future energy carrier, which can be easily obtained from methanol *via* several processes. Compared with thermochemical processes, photocatalytic reforming is an effective approach to produce hydrogen under ambient conditions and using sunlight [1].

Some of the previous studies are summarized in Table 1, [2–10]. In most of the cases listed in Table 1, TiO_2 is the most widely used semiconductor for various photocatalytic applications due to its relatively low cost, abundance, stability and low toxicity.

Noble metal-loaded TiO_2 has extensively been proved to be a good photocatalyst for hydrogen production, mainly because of its ability to enhance photoproduced electron–hole pair separation and photoinduced reduction processes [4, 11, 12]. However, non-precious metal-loaded TiO_2 may be cost-effective compared with noble metal-loaded TiO_2 since it also shows significant photocatalytic activity in hydrogen generation [13, 14].

In this study, we have investigated photoreforming of methanol (Eq. 1) with the use of different metal-loaded TiO_2 photocatalysts in a semi-continuous reactor set-up under UV light irradiation:

$$CH_3OH + H_2O \underset{\text{photocatalyst}}{\overset{E_{bg} \leq h\nu}{\longrightarrow}} CO_2 + 3\,H_2 \qquad (1)$$

The effects of several parameters on the hydrogen production were studied, including metal loading, H_2 pre-treatment and semiconductor material.

1 EXPERIMENTAL

1.1 Catalyst Preparation

All catalysts were prepared by using the incipient wetness impregnation method and TiO_2 (P25-Degussa) was used as the photoactive semiconductor. The appropriate mass of metal precursor was dissolved in deionized water to give the desired metal concentration. The volume added to the catalyst was just sufficient to fill the pores of the support. The impregnated support was then dried in an oven at 110°C overnight, and was calcined in a muffle furnace (*Nabertherm*- L15/12/P320). The calcined catalysts were lightly ground and sieved to < 53 μm before use. The metal content of the final catalyst was determined by the concentration of the metal precursor used in the impregnation, and was denoted by the weight percentage of metal in the catalyst: $Wt\% = (wt_X/wt_{X+TiO2}) \times 100$. The Au-$TiO_2$ catalyst reported in this study is the reference catalyst from the World Gold Council (WGC) prepared by a deposition precipitation method [15]. All the metal precursors and calcination conditions are summarized in Table 2, [16, 17].

1.2 Photocatalytic Activity Measurements

As shown in Figure 1, the photocatalytic activity measurements were performed in a custom made cylindrical quartz reactor (volume ~1.2 L) and kept inside the photoreaction chamber (*Luzchem* CCP- 4V). The chamber was equipped with 14 UVC lamps (*Luzchem* LZC-UVC: 15 W, 254 nm) and a magnetic stirrer. During irradiation, the temperature inside the photoreaction chamber was continuously monitored and always kept within 3°C above room temperature. Typically, 0.25 g powder of photocatalyst was dispersed in the reactor containing 500 mL of 50 vol.% methanol aqueous solution. Prior to the irradiation, the photocatalyst suspension in methanol solution was deaerated thoroughly for 30 min by a 100 mL/min flow of Ar. The reactor effluent was analyzed on-line with a Thermal Conductivity Detector (TCD) micro gas chromatograph (*Agilent Technologies* 3000) with a molecular sieve column (MolSieve 5 A). Ar was used as a carrier gas and the GC response was first calibrated by introducing a known flow of hydrogen into the system. All the gas lines were heated to prevent condensation of methanol and water.

In some experiments, the catalysts were reduced *in situ* prior to activity tests in a stream of 10 vol.% H_2/Ar mixture flowing at 100 mL/min at ambient temperature for 1 h to investigate the effect of H_2 pre-treatment.

1.3 Catalyst Characterization

X-Ray powder Diffraction (XRD) patterns were obtained using a *Bruker* D8 Advance X-Ray Diffractometer with Cu–Kα radiation (Kα = 1.54 Å) in the scanning angle (2θ) range of 20-80°. The specific surface areas of the catalysts were measured by N_2 physisorption using the Brunauer–Emmett–Teller (BET) method (*Micromeritics* Tristar II 3020). To remove traces of water and impurities from the catalyst surface, all samples were degassed (*Micromeritics* VacPrep 061) overnight at 200°C before the measurement.

2 RESULTS AND DISCUSSION

2.1 Photocatalytic Activity

The comparison of photocatalytic activity over different metal-loaded catalysts is shown in Figure 2. The photocatalytic activity decreased in the order of Pd > Au > Cu. However, after 10 h of reaction, the hydrogen evolution changed to Au > Pd > Cu. This is in agreement with the results of Sreethawong and Yoshikawa [18] on similar

TABLE 1
Previous studies on photocatalyst development for hydrogen production

Ref No.	Reactants	Catalyst and preparation method	Light source	Reaction conditions	Experiment	Results obtained
[2]	Methanol (liquid and gas phases)	TiO_2 and 1% Au/TiO_2 Flame spray pyrolysis	Iron halogenide Hg arc lamp (330-450 nm)	Plexiglass reactor 55±5°C	The performance of the suspension and vapor phase reactors was investigated	Intermediate product: formaldehyde Final products: H_2 and CO_2. Optimal results: A 30% higher reaction rate was attained with the vapor phase reactor
[3]	Ethanol and glycerol solutions	CuO_x-TiO_2 Water-in-oil microemulsion synthesis Wet impregnation (Cu loading was 2.5% by weight)	UV	20°C Pyrex batch reactor (250 mL) Cat. wt: 500 mg	The effect of the catalyst preparation method and the stability of the CuO_x-TiO_2 catalysts were studied with respect to carbonaceous deposits and copper leaching	The catalyst prepared by the embedded method showed the best performance
[4]	Methanol	Au/ TiO_2 Deposition-precipitation method	300 W Xe lamp	Pyrex reactor Cat. wt: 0.3 g	The effects of: 1. Au particle size, 2. the initial methanol concentration, 3. the initial pH value, on H_2 production rate and CO selectivity	– The H_2 production rate increased with decreasing Au particle size from 10 to smaller than 3 nm – The CO concentration decreased with reducing Au particle size – The rate varied as a function of the initial methanol concentration as a Langmuir-type curve – The molar ratio of CO/H_2 showed a weak increasing trend with increasing methanol concentration – Both acidic and basic solutions resulted in a considerable decrease in the rate of H_2 production and CO concentration
[5]	Methanol	Metal-titania (Pd, Au, Ir, Pt, Ru, Rh, Ni) Incipient wetness	400 W Xe arc lamp	Ambient temperature; Pyrex flask (100 mL) Cat. wt: 0.2 g	The model for the reaction was presented	– Pd was the best when comparing similar loading – In terms of molar loading of metal, Pt was a little better than Pd
[6]	Methane (methane hydrates)	Tungsten oxide doped with Cu, Pt, La/Cu	Hg vapor quartz lamp	1 MPa; 100°C; Quartz reactor (1 L)	The methanol production with various promoted tungsten oxide catalysts was investigated	Final products: methanol and H_2 By-products: ethane, oxygen, formic acid and CO_2 The catalysts doped with La showed the highest methane conversion and methanol yield

(continued)

TABLE 1 *(continued)*

Ref No.	Reactants	Catalyst and preparation method	Light source	Reaction conditions	Experiment	Results obtained
[7]	Methanol/ water solution	$CuO-TiO_2$ Impregnation method	400 W Hg lamp	Pyrex reactor (1.45 L) Cat. wt: 1 g	The effects of: 1. Cu content 2. Initial methanol concentration on H_2 generation activity The stability and recovery of catalysts over extended time was investigated	– The optimum Cu content was 9.1 mol% – The H_2 generation rate could be increased by increasing methanol concentration – The H_2 generation rate dropped under long irradiation time mainly due to the accumulation of byproducts, reduction of CuO and Cu leaching
[8]	Methane and water	Pt/TiO_2 Impregnation method and photo-deposition method	300 W Xe lamp (both UV and visible light)	50°C; 1 atm; Fixed bed Quartz flow reactor; Cat. wt: 0.5 g Cat. size: 400-600 μm	The effects of 1. Photocatalyst structure 2. Catalyst preparation method 3. Reaction conditions (CH_4 and H_2O concentration, light intensity) on the H_2 production rate were studied The reaction mechanism was also investigated	By-products: C_2H_6 and CO Final products: H_2 and CO_2 Optimal results: – the catalyst prepared by the photodeposition method showed much higher activity than the impregnation method; – the optimum concentration of CH_4 to sustain the high activity with moderate intermediates was around 50%; – the H_2 production rate was almost constant for moderate H_2O concentration (0.28-15%). With increasing light intensity, the H_2 production rate. increased first, and then the increase became small
[9]	Methanol and water	Au/TiO_2 Flame spray pyrolysis	Iron halogenide mercury arc lamp (330-450 nm)	45-mL cylindrical quartz vessel Reaction medium: Pure water or 6 vol.% methanol aqueous solution	The addition of methanol and the deposition of gold particles on titanium dioxide were investigated	The highest rate of hydrogen production was attained with gold-modified titanium dioxide Addition of methanol to the reactant suspension led to an increase in the hydrogen production rate
[10]	Methanol, ethanol, glycol, glycerin/ water	Lanthanide-doped titania (Gd, Eu, Yb, Ho) Incipient wetness method; cat. wt: 0.1 g	300 W Ultra-Vitalux (*Osram*) lamps	Reactor: Pyrex round-bottom flask (250 cc) Reaction medium: water and alcohol (50:1)	A series of platinized lanthanide-doped titania were tested for photocatalytic water splitting to find out the effect of lanthanide oxide type. Methanol, ethanol, glycol or glycerin were used to investigate the effect of sacrificial agents	– The best performance was obtained for the catalyst containing 0.5 mol% of Gd, Eu, Yb or Ho oxides as the dopant; – No oxygen was observed in the reaction products; – The best sacrificial agents were methanol and ethanol acting as the oxygen scavenger, and the worst was glucose or fructose.

catalysts. No significant hydrogen evolution is detected over the $Co-TiO_2$ catalysts under UV irradiation.

H_2 pre-treatment lead to a detrimental effect on the hydrogen evolution, as shown in Figure 3. This is in agreement with the results reported by Wu *et al.* [19] and Oros-Ruiz *et al.* [12] for TiO_2-supported metal photocatalysts, although the H_2 pre-treatment was performed at much higher temperatures in the previous reports.

TABLE 2

Metal precursors and calcination conditions for prepared catalysts

Catalyst	Metal precursor	Calcination conditions
Cu-TiO_2	$Cu(NO_3)_2.3H_2O$	200°C for 4 h
Pd-TiO_2	$PdCl_2$	500°C for 2 h [16]
Co-TiO_2	$Co(NO_3)_2.6H_2O$	400°C for 4 h [17]

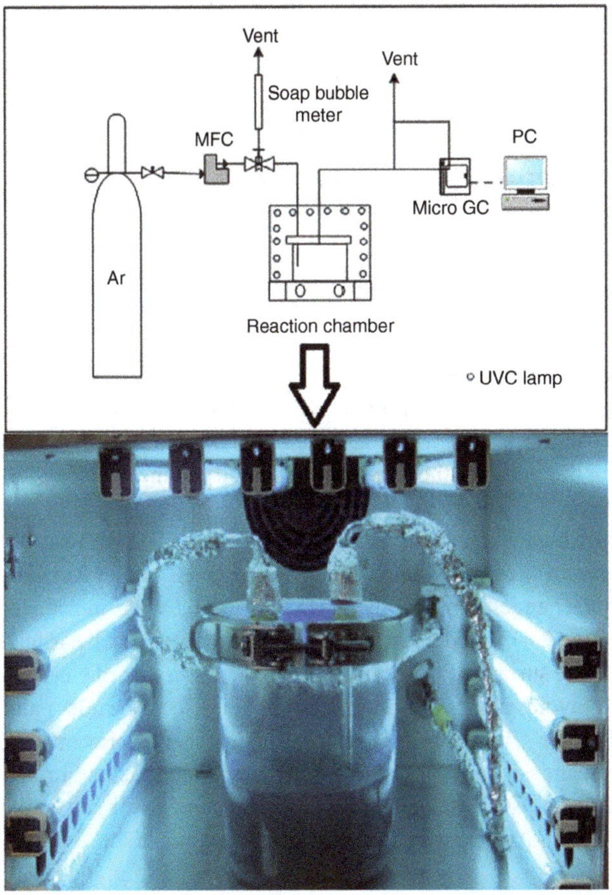

Figure 1

The photoreactor system utilized for the activity measurements.

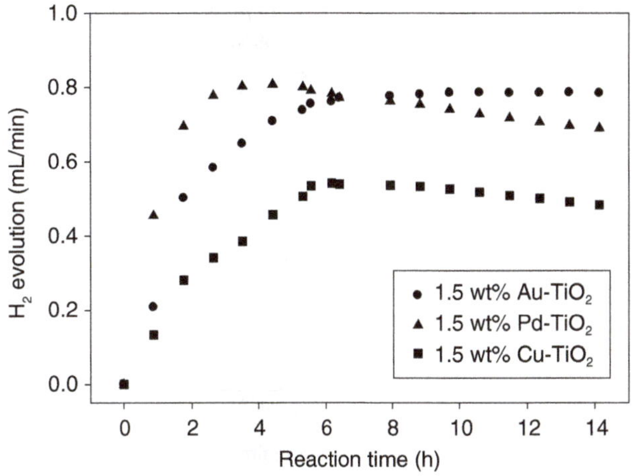

Figure 2

Comparison of photocatalytic activity over different metal-loaded catalysts. Reactor composition: 500 mL of 50 vol.% methanol aqueous solution. Catalyst weight: 0.25 g.

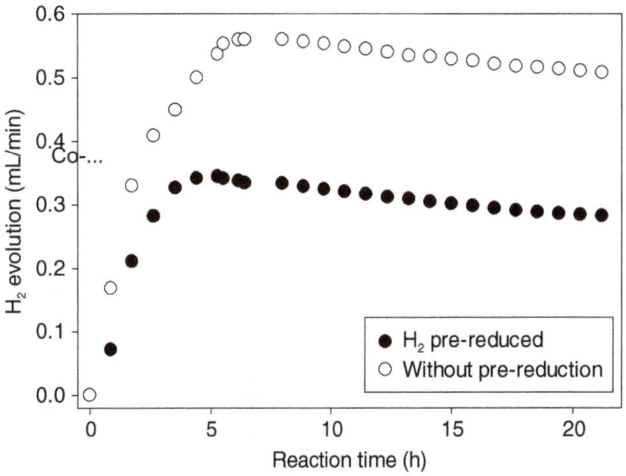

Figure 3

Effect of pre-reduction treatment in hydrogen over the 10 wt% Cu-TiO_2 catalysts. Feed composition: 500 mL of 50 vol.% methanol solution. Catalyst weight: 0.25 g.

The evolution of photocatalytic hydrogen over a 10 wt% Cu-TiO_2 was studied in the conditions of the initial methanol concentration being 50 vol.% at room temperature for 24 hours, as shown in Figure 4. The hydrogen evolution increases significantly at the beginning of the reaction and after approximately 5 hours, the hydrogen evolution reaches a maximum level and after that a small decrease in hydrogen evolution was observed. The hydrogen evolution can be recovered by refilling the reactor with fresh methanol solution to the initial concentration. The results indicate that the Cu-TiO_2 photocatalyst is a relatively stable catalyst for hydrogen production from methanol and that the decrease in hydrogen evolution is mainly due to the consumption of methanol from the solution.

2.2 Effect of Metal Loading

The effect of metal loading on the hydrogen evolution is shown in Figure 5. No hydrogen evolution is detected with

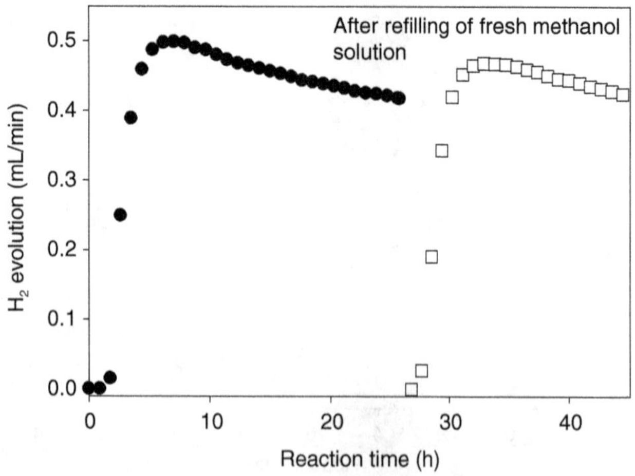

Figure 4

H_2 evolution *versus* time on stream in photoreforming of methanol. Feed composition: 500 mL of 50 vol.% methanol solution. Catalyst weight: 0.25 g.

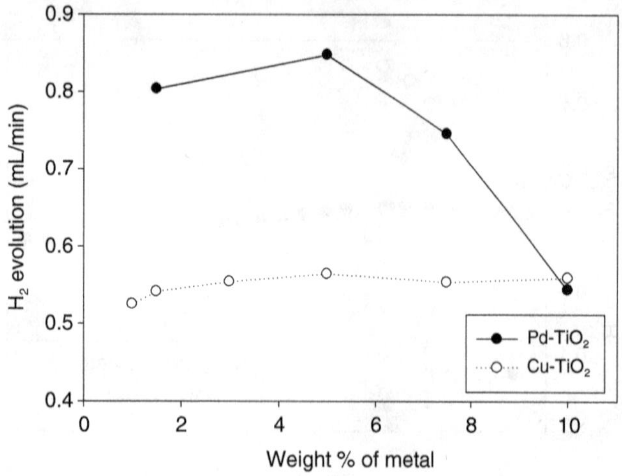

Figure 5

Effect of metal loading on hydrogen evolution. Reactor composition: 500 mL of 50 vol.% methanol aqueous solution. Catalyst weight: 0.25 g.

pure TiO_2 (P25-Degussa). The photocatalytic activity increased with Pd loading up to 5 wt% and then decreased with a further increase in palladium loading. In the case of copper loading, the change in the hydrogen evolution with copper loading is very small compared with Pd-TiO_2 catalysts. However, photocatalytic activity increased with copper loading up to 5 wt% and then showed a nearly constant hydrogen evolution with a further increase in copper loading. The photocatalytic activity decreased (Pd) or remained constant (Cu) above approximately 5 wt% metal loading.

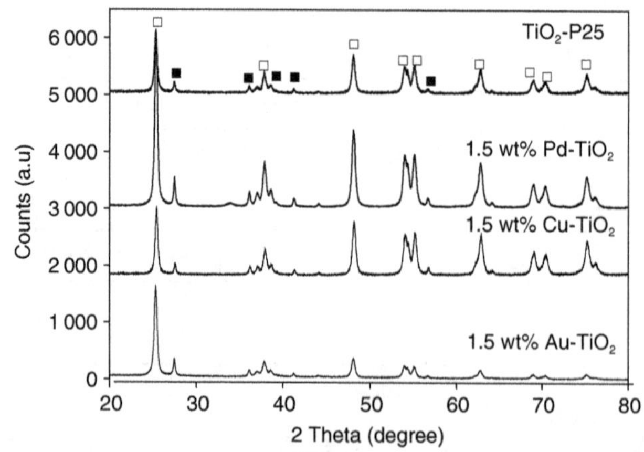

Figure 6

XRD patterns of different metal-loaded TiO_2 catalysts. Anatase (□) and Rutile (■).

The optimal loadings for Cu and Pd are somewhat higher than the reported values from Sreethawong and Yoshikawa [18], which are in the range 1-2 wt%. The reason for this could be the preparation method and metal dispersion. They also observed a relatively large change in the BET surface area when changing the metal loading. The BET surface area of the impregnated catalysts presented in this work did not change significantly when the metals were introduced (Sect. 2.4). The presence of a certain amount of metal particles on the surface of the semiconductor acts as a sink of electrons to avoid rapid recombination of the electron-hole pairs. Further increase in metal loading will prevent the transmission of light to the TiO_2 semiconductor surface. The Co-TiO_2 catalysts showed no detectable hydrogen evolution, and are hence not included in the comparison.

2.3 X-Ray Diffraction (XRD) Analysis

In the XRD patterns of the fresh catalysts, no detectable diffraction peaks of CuO, PdO or metallic Au crystallites could be detected for metal loadings of 1.5 wt%, as shown in Figure 6. This is most likely due to the relatively low content or because of the high dispersion of the particles on the TiO_2 surface.

No significant changes in the characteristic peaks of TiO_2 (P25) were observed for the fresh catalysts, indicating that there was no phase transformation in the TiO_2 support during the preparation. This is in agreement with the results of Raj and Viswanathan [20], where they concluded that the effect of rutilation of TiO_2 (P25) is not significant at temperatures below 500°C.

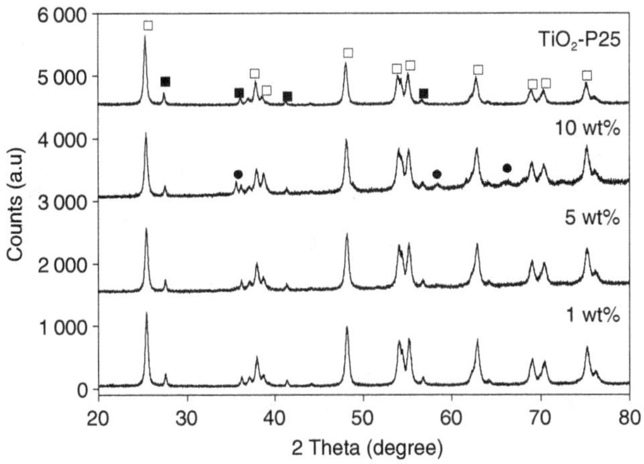

Figure 7

XRD patterns of fresh Cu-TiO_2 catalysts. Anatase (□), Rutile (■) and CuO (●).

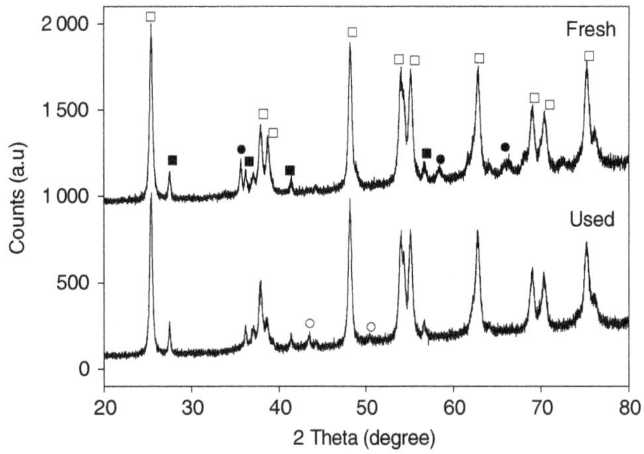

Figure 8

XRD patterns of fresh and used 10 wt% Cu-TiO_2 catalysts. Anatase (□), Rutile (■), Cu (○) and CuO (●).

Figure 7 shows XRD patterns of fresh Cu-TiO_2 catalysts. When the copper loading is more than 5 wt%, the XRD peaks corresponding to crystalline CuO were observed at 2θ of 35, 58 and 66° and the intensity increased with increasing copper loading, in agreement with the literature [7, 18]. CuO crystallite size analysis from XRD line broadening is difficult because of overlapping peaks between CuO and TiO_2. However, for the 10 wt% $CuTiO_2$ sample, the average crystallite size of CuO was estimated to be 32 nm from the Full Width at Half Maximum (FWHM) of the XRD main peak at $2\theta = 35°$ using the Scherrer equation.

The copper in all the fresh catalysts is present as CuO. However, a change in the state of the copper after reaction is observed in the diffractograms in Figure 8. The peaks corresponding to metallic copper are clearly visible after reaction. This indicates that Cu is reduced in the active state.

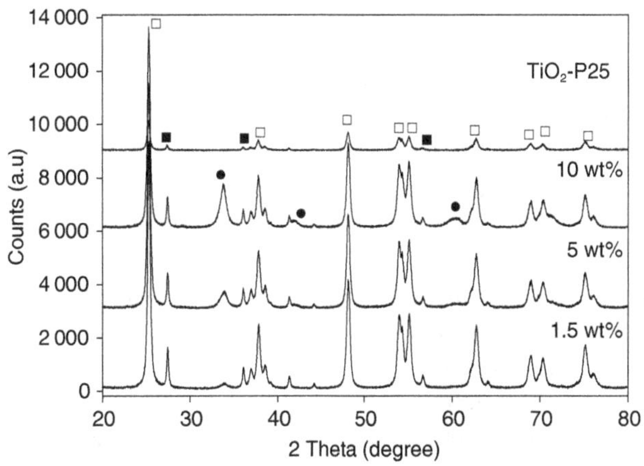

Figure 9

XRD patterns of fresh Pd-TiO_2 catalysts. Anatase (□), Rutile (■) and PdO (●).

As shown in Figure 9, the XRD peaks corresponding to crystalline PdO were observed at 2θ = 34 and 60, and the intensity increased with increasing palladium loading. The average crystallite sizes of PdO were calculated from the FWHM of the XRD main peak at $2\theta = 33°$ using the Scherrer equation. The particle size appears to be independent of the metal loading and is estimated to be 10 nm for all the samples.

2.4 BET Surface Areas of the Catalysts

The BET surface areas of fresh and used catalysts are presented in Table 3. The photocatalysts did not show any significant change in the surface area after being exposed to the reaction conditions. Bin Xu *et al.* [21] also observed a small change in the BET surface area with different amounts of CuO loading and reported that the loadings of CuO do not have a significant effect on the surface morphology of the support. As discussed in the previous section, photocatalytic activity increased with metal loading up to 5 wt% and then showed a decrease in hydrogen evolution or nearly constant hydrogen evolution with a further increase in metal loading. However, the BET surface area of the catalyst with high metal loading is only slightly lower than for the lower metal loading catalysts, indicating that the BET surface area may not be an important parameter in the photocatalytic activity. The BET surface areas of the 1.5 wt% catalysts are presented in Table 3. The values for the catalysts with other metal

TABLE 3

The BET surface areas of fresh and used catalysts (1.5 wt%)

Catalyst	BET surface area (m^2/g)	
	Fresh catalysts	Used catalysts*
Au-TiO_2	52	-
Pd-TiO_2	47	45
Cu-TiO_2	49	47

* After exposure to reaction conditions for 15 h.

loadings are not significantly different (all close to 50 m^2/g), and are hence not included in the table.

CONCLUSIONS

The photocatalytic activity decreased in the order of Pd > Au > Cu according to the results in the comparative study of Cu-TiO_2, Au-TiO_2 and Pd-TiO_2. Hydrogen evolution is not detected with the Co-TiO_2 catalyst. Optimum hydrogen evolution was achieved with 5 wt% Pd-TiO_2 and 5 wt% Cu-TiO_2 in the experimental conditions applied in this study. The BET surface area is not an important parameter for the photocatalytic at the applied conditions. No significant changes in the characteristic XRD peaks of TiO_2-P25 were observed for the fresh catalysts, indicating that there was no phase transformation in the TiO_2 support during the preparation. Finally, it can be concluded that methanol photoreforming may provide a potential low-cost method for the production of hydrogen from renewable sources and that TiO_2 loaded with non-precious metals such as copper is also a promising catalyst for producing hydrogen from methanol because of its photoactivity, stability and cost when compared with noble metal-loaded photocatalysts.

ACKNOWLEDGMENTS

The Norwegian University of Science and Technology (NTNU) is acknowledged for financial support through strategic funding.

REFERENCES

1 Rossetti I. (2012) Hydrogen Production by Photoreforming of Renewable Substrates, *ISRN Chemical Engineering* **2012**, 21.

2 Chiarello G.L., Forni L., Selli E. (2009) Photocatalytic hydrogen production by liquid- and gas-phase reforming of CH_3OH over flame-made TiO_2 and Au/TiO_2, *Catalysis Today* **144**, 69-74.

3 Gombac V., Sordelli L., Montini T., Delgado J.J., Adamski A., Adami G., Cargnello M., Bernal S., Fornasiero P. (2009) CuO_x–TiO_2 Photocatalysts for H_2 Production from Ethanol and Glycerol Solutions, *The Journal of Physical Chemistry A* **114**, 3916-3925.

4 Wu G., Chen T., Su W., Zhou G., Zong X., Lei Z., Li C. (2008) H_2 production with ultra-low CO selectivity *via* photocatalytic reforming of methanol on Au/TiO_2 catalyst, *International Journal of Hydrogen Energy* **33**, 1243-1251.

5 Al-Mazroai L.S., Bowker M., Davies P., Dickinson A., Greaves J., James D., Millard L. (2007) The photocatalytic reforming of methanol, *Catalysis Today* **122**, 46-50.

6 Taylor C. (2005) Photocatalytic conversion of methane contained in methane hydrates, *Top Catal.* **32**, 179-184.

7 Xu S., Sun D.D. (2009) Significant improvement of photocatalytic hydrogen generation rate over TiO_2 with deposited CuO, *International Journal of Hydrogen Energy* **34**, 6096-6104.

8 Yoshida H., Hirao K., Nishimoto J.-i., Shimura K., Kato S., Itoh H., Hattori T. (2008) Hydrogen Production from Methane and Water on Platinum Loaded Titanium Oxide Photocatalysts, *The Journal of Physical Chemistry C* **112**, 5542-5551.

9 Chiarello G.L., Selli E., Forni L. (2008) Photocatalytic hydrogen production over flame spray pyrolysis-synthesised TiO_2 and Au/TiO_2, *Applied Catalysis B: Environmental* **84**, 332-339.

10 Zalas M., Laniecki M. (2005) Photocatalytic hydrogen generation over lanthanides-doped titania, *Solar Energy Materials and Solar Cells* **89**, 287-296.

11 Lalitha K., Reddy J.K., Phanikrishna Sharma M.V., Kumari V.D., Subrahmanyam M. (2010) Continuous hydrogen production activity over finely dispersed Ag_2O/TiO_2 catalysts from methanol:water mixtures under solar irradiation: A structure–activity correlation, *International Journal of Hydrogen Energy* **35**, 3991-4001.

12 Oros-Ruiz S., Zanella R., López R., Hernández-Gordillo A., Gómez R. (2013) Photocatalytic hydrogen production by water/methanol decomposition using Au/TiO_2 prepared by deposition–precipitation with urea, *Journal of Hazardous Materials* **263**, 1, 2-10.

13 Chen W.-T., Jovic V., Sun-Waterhouse D., Idriss H., Waterhouse G.I.N. (2013) The role of CuO in promoting photocatalytic hydrogen production over TiO_2, *International Journal of Hydrogen Energy* **38**, 15036-15048.

14 Onsuratoom S., Puangpetch T., Chavadej S. (2011) Comparative investigation of hydrogen production over Ag-, Ni-, and Cu-loaded mesoporous-assembled TiO_2–ZrO_2 mixed oxide nanocrystal photocatalysts, *Chemical Engineering Journal* **173**, 667-675.

15 Holliday S.T.R. (2003) Gold Reference Catalysts, *Gold Bulletin* **36**, 1, 24.

16 Dickinson A., James D., Perkins N., Cassidy T., Bowker M. (1999) The photocatalytic reforming of methanol, *Journal of Molecular Catalysis A: Chemical* **146**, 211-221.

17 Takanabe K., Nagaoka K., Nariai K., Aika K.-i. (2005) Influence of reduction temperature on the catalytic behavior of Co/TiO_2 catalysts for CH_4/CO_2 reforming and its relation with titania bulk crystal structure, *Journal of Catalysis* **230**, 75-85.

18 Sreethawong T., Yoshikawa S. (2005) Comparative investigation on photocatalytic hydrogen evolution over Cu-, Pd-, and Au-loaded mesoporous TiO_2 photocatalysts, *Catalysis Communications* **6**, 661-668.

19 Wu N.-L., Lee M.-S., Pon Z.-J., Hsu J.-Z. (2004) Effect of calcination atmosphere on TiO_2 photocatalysis in hydrogen production from methanol/water solution, *Journal of Photochemistry and Photobiology A: Chemistry* **163**, 277-280.

20 Raj K.J.A., Viswanathan B. (2009) Effect of surface area pore volume and particle size of P25 titania on the phase transformation of anatase to rutile, *Indian J. Chem. A* **48**, 1378-1382.

21 Xu B., Dong L., Chen Y. (1998) Influence of CuO loading on dispersion and reduction behavior of CuO/TiO_2 (anatase) system, *Journal of the Chemical Society Faraday Transactions* **94**, 1905-1909.

18

New Fuzzy Model for Risk Assessment Based on Different Types of Consequences

K. Karimpour, R. Zarghami*, M.A. Moosavian and H. Bahmanyar

Process Design and Simulation Research Center, Oil and Gas Processing Centre of Excellence, School of Chemical Engineering, College of Engineering, University of Tehran, P.O. Box 11155/4563, Tehran - Iran
e-mail: rzarghami@ut.ac.ir

* Corresponding author

Abstract — *A new risk assessment methodology by using fuzzy logic is proposed in this paper. The new Fuzzy Inference System (FIS) was established by the Mamdani algorithm based on different consequences of an incident. A combination of two FIS formed the proposed fuzzy method. Human knowledge and brainstorming were the devices for making the rules and interdependencies between variables in the new model. Different types of consequences and effective parameters were considered as inputs for the first fuzzy inference system. The final consequence was the preliminary result of the first inference model. It added to the probability of failures, as inputs of the second inference model. The result of the second inference model was the risk factor, which was considered as the final output of the proposed new fuzzy model. This model makes risk assessment more convenient in the absence of suitable data. In addition, decision-making will be easier, since its results are more understandable than the results of classical methods. A case study and a comparison between the classic method and the new fuzzy model illustrated that the results of the proposed model are more accurate, reliable and convenient for use in decision-making.*

Résumé — Nouveau modèle à logique floue pour une évaluation de risque basée sur différents types de conséquences — Une nouvelle méthode d'évaluation des risques en utilisant une logique floue est proposée dans le présent article. Le nouveau système d'inférence flou a été établi en utilisant l'algorithme de Mamdani basé sur les différentes conséquences d'un incident. Une combinaison de deux systèmes d'inférence flous a donné naissance à la méthode floue proposée. Les connaissances humaines et un *brainstorming* ont été les outils qui ont permis d'établir les règles et interdépendances entre des variables du nouveau modèle. Différents types de conséquences et de paramètres effectifs ont été pris en considération comme entrées pour le premier système d'inférence flou. La conséquence finale qui est le résultat préliminaire du premier modèle d'inférence, est combinée avec la probabilité de risques, en tant qu'entrées du second modèle d'inférence. Cette combinaison conduit à considérer le facteur de risque comme le résultat final du nouveau modèle flou proposé. Ce modèle rend l'évaluation des risques plus facile, en l'absence de données adaptées. En outre, la prise de décision sera plus simple, puisque ses résultats sont plus compréhensibles que les résultats des méthodes classiques. Une étude de cas et une comparaison entre la méthode classique et le nouveau modèle flou ont démontré que les résultats du modèle proposé étaient plus précis, plus fiables et plus faciles à utiliser dans la prise de décision.

ABBREVIATIONS

DNV	Det Norske Veritas
FIS	Fuzzy Inference System
OGP	International Association of Oil and Gas Production
OREDA	Offshore Reliability Data
RMP	Risk Management Process
BLEVE	Boiling Liquid Expanding Vapor Explosion

NOMENCLATURE

A_n	Hole area
C_d	Release hole coefficient of discharge
CA_{inj}^{flam}	Final probability weighted personnel injury flammable consequence area
CA_{cmd}^{flam}	Final probability weighted component damage flammable consequence area
CA_{inj}^{AINL}	Personnel injury flammable consequence area for releases that are not likely to auto-ignition
CA_{cmd}^{AINL}	Component damage flammable consequence area for releases that are not likely to auto-ignition
CA_{cmd}^{Tox}	Final probability weighted personnel injury toxic consequence area
CA_{inj}^{Leak}	Personnel injury non-flammable, non-toxic consequence area for steam and acid leaks
CA_{cmd}^{Leak}	Component damage non-flammable, non-toxic consequence area for steam and acid leaks
CA_{inj}^{nfnt}	Final probability weighted personnel injury consequence area for non-flammable, non-toxic releases such as steam or acids
CA_{cmd}^{nfnt}	Final probability weighted component damage consequence area for non-flammable, non-toxic releases such as steam or acids
CA_{inj}^{INST}	Personnel injury flammable consequence area for an instantaneous release
CA_{cmd}^{INST}	Component damage flammable consequence area for an instantaneous release
CA_{inj}^{CONT}	Personnel injury flammable consequence area for a continuous release
CA_{cmd}^{CONT}	Component damage flammable consequence area for a continuous release
$fact^{AIT}$	Auto-ignition temperature consequence area blending factor
$fact^{IC}$	Continuous/instantaneous consequence area blending factor determined for each release hole size
gff	Generic failure frequency
g_c	Gravitational constant
W_n	Theoretical release rate
$K_{v,n}$	Liquid flow viscosity correction factor
P_{atm}	Atmospheric pressure
P_s	Storage or normal operating pressure
ρ_L	Liquid density in storage or normal operating conditions

INTRODUCTION

There are many different definitions of risk in the literature, based on various aspects. The simplest definition of risk was introduced by Hertz and Thomas (1994): a barrier for a successful operation. Jafari (2001) defined risk as the possibility of losses in a project. More complete and deeper definitions were provided by Varnes (1984), Perry and Hayes (1985), Chapman and Ward (1997), Baloi and Price (2003), Pokoradi (2002), Mark *et al.* (2004), Kirchsteiger (2005) and Darbra *et al.* (2008a). In spite of various definitions of risk, there are some common features between different sources of risks, which were highlighted by Chia (2006) and Nieto-Morote and Ruz-Vila (2011) as follows:

- risk is a feature concern and event;
- there is no confidence in occurrence of a risk;
- risk shall be an uncertain condition or event that can affect projects' conditions, operations, or goals, if it occurs;
- the probability of a risk must be between 0-100%;
- the impact, consequences, losses, or any other effects of risk must be unplanned.

Hence, risk can be introduced as the existence of inherent uncertainties in systems that affect projects' goals and decrease success due to unpredictable and unplanned probabilities and possibilities. This definition is very useful for being applied in risk assessment by concentrating on uncertainty and probability. In addition, it makes risk a probable factor to be calculated by the study of probabilities and possibilities. In fact, this definition showed that probability and possibility were the basic steps of risk introduction.

To reduce unplanned and undesirable effects of risk during projects and control its level, risk must be managed (Naderi, 2008) by a Risk Management Process (RMP). Like risk sources and aspects, risk management processes consist of different steps, activities and directions. Commonly, risk management includes risk assessment, risk mitigation, risk acceptance and risk communication (API, 2008). Risk assessment is the most important step in risk management (Darbra *et al.*, 2008a), which includes hazard identification, frequency analysis and consequence evaluation. There are many probability-based methods by which risk is assessed but new techniques based on possibility methods were developed since mathematical relations and parameters for risk assessment were very difficult to model (Lees, 2001).

Pokoradi (2002) defined the preliminary basis of risk assessments by possibility methods and fuzzy logic. Bowles and Pelaez (1995) evolved the use of fuzzy arithmetic and linguistic variables in risk assessments while characterizing the system reliability. For a period, nuclear engineering systems were assessed through Fuzzy Inference System (FIS), particularly by Guimaraes and Lapa (2007). Karimi and Hüllermeier (2007) suggested a modular framework for risk assessment by fuzzy logic. They used possibility-probability distribution as a new approach for analyzing risks. Markowski and Mannan (2008, 2009a) and Markowski *et al.* (2009b) developed a risk matrix based on fuzzy thinking and described fuzzification of the frequency and severity of the consequences of an incident scenario as basic inputs for fuzzy risk assessment. In addition, many attempts were made in the models at developing possibility-based risk assessments in environmental issues since there were many uncertainties and lack of information in environmental risk analysis. During the last decade, different applications of fuzzy logic in environmental risk assessment have been discussed in papers. Ma (2002), Dahab *et al.* (1994), Uricchio *et al.* (2004), McKone and Deshpande (2005), and Darbra *et al.* (2008b) established different FIS for evaluating risk in environmental issues. Finally, Kentel and Aral (2007) and Vemula *et al.* (2004) tried to use a hybrid of probabilistic and fuzzy methods to analyze environmental risks.

Most of the studies mentioned applied a general definition of risk for analyzing by considering the frequency of failure and consequence as risk factors (Lees, 2001; Modarres, 2006). In real situations, each of the frequency and consequence factors is affected by different variables so that inattention to them has major impacts on risk assessment. A study was conducted on the details of consequences in this paper by developing new fuzzy inference systems. Different types of consequences and effective variables on severities were considered as inputs to a consequence model FIS and the final consequence was calculated as output. Then, the final consequence and frequency were entered in the risk model FIS and the final risk factor was calculated by the second FIS. The main objective of this paper is to show the effects of different variables in consequence calculation and decrease uncertainties in risk calculation by fuzzy logic.

1 THEORY

1.1 Fuzzy Logic

Fuzzy logic is a set of collective functions and relations to assess fuzzy sets (Zadeh, 1965). Fuzzy logic methods have been developed in risk assessment for a decade.

The principle of fuzzy thinking and multi-valued logics has a very long history. Jan Lukasiewicz (Duboise *et al.*, 2007) established the principle of multi-valued logics for the first time in 1920, and Black (1937) developed vagueness and multi-valued logic application to set objects. Finally, Zadeh introduced the first fuzzy sets in 1965 and defined fuzzy sets as sets with a membership function (Zadeh, 1965). Based on Zadeh's description, if U is the universe of discourse (U = {u}), a fuzzy set (F of U) is defined by membership functions as presented by:

$$\mu_F : U \rightarrow [0, 1] \quad (1)$$

$$F = \{\langle u, \mu_F(u) | u \in U \rangle\} \quad (2)$$

where $\mu_F(u)$ is the degree of membership of u in F, and $\mu_F(u)$ will be between 0 and 1. This description changed the arithmetic methods in different branches of science. Following fuzzy set description, risk studies and assessment methods changed as well. Fuzzy logic was established based on possibility theories (Duboise and Prade, 1990).

In risk studies of a project, fuzzy logic methods are used to answer "How safe is the project?", while classical risk methods answer a different question, "Is the activity safe?" (Markowski *et al.*, 2009b). In addition, some other characteristics make risk assessment by fuzzy logic more applicable than classical methods. The main advantages of fuzzy system usage in risk assessment are considered as follows:

- fuzzy logic risk assessment expresses the possibility of an outcome, but classical risk assessment methods estimate the likelihood of an outcome (Darbra *et al.*, 2008a);
- input and output relationships in fuzzy logic risk assessment are determined not by complicated equations, but by sets of logical rules (Blair *et al.*, 2001);
- fuzzy logic risk assessment can be used for cases with input data that is vague, imprecise and insufficient (Darbra and Casal, 2009);
- results of risk assessment by fuzzy logic are easy for decision-making. In fact, managers can understand results and outgoings better and more precisely;
- partial failures can be studied by fuzzy sets and make it possible to conduct study of risk in more detail and more real conditions.

In spite of the mentioned advantages of fuzzy logic application in risk assessment, there are some disadvantages, on the other hand:

- reliance on subjective inputs (Darbra *et al.*, 2008a);
- complete data sets and complex parameter correlations cause failure (Ferson, 2002);
- the experimental basis of input preparation.

1.2 Fuzzy Logic Inference Description

Fuzzy models and calculating processes include different steps (Fig. 1). As shown, there are three main steps to define a fuzzy system: fuzzification, fuzzy inference and defuzzification (Zadeh, 1965).

1.2.1 Fuzzification

Crisp and real values are changed to linguistic functions and fuzzy sets in fuzzification. In fact, the linguistic terms and membership functions are defined based on actual and real values to determine the degree of truth for each assumption. As shown in Figure 1, the output of this step is a fuzzy value. In other words, membership functions are defined in this step to transfer data and values from the classical world to the fuzzy systems (Belohlavek and Klir, 2011).

There are five different types of membership functions: triangular, trapezoidal, Gaussian, singleton and piecewise linear

The most common and popular membership functions were categorized as triangular, trapezoidal and Gaussian types (Markowski and Mannan, 2008; Jamshidi *et al.*, 2012; Wulan and Petrovic, 2012). Suitable types of functions are selected based on experience, knowledge, problem, statements (Grima *et al.*, 2000) and characteristics of variables (Xie, 2003). Triangular and trapezoidal membership functions have the advantage of simplicity. On the other hand, Gaussian membership functions are the most natural (Grima *et al.*, 2000), smooth, and non-zero at all points (Jang and Sun, 1997). Although Gaussian membership functions enjoy the advantages mentioned they are unable to specify asymmetric membership functions (Mamdani, 1976) and it should be considered in type selections. In spite of the all facts mentioned, the types of membership function do not affect the result essentially (Markowski *et al.*, 2009b).

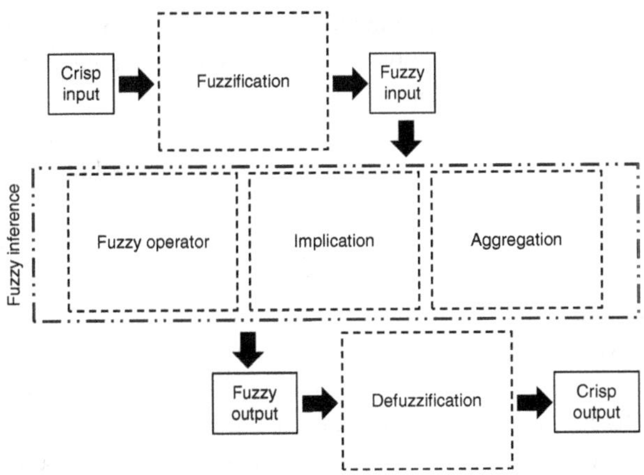

Figure 1

Fuzzy methodology steps.

1.2.2 Fuzzy Inference

The fuzzy inference step is the most important step in a fuzzy logic process (Mamdani, 1976). The fuzzy inference models consist of three sub-processes: fuzzy operator, implication and aggregation (Li, 2006). These sub-processes use different composition methods to establish a fuzzy model. These compositions are combined by sets of if-then rules, which are obtained from human experience, and synthesize fuzzy inputs into fuzzy outputs.

There are three main different fuzzy inference systems: Mamdani fuzzy models, Sugeno fuzzy models and the Tsukamoto fuzzy inference model (Gentile *et al.*, 2003), that have been used in different studies. Fuzzy operators make different models; hence, aggregation and defuzzifications are different in each model. In fact, these fuzzy models vary in the way outputs are determined (Jang and Sun, 1997).

Mamdani's fuzzy model is the most commonly seen fuzzy inference system (Jamshidi *et al.*, 2012). It uses the fuzzy sets and fuzzy logic concepts to translate linguistic and subjective terms into an algorithm (Mamdani and Assilian, 1975). Different composition methods are used to establish a Mamdani fuzzy model. Max-min composition was used in this paper. It is defined arithmetically as follows:

$$\mu_{Ck}(z) = \max\left[\min\left[\mu_{Ak}(\text{input}(x)), \mu_{Bk}(\text{input}(y))\right]\right] = 1, 2, \ldots, r \quad (3)$$

where μ_{Ck}, μ_{Ak}, and μ_{Bk} are the membership functions of output "z" for the rule "k", input "x", and input "y", respectively.

1.2.3 Defuzzification

The fuzzy set, which was calculated in previous steps, changes to a single number in defuzzification. There are different methods for defuzzifying: center of area, bisector, middle of maximum, largest of maximum and smallest of maximum. Different defuzzification methods are shown in Figure 2, where μ, is the membership function and Z is the universe of discourse.

Perhaps the most popular defuzzification method is the center of area method, which returns the center of the area under the curve (Chiu, 1994). The center of area method uses Equation (4) for defuzzification (Iphar and Goktan, 2006).

$$Z_{COA} = \frac{\int_z \mu_A(z)z.dz}{\int_z \mu_A(z).dz} \quad (4)$$

where Z_{COA} is the crisp value for the "z" output and $\mu_A(z)$ is the output membership function.

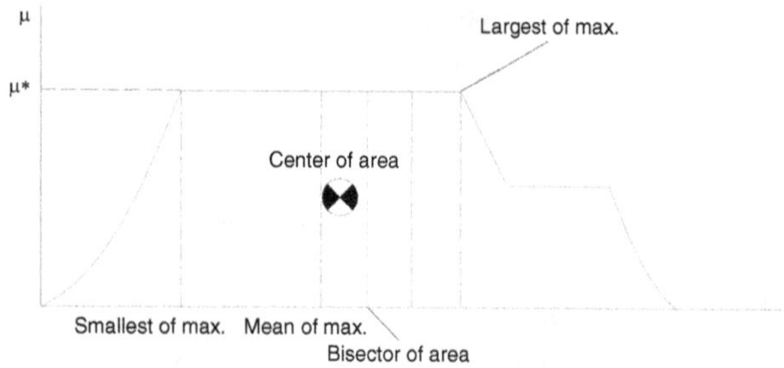

Figure 2

Different methods of defuzzification.

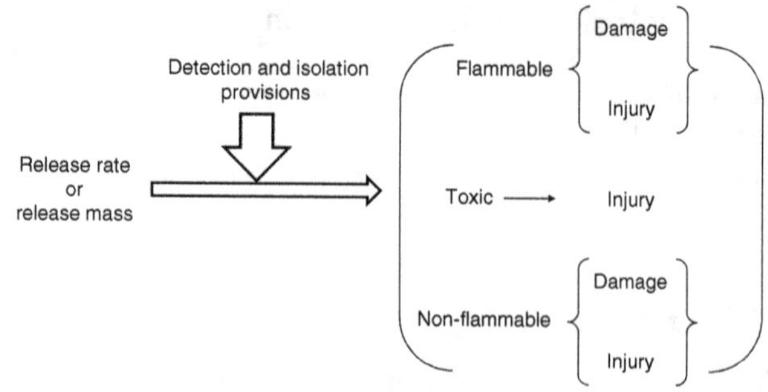

Figure 3

Various consequences for a release/rupture scenario.

2 RISK ASSESSMENT BY FUZZY LOGIC

Risk of releases and their probable consequences in a chemical plant were studied by a new method based on fuzzy logic systems. In this paper, a general definition of risk is selected: risk is a combination of the probability of an event and its consequences (Lees, 2001; Markowski *et al.*, 2009b). Indeed, the risk factor is calculated based on the probability of failure (frequency of an event occurrence) and its probable consequences. Most of the methods that were developed recently worked on the probability of occurrences. Unlike previous studies, the consequences and impacts of incidents, which were raised by release of chemical materials in industries, were studied in this paper.

Probabilities were extracted from handbooks and guidelines, such as OREDA, DNV Note 14 and OGP guidelines (OREDA, 2009; DNV, 2006; OGP, 2010).

2.1 Basis of Fuzzy Modeling

There are many consequences associated with any release in a plant. Cost impacts, time delay, injury and fatalities, and environmental effects are the main consequences in the case of equipment/pipeline failure (Lees, 2001; Markowski and Mannan, 2008; AIChE/CCPS, 1999; Kletz, 1977).

The type of material released specifies the types of consequences. As shown in Figure 3, release of flammable, toxic and non-flammable material has various consequences. Release of flammable and explosive materials has potential for damaging equipment and personnel injury from thermal radiation and explosion overpressure. Release of non-flammable materials was also considered. Consequences such as chemical splashes and high temperature steam affect personnel health. Moreover, physical explosions and Boiling Liquid Expanding Vapor Explosion (BLEVE) can damage equipment and have personnel injury effects. Toxic material releases cause personnel injury and fatalities due to overexposure of personnel to toxic components and concentrations.

Different equations for calculating consequences are presented in Table 1. As mentioned in this table, complicated mathematical operations shall be done to evaluate different consequences. In addition, many assumptions and data are required to solve any of the provided equations. The required data for calculating the consequences of an event are not ready or easy to access most of the time. These limitations cause many restrictions in the usage of arithmetic and classical methods.

The fuzzy logic model was established based on the Mamdani algorithm as the most popular algorithm (Jamshidi *et al.*, 2012) and the max-min composition method as the most popular method (Ross, 2010), which was used for composite relations. The inputs and outputs which were used to make this FIS are detailed in Figure 4. As shown, two FIS models were established to calculate the risk factor: a consequence model FIS and risk model FIS. As mentioned in Figure 4, the release rate was the first required data with which consequence modeling started. Magnitudes of releases have very important effects on sequential events and final consequences. Different parameters determine release rates: the physical properties of the material, the initial phases, the operating conditions and release hole sizes (API, 2008).

Detection and isolation provisions that decrease the consequence of releasing are considered as one of the FIS inputs (Mamdani and Assilian, 1975). Detection and isolation systems are provided to reduce the effects of a release of hazardous materials. These systems affect the releases in two different ways: detection and isolation systems to reduce the rate and duration, and mitigation systems to reduce the consequences of a release by minimizing ignition possibility or limiting spread of materials. Different types of consequences considered as inputs to the first FIS and final

TABLE 1
Different equations for calculating consequences

Type of consequence	Equations	Remark
Flammable	$CA^{flam}_{inj,n} = CA^{flam-AIL}_{inj,n} \cdot fact^{AIT} + CA^{AINL}_{inj,n}\left(1 - fact^{AIT}\right)$	
	$CA^{flam}_{cmd,n} = CA^{flam-AIL}_{cmd,n} \cdot fact^{AIT} + CA^{AINL}_{cmd,n}\left(1 - fact^{AIT}\right)$	
	$CA^{flam}_{inj} = \frac{\sum_{n=1}^{4} gff_{n} \cdot CA^{flam}_{inj,n}}{gff_{total}}$	
	$CA^{flam}_{cmd} = \frac{\sum_{n=1}^{4} gff_{n} \cdot CA^{flam}_{icmd,n}}{gff_{total}}$	
Toxic	$CA^{tox}_{inj,n} = C_8 \cdot 10^{\left(c.\log_{10}\left[c_4.rate^{tox}_{n}\right] + d\right)}$	For HF or H_2S only
	$CA^{tox}_{inj,n} = C_8 \cdot 10^{\left(c.\log_{10}\left[c_4.mass^{tox}_{n}\right] + d\right)}$	C_8, c, C_4 and d are constants
	$CA^{tox}_{inj,n} = e\left(rate^{tox}_{n}\right)^{f}$	For ammonia or chlorine
	$CA^{tox}_{inj,n} = e\left(mass^{tox}_{n}\right)^{f}$	e and f are constants
	$CA^{tox}_{inj} = \frac{\sum_{n=1}^{4} gff_{n} \cdot CA^{tox}_{inj,n}}{gff_{total}}$	
Non-flammable-non-toxic	$CA^{leak}_{inj,n} = CA^{INST}_{inj,n} \cdot fact^{IC} + CA^{CONT}_{inj,n}\left(1 - fact^{IC}_{n}\right)$	
	$CA^{leak}_{cmd,n} = CA^{INST}_{cmd,n} \cdot fact^{IC} + CA^{CONT}_{cmd,n}\left(1 - fact^{IC}_{n}\right)$	
	$CA^{nfnt}_{inj} = \frac{\sum_{n=1}^{4} gff_{n} \cdot CA^{leak}_{inj,n}}{gff_{total}}$	
	$CA^{flam}_{cmd} = \frac{\sum_{n=1}^{4} gff_{n} \cdot CA^{leak}_{cmd,n}}{gff_{total}}$	

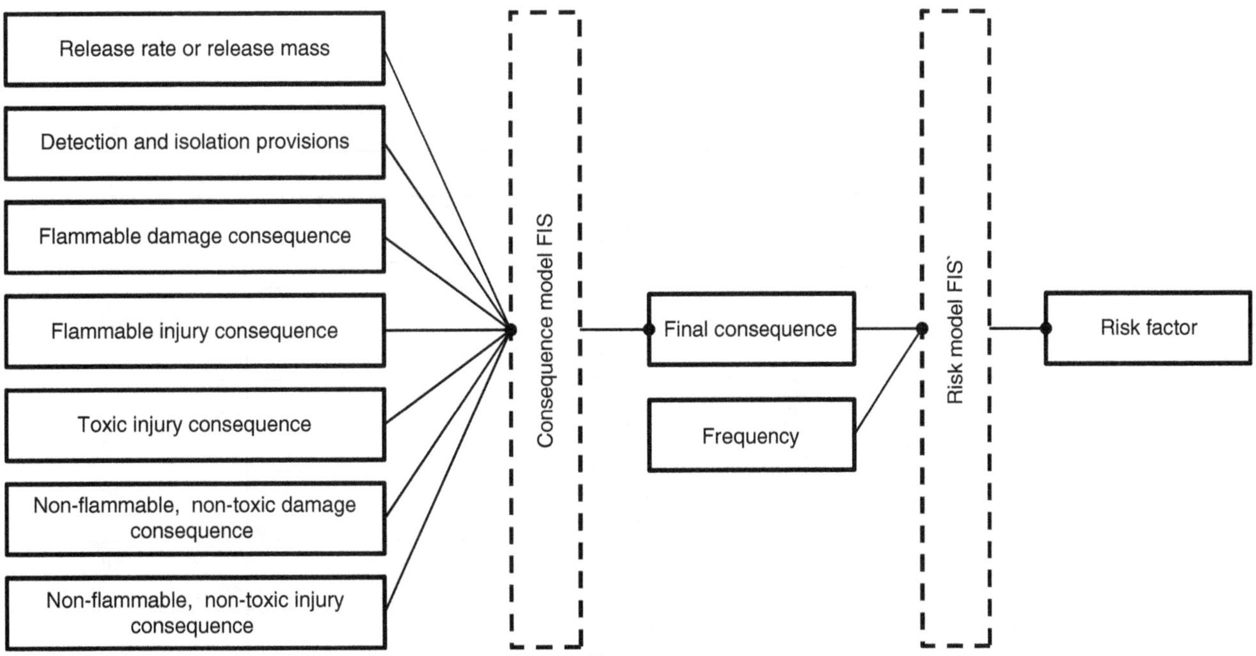

Figure 4

FIS configuration.

consequence were the results of the consequence model FIS. A combination of the final consequence and frequency in the second FIS resulted in a risk factor for each release incident.

2.2 Establishment of Fuzzy Model

To create FIS, fuzzification was the first step. The Gaussian type of membership functions were used for consequence inputs and outputs, since they were the most natural (Markowski and Mannan, 2009b), non-zero and smooth functions at all points (Xie, 2003). The Gaussian membership functions are based on the following equation:

$$Gaussian(x : c, \sigma) = e^{-\frac{1}{2}\left(\frac{x-c}{\sigma}\right)^2} \quad (5)$$

where c is the center and σ is the width of the membership function.

The Gaussian functions have between 45-55% overlapping to make sure that there is no hole (Jamshidi *et al.*, 1997).

Frequencies were fuzzified by trapezoidal membership functions since they reflect the best failure data based on experience (Markowski and Mannan, 2008). The trapezoidal membership function depends on four parameters a, b, c and d, as given by Equation (6). The parameters a and d show the "feet" and the trapezoid; the parameters b and c locate the "shoulders".

$$Trapezoidal(x, a, b, c, d) = \begin{cases} 0 & x \leq a \\ \frac{(x-a)}{(b-a)} & a \leq x \leq b \\ 1 & b \leq x \leq c \\ \frac{(d-x)}{(d-c)} & c \leq x \leq d \\ 0 & d \leq x \end{cases} \quad (6)$$

The values of fuzzifying membership functions for inputs and outputs of both FIS are presented in Table 2.

As presented in Table 2, the universe of discourse for consequence variables was between zero and 100 to cover the percentage of impacts. Various membership functions are entitled and characterized in Table 2 based on types of consequence variables. The universe of discourse for frequencies was selected based on experience, general guidelines and standards (OREDA, 2009; DNV, 2006; OGP, 2010). It is detailed in membership functions to cover very rare circumstances and most credible events.

TABLE 2
Values of membership functions

Gaussian membership function				
Variable	Universe of discourse	Membership functions	C	σ
Release rate or release mass	[0 100]	Very small	0	10
		Small	25	10
		Medium	50	10
		Large	75	10
		Very large	100	10
Detection and isolation provisions	[0 100]	Perfect operation	0	10
		Good operation	25	10
		Medium operation	50	10
		Weak operation	75	10
		Very weak operation	100	10
Flammable damage consequence	[0 100]	No damage	0	10
		Self equipment devastation/damage	25	10
		Beside equipment devastation/damage	50	10
		Unit devastation/damage	75	10
		Plant devastation/damage	100	10

(continued)

TABLE 2 *(continued)*

Gaussian membership function

Variable	Universe of discourse	Membership functions	C	σ
Flammable injury consequence	[0 100]	No injury consequence	0	21
		Injury of personnel	50	21
		Death of personnel	100	21
Toxic injury consequence	[0 100]	No injury consequence	0	21
		Injury of personnel	50	21
		Death of personnel	100	21
Non-flammable-non-toxic damage consequence	[0 100]	No damage	0	10
		Self equipment devastation/damage	25	10
		Beside equipment devastation/damage	50	10
		Unit devastation/damage	75	10
		Plant devastation/damage	100	10
Non-flammable-non-toxic injury consequence	[0 100]	No injury consequence	0	21
		Injury of personnel	50	21
		Death of personnel	100	21
Final consequence	[0 100]	Insignificant	0	10
		Minor	25	10
		Moderate	50	10
		Major	75	10
		Catastrophic	100	10
Risk factor	[−9 0]	Low risk	−9	0.76
		Negligible risk	−7.2	0.76
		Tolerable risk	−5.4	0.76
		Undesirable risk	−3.6	0.76
		Intolerable risk	−1.8	0.76
		High risk	0	0.76

Trapezoidal membership function

Variable	Universe of discourse	Membership functions	a	b	d	c
Frequency	[−8 0]	Impossible	−9.43	−8.15	−7.83	−6.55
		Improbable	−7.86	−6.58	−6.26	−4.98
		Remote	−6.24	−4.96	−4.64	−3.36
		Occasional	−4.64	−3.36	−3.04	−1.76
		Probable	−3.04	−1.76	−1.44	−0.16
		Frequent	−1.45	−0.17	−0.15	1.43

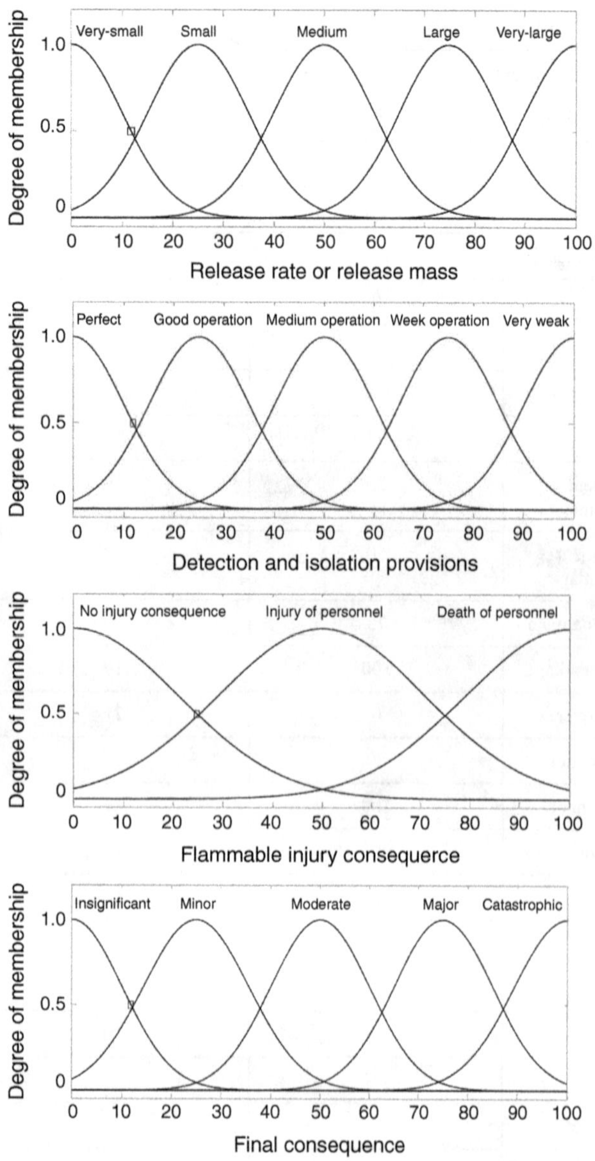

Figure 5

Example of membership functions for the consequence model FIS.

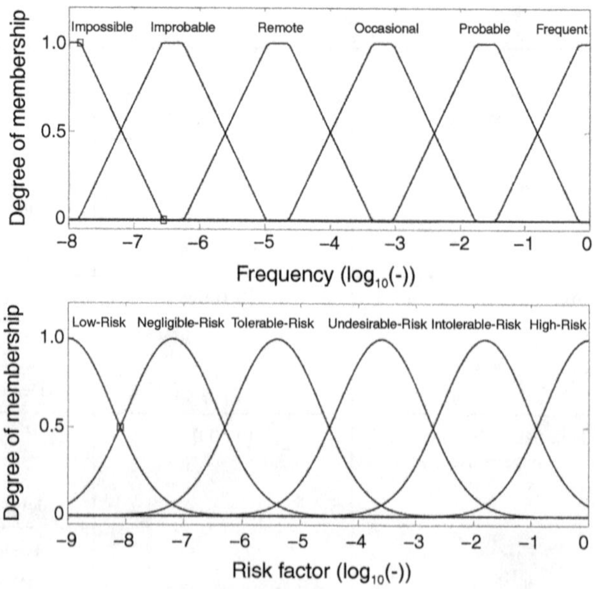

Figure 6

Membership functions for the risk model FIS.

Inputs, outputs curves and examples of the membership functions of the consequence model FIS and risk model FIS are shown in Figures 5 and 6. As shown in Figure 5, all consequence variables were Gaussian type and covered all the universe of discourse based on the values presented in Table 2. In addition, as shown, both inputs and outputs were fuzzy statements based on the Mamdani model. On the other hand, frequency membership functions were selected as logarithmic values in trapezoidal curves, as shown in Figure 6.

The second step to establish a FIS was making if-then rules based on human knowledge. These rules, which were made by expert knowledge during brainstorming meetings, defined the relations between fuzzy inputs and outputs. Brainstorming meetings were held with the participating operating HSE engineer, operating process engineer, operating manager, process-safety design engineer, process design engineer, piping design engineer, instrument design engineer and project design manager. About 116 rules connected inputs and outputs in the first FIS. For instance, one of the rules that was made in the first FIS was: if 'release rate' is small and detection and isolation is very weak and 'flammable damage' is no damage then 'final consequence' is minor.

In the first FIS, the final consequence was the target point resulting from different logics and linguistic terms, originated by human knowledge and experiments. The validation and accuracy of the consequence model FIS was much better than existing FIS for assessing consequences. The new FIS analyzes consequences by concentrating on details of reasons for developing or decreasing release circumstances. In addition, different types of damage and injuries, which may occur during a release, were considered in rule-making. The items mentioned made the consequence model FIS more accurate for calculating consequence factors in comparison with the recommended general risk assessment methods (Markowski and Mannan, 2008; Ma, 2002) and suggested piping risk assessment method (Jamshidi *et al.*, 2012). More than 30 rules connected the variables of the risk model FIS together. An example of the rules that were provided in

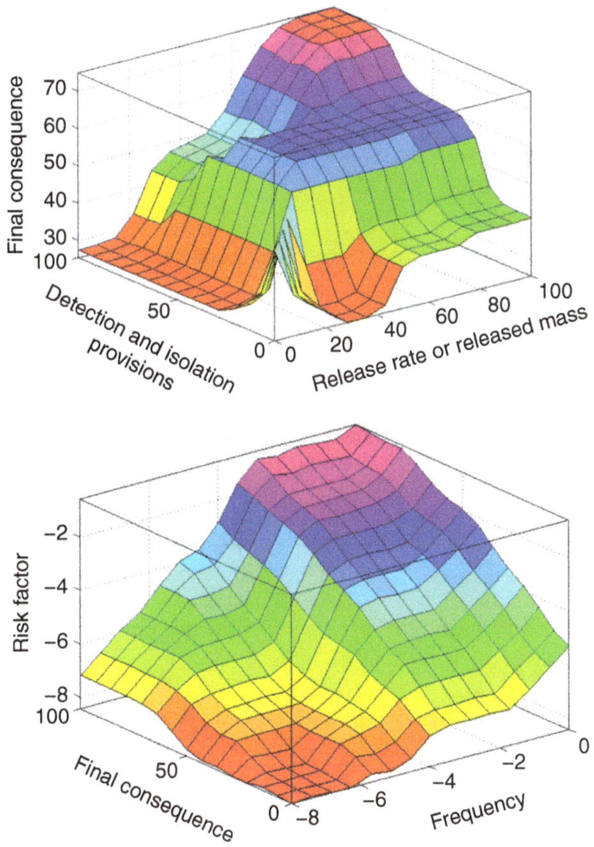

Figure 7

Interdependency of variables for the consequence and risk models' FIS.

the risk model FIS was: if 'final consequence' is insignificant and 'frequency' is impossible then 'risk factor' is low risk.

Different rules studied various states of consequences and probability of occurrences for assessing risk factors in a circumstance. It was implemented by using human knowledge and very simple linguistic terms which made fuzzy inference more convenient to understand and analyze. The final risk factor was calculated by considering both FIS in a series run.

The interdependency of different variables for both FIS is shown in Figure 7. There was different interdependency between different variables in the FIS, some of which are shown in Figure 7. This interdependency showed that any change in variables, even minor, would affect the final graph and risk factor. In addition, it illustrated that rule configuration was noticeably effective to provide an acceptable risk assessment method.

In the last step, the defuzzification process was carried out to transfer data from fuzzy values to crisp ones. The center of area method, which is one of the most common ways, was used in this paper for defuzzification (Gottwald, 2006; Crowl and Louvar, 2001).

3 CASE STUDY

Application of the proposed fuzzy model for calculating risk was checked in this section by studying a gas pipeline between a gas plant and an export jetty. The pipeline, 7 km length, was routed to the jetty through a general unmanned area and the last 1 km of this pipeline crossed next to a general warehouse where non-flammable materials were stored. Operators were working 12 h a day in the warehouse. The general specification of the gas pipeline is presented in Table 3.

As this table shows, liquid propane flows through 24″- diameter pipe at high pressure and in low temperature conditions. Risk of personnel injury due to flammable material releases from the gas pipeline was studied here by two different methods: the new fuzzy method proposed in this paper and the classical method normally used in such studies. Risk calculations were carried out for a 25-mm leak hole and full-bore rupture as examples. The procedure of the classical method for risk calculation is shown in Figure 8. As shown, the first step for starting risk calculation was selecting the release hole size. Among common leak holes, 25-mm and full-bore release holes were selected as examples of risk calculations (API, 2008). Calculation of the release rate was the next step, which was done based in the equation mentioned in Figure 8. Since there were different hole sizes, various release rates were calculated, as mentioned in Table 4 (W_n). As the third step, the impacts of detection and isolation systems on reduction of the release magnitude were studied. The reduction factor was extracted from standard tables based on the types of detection and isolation systems. The detection type for the gas pipeline was visual detection,

TABLE 3

Pipeline operational conditions

Composition	Propane
Size	24″
Phase	L
Temperature (°C)	−31.7
Pressure (bar)	25.1
Vol. flow (m^3/h)	54.7
Density (kg/m^3)	577.7
MW	44.2

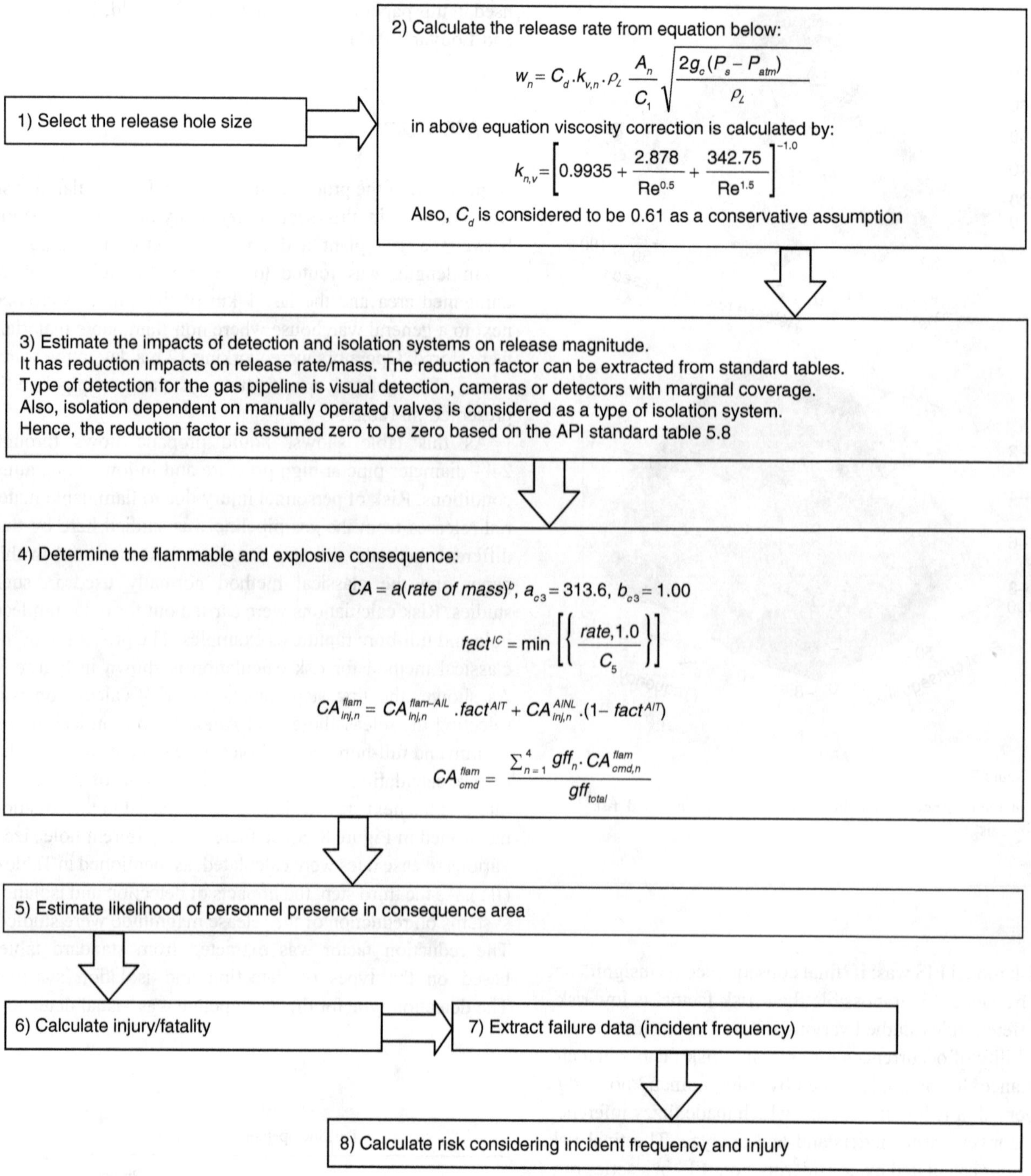

Figure 8

Procedure of risk calculation by the classical method.

TABLE 4
Results of risk calculation by the classical method and proposed new FIS

Classical method		
Release hole size	25 mm	Full bore
Wn (kg/s)	13.2	4 182.35
$CA^{Inj}(m^2)$	1 600	270 400
Likelihood of presence of personnel	0.5	0.5
Injury/fatality probability	3.4×10^{-2}	0.8
Failure data (per year)	2.4×10^{-6}	6.5×10^{-5}
Risk (per year)	8.3×10^{-8}	1.2×10^{-4}
Proposed new fuzzy system		
Release mass/release rate	45	88
Detection and isolation	90	90
Flammable injury consequence	30	80
Final consequence	30	80
Frequency	−5.62	−4.187
Risk factor (per year)	−7.12	−4

cameras or detectors with marginal coverage. Isolation of the pipeline depended on manually operated valves that were considered as types of isolating systems. Hence, the reduction factor was assumed to be zero for the gas pipeline based on the API standard table 5.8 (API, 2008). The fourth step for classical risk calculation was determining consequences. The area involved where personnel may be affected by sequential events of a release was specified in this step (CA^{Inj}).

The fifth step was estimating the likelihood of the presence of personnel in the consequence area. Based on the storage operator's report, the likelihood of personnel presence in the area of the pipeline was 50% (POGC, 2010). Hence, fatality and injury calculations were done based on the reported values as the sixth step. Incident frequency calculation was done by extracting failure data from guidelines, and finally, risk was calculated considering incident frequency and injury.

Furthermore, risk of leak/release from the gas pipeline was calculated by the proposed fuzzy system. The graphical rules for calculating risk by the risk model FIS are shown in Figure 9. As shown, the final consequence value was calculated by the first fuzzy inference system as an input for the second fuzzy inference system. A combination of the frequency and final consequence formed the risk factor in the second fuzzy inference system.

The input data, calculated values, and results for both the classical method and the proposed new fuzzy system are presented in Table 4. This table contains two independent parts. The first part shows the calculated parameters and values based on the classical procedure for calculating risk. As shown in Table 4, the fundamental values for calculating risk in the classical procedure were quantitative values, which were calculated based on the complicated equations described in Figure 8 and Table 1. Moreover, the second part of Table 4 shows the input variables for calculating risk by the proposed fuzzy inference system. As mentioned, the input data for the new fuzzy system were linguistic terms that were produced by human knowledge and experiments that were easy to understand and easy to use in decision-making. For instance, based on decisions that were made during meetings of experts, the following inputs were considered for calculating the risk factor by the proposed new fuzzy system in the case of a 25-mm leak hole: release mass/release rate: medium; detection and isolation: weak operation or very weak; flammable injury consequence by considering the results of simulations and magnitude of the release rate: no injury or fatalities of personnel; frequency based on guidelines: remote and improbable. Considering the inputs mentioned, the final consequence was calculated as a minor or moderate consequence. Eventually, the risk factor was calculated

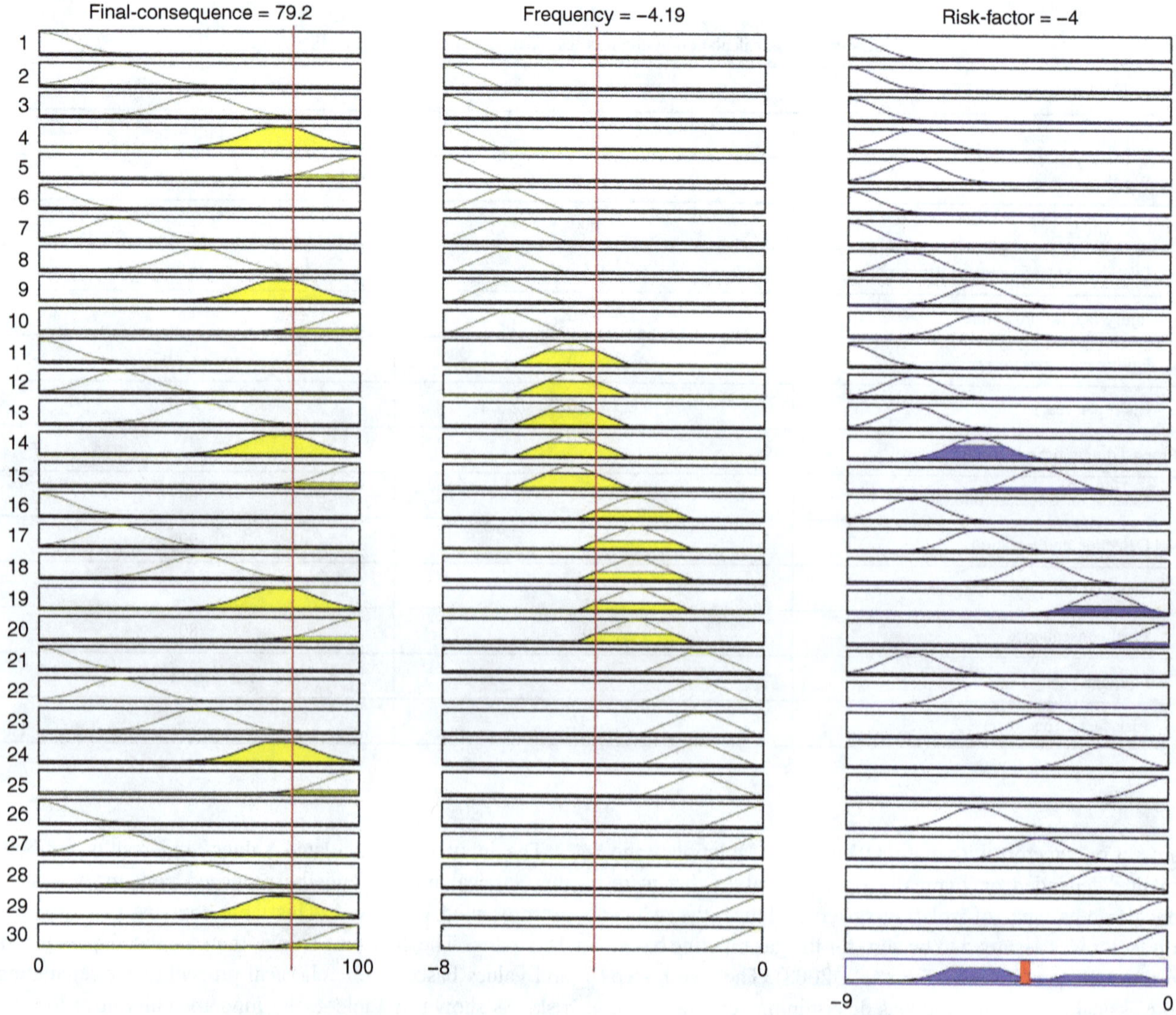

Figure 9

Graphical rules for the risk model FIS.

based on the final consequence and frequency as a negligible risk.

All values that were used for calculating the risk are presented in Table 4 for both the classical and new fuzzy methods. As presented, the proposed fuzzy inference system demonstrated that risk of release from a 25-mm hole was positioned in a negligible risk area and was not considerable. It was certified by the classical method calculation for a leak hole of small diameter, since the calculated risk was very low in comparison with the standard risk criteria (IP 323, 1998). On the other hand, based on the calculated value by the new fuzzy inference system, the risk of full-bore rupture was 80% undesirable and 20% tolerable. It was certified again by the classical method, since the calculated risk was considerable and high in comparison with the risk criteria.

A comparison between the calculated risks demonstrated that there was 15% difference between the result of the proposed new FIS and the classical method.

CONCLUSION

The risk factor was calculated by a combination of two FIS in this work. A combination of a consequence model FIS and risk model FIS made up the new FIS based on different possible consequences of an incident. The most important advantage of the proposed fuzzy system was its comprehensiveness for considering the effects of incidents. In addition, the proposed system has other advantages, presented as follows:

- the relation between inputs and outputs were linguistic terms. It made the model more flexible for changing in different conditions;
- it made the decision-making process simpler and faster, since its results were more understandable and reliable;
- the case study showed that the proposed model has more realistic results than the classical methods. A study of different types of consequences and various variables that affect the consequences of an event showed that risk assessment by the new model was more reliable than classical methods, in which the effects of very important items such as detection and isolation are normally ignored during risk assessments. In addition, brainstorming of experts to estimate parameters gave results based on experimental outcomes. A fuzzy mathematical method also decreases uncertainties and increases accuracy;
- simplicity in the use of this model made it convenient for users throughout complicated risk assessments. In addition, it was a very useful method while suitable data for assessing risk were not accessible.

REFERENCES

AIChE/CCPS (1999) *Guidelines for Consequence Analysis of Chemical Releases*, The Center for Chemical Process Safety of the American Institute of Chemical Engineers Publishers, New York.

API (2008) Recommended Practice 581, *Risk based inspection technology, Part 3: Consequence analysis in an API RBI assessment*, 2nd ed., American Petroleum Institute, Washington DC.

Baloi D., Price A.D.F. (2003) Modeling global risk factors affecting construction cost performance, *International Journal of Project Management* **21**, 261-269.

Belohlavek R., Klir G. (2011) *Concepts and fuzzy logic*, The MIT Press, Massachusetts.

Black M. (1937) Vagueness: An exercise in logical analysis, *Philosophy of Science* **4**, 427-455.

Blair A.N., Ayyub B.M., Bender W.J. (2001) Fuzzy stochastic risk-based decision analysis with the mobile offshore base as a case study, *Marine Structures* **14**, 69-88.

Bowles J.B., Pelaez C.E. (1995) Application of fuzzy logic to reliability engineering, *Proceeding of IEEE* **83**, 3, 435-449.

Chapman C.B., Ward S.C. (1997) *Project risk management: process techniques and insights*, John Wiley and Sons, New York.

Chia S.E. (2006) Risk assessment framework for project management, *2006 IEEE International Engineering Management Conference*, 17-20 Sept, Bahia, pp. 376-379.

Chiu S. (1994) Fuzzy model identification based on cluster estimation, *Journal of Intelligent and Fuzzy Systems* **2**, 3-10.

Crowl D.A., Louvar J.F. (2001) *Chemical Process Safety: Fundamentals with Applications*, 2nd ed., Prentice Hall, Englewood Cliffs, New Jersey.

Dahab M.F., Lee Y.W., Bogardi I. (1994) A rule based fuzzy set approach to risk analysis of nitrate contaminated groundwater, *Water Science and Technology* **7**, 45-52.

Darbra R.M., Eljarrat E., Barcelo D. (2008a) How to measure uncertanties in environmental risk assessment, *Trends in Analytical Chemistray* **27**, 377-387.

Darbra R.M., Demichela M., Mure S. (2008b) Preliminary risk assesment of ecotoxic substances accidental releases in major risk installations through fuzzy logic, *Process Safety and Environmental Protection* **86**, 103-111.

Darbra R.M., Casal J. (2009) Environmental risk assessment of accidental releases in chemical plants through fuzzy logic, *Chemical Engineering Transactions* **17**, 287-292.

DNV (2006) Report No. T14: Technical note process equipment failure frequencies.

Duboise D., Esteva F., Godo L., Prade H. (2007) *Fuzzy set based logics- an history oriented presentation of their main development. Handbook of the History of Logic*, **Vol. 8**, Elsevier, New York.

Duboise D., Prade H. (1990) Resolution principles in possibilistic logic, *International Journal of Approximate Reasoning* **4**, 1-21.

Ferson S. (2002) *RAMAS Risk calc 4.0 software: risk assessment with uncertain numbers*, Lewis publishers, Boca Raton.

Gentile M., Rogers W.J., Mannan M.S. (2003) Development of a fuzzy logic-based inherent safety index, *ICHEM Journal* **81**, 444-456.

Gottwald S. (2006) Universes of fuzzy sets and axiomatizations of fuzzy set theory. Part I: Model-based and axiomatic approaches, *Studia Logica* **82**, 211-244.

Grima M.A., Bruines P.A., Verhoef P.N.W. (2000) Modeling tunnel boring machine performance by nuero fuzzy methods, *Tunneling and Underground Space Technology* **15**, 259-269.

Guimaraes A.C.F., Lapa C.M.F. (2007) Fuzzy inference to risk assessment on nuclear engineering systems, *Applied Soft Computing* **7**, 17-28.

Hertz D.B., Thomas H. (1994) *Risk analysis and its applications*, 2nd ed., John Wiley and Sons, New York.

IP 323 (1998) *A Risk-Based Approach to Hazardous Area Classification*, The Institute of Petroleum, London.

Iphar M., Goktan R.M. (2006) An application of fuzzy sets to the diaggability index rating method for surface mine equipment selection, *International Journal of Rock Mechanicas and Mining Science* **43**, 253-266.

Jafari A. (2001) Management of risk, uncertainties and opportunities on projects: time for fundamental shift, *International Journal of Project Management* **19**, 89-101.

Jamshidi A., Jang J.S.R., Sun C.T., Mizutani E. (1997) *Neural fuzzy and soft computing*, Prentice Hall, Englewood Cliff, New Jersey.

Jamshidi A., Yazdani A., Yakhchali S., Khaleghi S. (2012) Developing a new fuzzy inference system for pipeline risk assessment, *Journal of Loss Prevention in the Process Industries*, In Press.

Jang J.S., Sun C.T. (1997) *Neuro fuzzy and soft computing: a computational approach to learning and machine intelligence*, Prentice Hall, New Jersey, Englewood Cliffs.

Karimi I., Hüllermeier E. (2007) Risk assessment system of natural hazards: A new approach based on fuzzy probability, *Fuzzy Sets and Systems* **158**, 987-999.

Kentel E., Aral M.M. (2007) Risk Tolerance Measure for Fuzzy Health Risk Assessment, *International Journal for Stochastic Environmental Research and Risk Assessment* **21**, 405-417.

Kirchsteiger C. (2005) Review of industrial safety management by international agreements and institutions, *Journal of Risk Research* **8**, 31-51.

Kletz T.A. (1977) Unconfined Vapor cloud explosions, *AICHE Loss Prevention* **11**, 50-61.

Lees F.P. (2001) *Loss prevention in the process industries, hazard identification, assessments and control*, 2nd ed., Butter Worth-Heinemann, Oxford.

Li Z. (2006) *Fuzzy chaotic systems: modeling, control, and applications*, Springer, Germany.

Ma H.W. (2002) Stochastic multimedia risk assessment for a site with contaminated groundwater, *Stochastic Environmental Research and Risk Assessment* **16**, 464-478.

Mamdani E.H. (1976) Advances in the linguistic synthesis of fuzzy controllers, *International Journal of Man-Machine Studies* **8**, 669-678.

Mamdani E.H., Assilian S. (1975) An experiment in linguistic synthesis with a fuzzy logic controller, *International Journal of Man-Machine Studies* **7**, 1-13.

Mark W., Cohen P.E., Glen R.P. (2004) Project risk identification and management, *AACE International Transaction* **01**, 1-5.

Markowski A.S., Mannan M.S. (2008) Fuzzy risk matrix, *Journal of Hazardous Materials* **159**, 152-157.

Markowski A.S., Mannan M.S. (2009a) Fuzzy logic for piping risk assessment, *Journal of Loss Prevention in the Process Industries* **22**, 921-927.

Markowski A.S., Mannan M.S., Bigoszewska A. (2009b) Fuzzy logic for process safety analysis, *Journal of Loss Prevention in the Process Industries* **22**, 695-702.

McKone T.E., Deshpande A.W. (2005) Can fuzzy logic bring complex environmental problems into focus? *Environmental Science Technology* **39**, 42-47.

Modarres M. (2006) *Risk analysis techniques, tools and trends*, CRC Press, Florida.

Naderi M. (2008) Fuzzy logic application in risk analysis of construction management, *M.S. Thesis*, Dep. of Civil and Environmental Engineering, Alberta

Nieto-Morote A., Ruz-Vila F. (2011) A fuzzy approach to construction project risk assessment, *International Journal of Project Management* **29**, 220-231.

OREDA (2009) *Offshore Reliability Data Handbook*, 5th ed., SINTEF Industrial Management, Norway.

OGP (2010) *Report No. 434-1: Risk Assessment Data Directory: Process Release Frequencies, International Association of Oil and Gas Procedures.*

Perry J.G., Hayes R.W. (1985) Risk and its management in construction projects, *Proceeding of Institution Civil Engineers* **78**, 499-521.

POGC (2010) *Pars Oil and Gas Company (POGC), report on general storage area in Assaluye and Tobmak in South Pars Industrial area*, www.POGC.ir, Tehran.

Pokoradi L. (2002) Fuzzy logic based risk assessment, *Academic and Applied Research in Military Science* **1**, 63-73.

Ross T.J. (2010) *Fuzzy logic with engineering applications*, 2nd ed., John Wiley and Sons, New York.

Uricchio V.F., Giordano R., Lopez N. (2004) A fuzzy knowledge based decision support for groundwater pollution risk evaluation, *Journal of Environmental Management* **73**, 189-197.

Vemula V.R.S., Mujumdar P.P., Ghosh S. (2004) Risk evaluation in water quality management of a river system, *Journal of Water Resources Planning and Management* **130**, 411-420.

Wulan M., Petrovic D. (2012) A fuzzy logic based system for risk analysis and elevation within enterprose collaborations, *Computers in Industry* **63**, 739-748.

Xie M. (2003) *Fundamentals of robotics: linking perception to action*, World Scientific Publishing Co., London.

Zadeh L. (1965) Fuzzy sets, *Information and Control* **8**, 338-353.

Varnes D.J. (1984) *Commissioning on landslides and other mass movements, IAEG. Landslide hazard zonation: a review of principles and practices*, The UNESCO Press, Paris.

19

Prediction of the Gas Injection Effect on the Asphaltene Phase Envelope

Peyman Bahrami[1]*, Riyaz Kharrat[2], Sedigheh Mahdavi[3] and Hamed Firoozinia[4]

[1] *Department of Oil and Gas Engineering, Science and Research Branch, Islamic Azad University, Tehran - Iran*
[2] *Petroleum University of Technology, Tehran - Iran*
[3] *Memorial University of Newfoundland - Canada*
[4] *Research Institute of Petroleum Industry, Tehran - Iran*
e-mail: pymnbahrami@yahoo.com - kharrat@put.ac.ir - s.mahdavi@mun.ca - firooziniah@ripi.ir

* Corresponding author

Abstract — *Asphaltene instability may occur when pressure, temperature and compositional variations affect the reservoir oil. Permeability reduction, wettability alteration, and plugging of wells and flow lines are the consequences of this phenomenon. Therefore, it is crucial to investigate the asphaltene behavior in different thermodynamic conditions by knowing the Asphaltene Precipitation Envelope (APE) in a preventive way rather than the costly clean-up procedures. The selected reservoir oil has faced a remarkable decline in production due to several years of extraction, and Enhanced Oil Recovery (EOR) has been considered as a solution. Therefore, in this paper, a comprehensive study was carried out to predict the effects of different injected gases on asphaltene onset and to prevent future asphaltene precipitation based on the laboratory data. The Advanced Redlich-Kwong-Soave (RKSA) equation of state was considered to develop APE using Multiflash (*Infochem Co.*). For the selected reservoir oil, with temperature reduction at low temperatures, asphaltene precipitation weakened and made the onset pressure decrease, so this behavior is different from the results obtained in other published reports. On the basis of this model, several sensitivity analyses were performed with different gases (*i.e.*, methane, CO_2, N_2 and associated gases) to compare the risk of each gas for future EOR strategies. APE tend to expand as the amount of injected gases increases, except for CO_2 gas injection, that showed another unconventional behavior for this crude oil. It was observed that for CO_2 gas injection below a certain temperature, asphaltene stability increased, which can be considered as a good inhibitor of asphaltene precipitation.*

Résumé — **Prévision de l'effet d'injection de gaz sur l'enveloppe de phase des asphaltènes** — L'instabilité des asphaltènes peut se produire lorsque des variations de pression, de température et de composition affectent les réservoirs d'huile conduisant à une réduction de la perméabilité, à l'altération de la mouillabilité et au blocage de l'écoulement dans les puits et les conduites. Par conséquent, il est crucial d'étudier le comportement des asphaltènes pour différentes conditions thermodynamiques en connaissant leur domaine de précipitation (APE, *Asphaltene Precipitation Envelope*) de manière préventive pour limiter les opérations coûteuses de nettoyage. Le réservoir d'huile choisi pour cette étude a subi un déclin remarquable de sa production du fait de plusieurs années d'extraction et la récupération assistée (EOR, *Enhanced Oil Recovery*) a été considérée comme une solution prometteuse. Par conséquent, dans ce document, une étude détaillée basée sur des données de laboratoire est réalisée afin de prévoir les effets d'injection de différents gaz lors du début de la précipitation des asphaltènes et pour empêcher toute précipitation ultérieure. L'équation d'état avancée Redlich-Kwong-Soave (RKSA) a été considérée pour établir l'enveloppe APE en utilisant le logiciel Multiflash (*Infochem Co.*). Pour le réservoir de pétrole sélectionné, une réduction de la température diminua la précipitation d'asphaltènes et fit chuter la pression de départ. Ce comportement diffère des résultats précédemment publiés. De nombreuses analyses de sensibilité ont été réalisées pour différents gaz (méthane, CO_2, N_2 et gaz associé) dans le but de comparer le taux de risque de chaque gaz pour différentes stratégies de récupération EOR. Les enveloppes APE ont tendance à croître lorsque la quantité de gaz injecté augmente, sauf pour de l'injection de CO_2 qui a révélé un comportement non conventionnel avec ce pétrole brut. Il a été constaté qu'avec l'injection de CO_2, en-dessous d'une certaine température, la stabilité des asphaltènes augmentait, ce qui peut être considéré comme un bon système d'inhibition de la précipitation des asphaltènes.

INTRODUCTION

Asphaltene deposition in near-wellbore formations, production tubing and surface facilities can occur upon pressure, temperature and composition variations. The latter may happen when reservoir oil blends with various gases during enhanced oil recovery (CO_2, N_2, dry and enriched gas injection), gas lift and acid stimulation. The effects of different gases on asphaltene stability have previously been studied in various fluids. Experimental investigation of asphaltene aggregation due to N_2 injection showed asphaltene particles tend to precipitate as the amount of injected N_2 increases (Jamaluddin *et al.*, 2002), and other gases such as CO_2, methane and propane could increase the asphaltene instability (Vafaie-Sefti *et al.*, 2003). In addition to the fact that these parameters affect asphaltene stability, they could also change the Asphaltene Precipitation Envelope (APE). Observations showed that as the temperature increases, the onset pressure decreases (Andersen, 1994), whereas it might cause an increase at onset pressure in some cases (Gonzalez *et al.*, 2012). Some authors reported that CO_2 gas injection increases the Asphaltene Onset Pressure (AOP) at any temperature (Novosad and Costain, 1990; Srivastava and Huang, 1997; Takahashi *et al.*, 2003; Vazquez and Mansoori, 2000), while other studies found that at some temperatures asphaltene particles could be more stable (Gonzalez *et al.*, 2007). An oil production system can also experience a significant temperature change from the reservoir to the separator. This effect is even more significant in production of deep-sea reservoirs where seafloor temperatures are about 4°C. The consequences generally include changes in reservoir wettability, formation damage, downhole facility plugging and finally, affect the Enhanced Oil Recovery (EOR) project due to unfavorable economic consequences such as production delays and costly clean-up procedures (Akbarzadeh *et al.*, 2007; Creek, 2005; Escobedo and Mansoori, 1997; Mansoori *et al.*, 2007; Zekri *et al.*, 2007).

The reservoir under study faced asphaltene deposition problems during early production of oil and the problem has become even more severe with further production. The producing interval is in the range of 2 692-2 725 m and the oil is transferred to an operation plant near the well site. After the first appearance of asphaltene deposition in surface facilities, a comprehensive study was performed in the laboratory to evaluate the AOP and contents of Saturates, Aromatics, Resins and Asphaltenes (SARA). In order to maintain the reservoir pressure and also prevent more asphaltene deposition and further problems in the future, it is necessary to investigate different strategies to predict asphaltene precipitation behavior. The amount of gas that can be dissolved in the oil before the asphaltene precipitation is an important parameter to consider in the design of the optimal gas injection scheme. The prediction of the phase behavior of asphaltene is essential in the design of well completion and/or surface facilities.

Although experimental determination of the full set of thermodynamic conditions in which asphaltenes begin to flocculate might be desirable for each specific crude oil, the cost of such a detailed characterization is prohibitive. An alternative is to obtain a minimum amount of laboratory data and incorporate that information into a thermodynamic model to predict the onset pressure over a wide range of conditions. To mimic experimental observations, current thermodynamic models generally require the use of many adjustable parameters, some of which cannot be verified experimentally (Leontaritis and Mansoori, 1987; Nghiem *et al.*, 1993; Pan and Firoozabadi, 1998).

One of the first effective models for asphaltene flocculation was a colloidal model that recognized asphaltene particles in crude oil as a colloid (Hirschberg *et al.*, 1984). They treated the oil phase as a homogeneous liquid and modeled flocculation as a new phase of pure asphaltene. In some models, a solid phase was formed when flocculation occurred (Chung, 1992). Both approaches could result in a model that requires too many adjustable parameters and was very sensitive to small changes in parameter values. This means that the model must be refitted to a large quantity of experimental data for each fluid. Leontaritis and Mansoori (1987) proposed a thermodynamic colloidal model that is able to predict the onset of asphaltene flocculation. According to this model, asphaltene solid particles are stabilized by resin on their surface in a colloidal suspension system, and the state of crude oil and molecular weight of asphaltene can be determined by this model. Single-component solid model is one of the simplest models for asphaltene precipitation that was proposed by Nghiem *et al.* (1993), who used the model along with a flash calculation procedure for vapor/liquid/asphalt systems to predict asphaltene precipitation data under pressure depletion, gas or solvent injection conditions. They discovered that the model could not depict the true behavior of asphaltene precipitation with increasing solvent concentration under gas/solvent injection conditions. The micellization model of asphaltene precipitation is another model and it is assumed that one of the key factors in asphaltene deposition is micelle formation (Victorov and Firoozabadi, 1996). This model has shown good agreement with most experimentally observed results. However, this model could not predict asphaltene behavior under gas titration conditions. Later, another model was proposed to understand asphaltene precipitation with a detailed molecular-thermodynamic framework (Wu *et al.*, 2000). This model considered strong interactions between asphaltene and resin molecules and between asphaltene molecules using the SAFT association theory, where asphaltene precipitation is treated as a liquid-liquid equilibrium with the oil. An effective model that describes solid deposition allows the engineer to carry out a variety of tasks from screening for potential problems through to examination of possible remediation strategies. In this work, the Multiflash commercial software was used for predicting the APE based on the RKSA Equation Of State (EOS). The purpose of this paper is to investigate the AOP for a live oil during pressure reduction, blending with different gases and to demonstrate the future strategies for the selected reservoir.

1 CHARACTERIZATION METHODOLOGY

Asphaltenes may destabilize, and precipitate out of the reservoir fluid when subjected to pressure, temperature or compositional changes. It was used screening criteria to investigate the tendency of asphaltenes particles to precipitate. The crude oil under study was a 20.32 °API fluid with the characteristics shown in Table 1, and the total content of asphaltene was 12.8 wt%.

TABLE 1
Reservoir fluid characterization

Reservoir temperature (°C)	96.1
Reservoir pressure (MPa)	30.82
Bubble point pressure (MPa)	9.87
Gas/oil ratio (m^3/m^3)	56.81
Oil API gravity (deg)	20.32
Molecular weight of residual oil	269
Molecular weight of reservoir oil	169
SARA contents (ASTM D4124-97)	
Saturates (wt%)	24.8
Aromatics (wt%)	45.6
Resins (wt%)	16.8
Asphaltenes (wt%)	12.8

The data required for the screening were PVT properties including the oil density at the bubble point pressure, the reservoir pressure, and the SARA contents of the Stock-Tank Oil (STO). Three plots are used for the screening; namely, the de Boer plot, the Asphaltene-Resin ratio and the ratio of Saturate/Aromatic (S/A) to Asphaltene/Resin (A/R). The preliminary screening was conducted on the selected sample. The de Boer plot was developed based on field observations (de Boer *et al.*, 1995), Figure 1 indicates that the reservoir fluid is in a moderate problem region. Moreover, the Asphaltene-Resin ratio screening method (Jamaluddin *et al.*, 2001), suggests that the asphaltenes are unstable in the given fluid (Fig. 2). The last criterion demonstrated in Figure 3 also shows the reservoir fluid is in unstable zone (Stankiewicz *et al.*, 2002). Thus, the results from the stability screening suggested that there may be potential for asphaltene problems in reservoir conditions. Hence, detailed measurements were conducted to determine the onset of asphaltene destabilization at various temperatures.

The reservoir fluid was flashed from reservoir conditions to ambient conditions to measure the Gas-Oil Ratio (GOR). Thus, it was anticipated that the STO would keep all the solids present in the reservoir fluid.

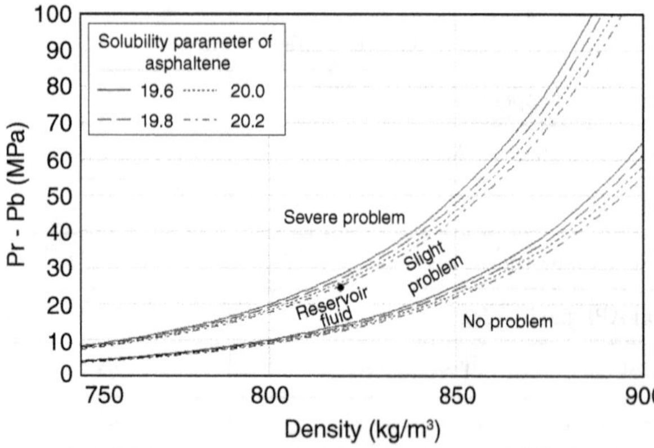

Figure 1
De Boer plot for reservoir sample.

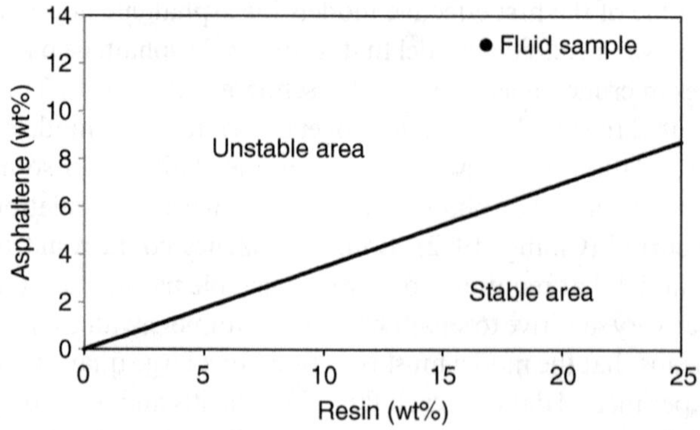

Figure 2
Asphaltene *vs* resin for reservoir sample.

TABLE 2
Reservoir fluid composition

Components	MW (g/mol)	Density (g/cm^3)	STO		Flashed gas		Monophasic fluid	
			(mol%)	(wt%)	(mol%)	(wt%)	(mol%)	(wt%)
N_2	28.04	0.809	0	0	0.91	0.792	0.39	0.065
CO_2	44.01	0.817	0	0	4.10	5.603	1.74	0.458
C1	16.04	0.300	0	0	48.38	24.094	20.55	1.97
C2	30.07	0.356	0	0	17.22	16.078	7.31	1.314
C3	44.1	0.508	0.10	0.016	12.43	17.02	5.34	1.409
i-C4	58.12	0.567	0.13	0.029	2.17	3.917	1.00	0.347
n-C4	58.12	0.586	1.67	0.364	6.34	11.441	3.65	1.269
i-C5	72.15	0.625	3.85	1.041	2.10	4.704	3.10	1.337
n-C5	72.15	0.631	6.44	1.741	2.45	5.489	4.75	2.049
C6	84	0.690	7.73	2.432	2.44	6.364	5.48	2.753
C7	96	0.727	4.82	1.733	1.07	3.189	3.23	1.853
C8	107	0.749	2.03	0.813	0.36	1.197	1.32	0.844
C9	121	0.768	3.93	1.781	0.03	0.112	2.27	1.642
C10	134	0.782	3.80	1.907	0	0	2.19	1.754
C11	147	0.793	3.15	1.734	0	0	1.81	1.59
C12+	370	0.9769	62.35	86.409	0	0	35.87	79.346
Total			100	100	100	100	100	100

Also, the flashed gas and STO were used to measure their molecular composition using gas chromatography and microdistillation, respectively. These compositions were then recombined numerically to obtain the reservoir fluid composition using the measured GOR. In addition, the molecular weight and density of the STO were measured. After measuring the density and the molecular weight of the STO, these properties can be calculated for the plus fraction. The measured and calculated data for modeling purposes are presented in Table 2. It has

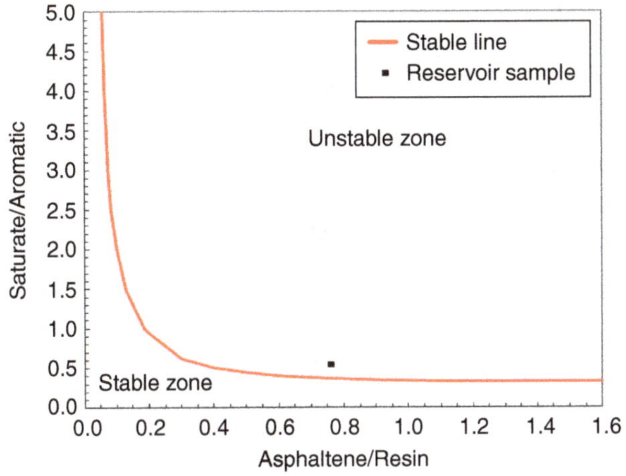

Figure 3

Saturate/Aromatic *vs* Asphaltene/Resin plot for reservoir sample.

been observed that the light components in the oil affect both the bubble pressure and AOP significantly (Gonzalez *et al.*, 2005). Additionally, gas injection typically used for EOR purposes is generally rich in lighter hydrocarbons, and therefore the methodology proposed in the present work will enable good predictions for this type of injection.

2 PRESSURE DEPLETION AND GAS INJECTION EXPERIMENTS

A filtration method was used to measure the asphaltene content in natural depletion and gas injection tests. The reservoir fluid was transferred into the PVT cell to apply the desired pressure and temperature in contact with different ratios of gases. A 0.5-micron metal filter made by the *Swagelok* company was used during the experiments. After the reservoir oil is stabilized in the PVT cell (in the pressure depletion scenario and/or in the gas injection scenario), the fluid passes through the filter during sampling at each stage. All asphaltene particles bigger than 0.5 microns do not pass through the filter, and the asphaltene content of the filtered fluid was determined by the IP-143 method (ASTM D6560-00, 2000). Finally, the differences between the asphaltene content of the original reservoir oil and the filtered samples can determine the amount of deposited asphaltene at each step. The asphaltene that this technique gives is physically extracted from oil, and by using some other techniques such as SARA analysis and molecular weight measurement, it can be further studied. The asphaltene onset points at different temperatures were achieved by extrapolation on asphaltene wt% *versus* the pressure plot by using the Peng-Robinson (PR) EOS (Nghiem *et al.*, 1993; Nghiem and Coombe, 1997), and are summarized in Table 3. Figure 4 is an example of achieving the asphaltene onset pressure at 80°C.

TABLE 3

Upper onset points at different temperatures

Temperature (°C)	96.1	80	60
Upper onset point (MPa)	32.34	34.96	31.36

3 MODELING OF ASPHALTENE PRECIPITATION

In this work, the plus fraction splits into pseudocomponents with different ranges of molecular weight, and single carbon cuts can also be grouped into pseudocomponents (Whitson, 1983). The molecular weight and specific gravity of each component can be obtained by correlation (Riazi and Al-Sahhaf, 1996), and they are adjusted to match the measured values for the whole liquid sample. The mole fraction of each component is obtained by converting the weight fraction using the molecular weight. Then the mole fraction of the monophasic fluid can be calculated. The critical properties and acentric factors of each single carbon cut and pseudocomponents can be predicted by correlations (Kesler and Lee, 1976).

Here, the experimental results are used in the commercial software Multiflash to generate and estimate APE. The models that have been proposed in the past half century have some limitations. Poor prediction of the liquid phase density, lack of asphaltene-asphaltene and asphaltene-resin interactions, and a comprehensive correlation for vapor pressure of polar compounds are some common problems in these models (Edmonds *et al.*, 1999a, 1999b).

The RKSA EOS is able to predict the properties of non-polar molecules in the vapor phase, and it is given by Equation (1) (Soave, 1972):

$$P = \frac{NRT}{V-b} + \frac{a}{V(V+b)} \tag{1}$$

$$a = \sum_{ij} \sqrt{a_i a_j}(1-K_{ij})n_i n_j \tag{2}$$

$$b = \sum_i b_i n_i \tag{3}$$

$$a_i = a_{ci}\left(1+\kappa_i\left(1-\sqrt{T/T_{ci}}\right)\right)^2 \tag{4}$$

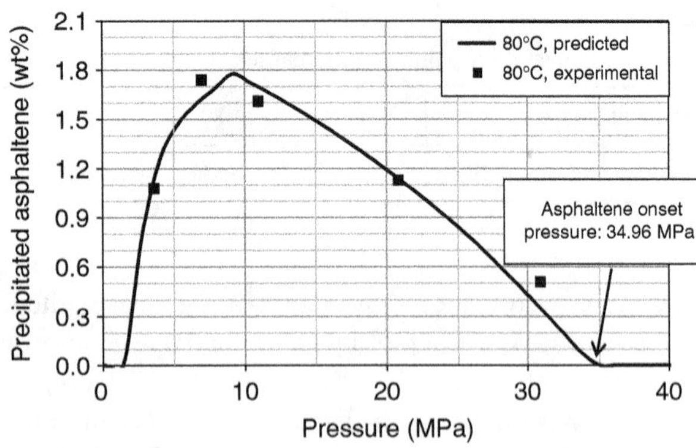

Figure 4

Precipitated asphaltene wt% in natural depletion at 80°C.

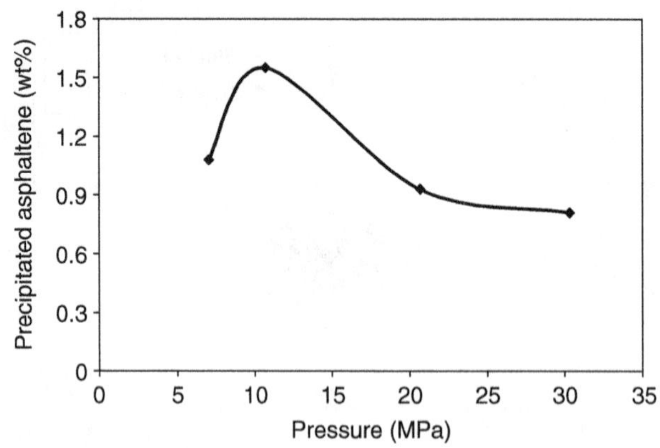

Figure 5

Experimental amount of asphaltene precipitation for reservoir oil at the reservoir temperature during a natural depletion test.

$$a_{ci} = 0.42748 \frac{R^2 T_{ci}^2}{P_{ci}} \tag{5}$$

$$\kappa_i = c_0 + c_1 \omega_i - c_2 \omega_i^2 \tag{6}$$

$$b_i = 0.08664 \frac{R T_{ci}}{P_{ci}} \tag{7}$$

where $n_{i,j}$ are the molar fractions of components, ω is the acentric factor, P_{ci} and T_{ci} are the critical pressure and temperature, R is the gas constant, and K_{ij} is the binary interaction between components i and j. Equation (6) is a correlation for vapor pressure and c_0, c_1 and c_2 were determined based on vapor pressure data. Poor prediction of vapor pressure correlation at low temperatures, large errors in calculating liquid density and large density deviation at the critical point are some problematic issues in this model.

Pedrosa *et al.* (2013) applied some modifications to the EOS and presented the RKSA EOS to eliminate these issues. For an improved phase behavior prediction, K_{ij} coefficients are fitted to vapor pressure data from the triple point to the critical point for each component. Furthermore, for a better prediction of liquid volume and density, Equation (8) is used (Peneloux *et al.*, 1982).

$$V = V^{\mathrm{RKS}} - \sum c_i n_i \tag{8}$$

where c_i is the temperature-dependent volume shift parameter, and it is fitted to the experimental saturation liquid density. The association and solvation of asphaltene molecules are described by the RKSA cubic EOS. In this model, asphaltene-asphaltene and asphaltene-resin interaction coefficients (K_{AA} and K_{AR}), the temperature-dependent association constants, are used to characterize the interactions between asphaltenes and asphaltene-resin particles (Zhang *et al.*, 2012). This model is a computationally efficient way of considering the complex compounds and it is appropriate for engineering studies.

4 RESULTS AND DISCUSSIONS

4.1 Effect of Pressure and Temperature

The asphaltene phase behavior during a natural depletion test at the reservoir temperature is demonstrated in Figure 5, based on the filtration method (Nakhli *et al.*, 2011). Figure 6 also shows the experimental oil density and calculated solubility parameter at different pressures. In this figure, the solubility is calculated based on the Asphaltene Instability Trend (ASIST) (Wang *et al.*, 2006). As the pressure drops from the reservoir pressure to the bubble point pressure, the density and solubility parameters of the oil will decrease and it makes the asphaltene less stable. Further reduction in pressure below the bubble point pressure causes the solution gas to evolve from the liquid and as a result, the oil becomes richer in heavy-end components. Therefore, the density and solubility parameters of the oil increase and again the oil becomes a better solvent for precipitated asphaltene. The calculated solubility parameter of the oil supports this speculation. The minimum amount of solubility and oil phase density occurs at the bubble

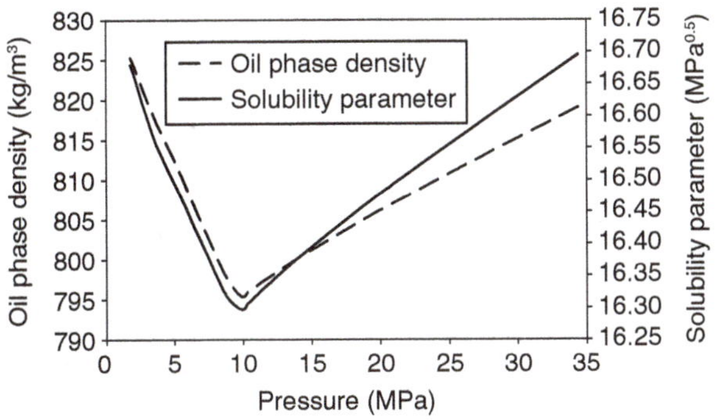

Figure 6

Calculated oil phase density and solubility parameters during a natural depletion test at the reservoir temperature.

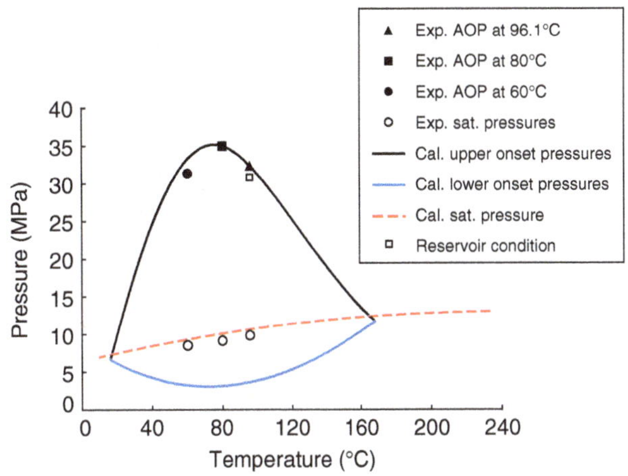

Figure 7

Asphaltene precipitation envelope for reservoir sample.

point pressure where the asphaltene precipitation reaches its maximum.

When the pressure of the reservoir fluid drops at a specified temperature, the asphaltene is in the liquid phase region as the pressure is higher than the upper onset point of precipitation and there is no appearance of asphaltene flocculation. Asphaltene instability occurs when the pressure is between the upper onset and bubble point and between the bubble point and lower onset. The fluid is in oil-asphaltene two-phase and gas-oil-asphaltene three-phase equilibrium respectively. As the reservoir fluid depressurizes below the lower onset point, the asphaltene particles redissolve completely in the liquid phase and the fluid is in the gas-oil two-phase region (Panuganti *et al.*, 2012). Figure 7 presents the APE for the reservoir fluid with the plots of the bubble point and upper and lower onset pressures as a function of temperature and as can be clearly seen, there is good agreement between the calculated and experimental results. The oil is originally unstable, probably due to its initial high content of light components, which has a strong effect on asphaltene stability.

Asphaltenes are polar compounds which are stabilized in crude oil by the presence of resins. When the oil is diluted by light hydrocarbons, the concentration of resins decreases and a point may be reached where the asphaltenes are no longer stabilized. At this point, the asphaltene particles start to grow in size, and finally they flocculate and form a deposit. Since resins, as asphaltene stabilizers, work through the mechanism of polar interactions, their effect becomes weaker as the temperature rises, *i.e.* flocculation will occur as the temperature increases. Also, a decrease in the density of the fluid is another reason for asphaltene precipitation. However, at a specified temperature, the solution entropy increases and the asphaltene melts and redissolves in the oil as the temperature increases. It causes a counterbalancing effect, and make the asphaltene precipitation either stronger or weaker when the temperature increases (Ting, 2003); while in other work, it was proposed that an increase in temperature will redissolve asphaltene particles (Hassanvand *et al.*, 2012; Verdier *et al.*, 2006). When the temperature is below 73.9°C, the density of the oil decreases and causes asphaltene particles to flocculate as the temperature increases, and above this temperature the redissolution phenomenon is stronger. According to Figure 7, the temperature plays a more significant role in the upper onset pressure than lower onset pressure. At higher pressures the fluid non-ideality is stronger than at lower pressures, and non-ideality relates to interaction between molecules. At high pressures, molecules have lower mobility, but an increase in temperature changes the mobility significantly. However, at low pressure the non-ideality is weak and the fluid mobility is higher, therefore temperature plays a less important role. That is why at higher pressures and temperatures the change in the upper onset is greater than in the lower onset.

4.2 Effect of Gas Injection

Based on the model in the natural depletion test, the effect of different light gases was observed on the onset of the asphaltene precipitation. The possibility of asphaltene flocculation increases when light gases are exposed to the oil during miscible gas flooding processes.

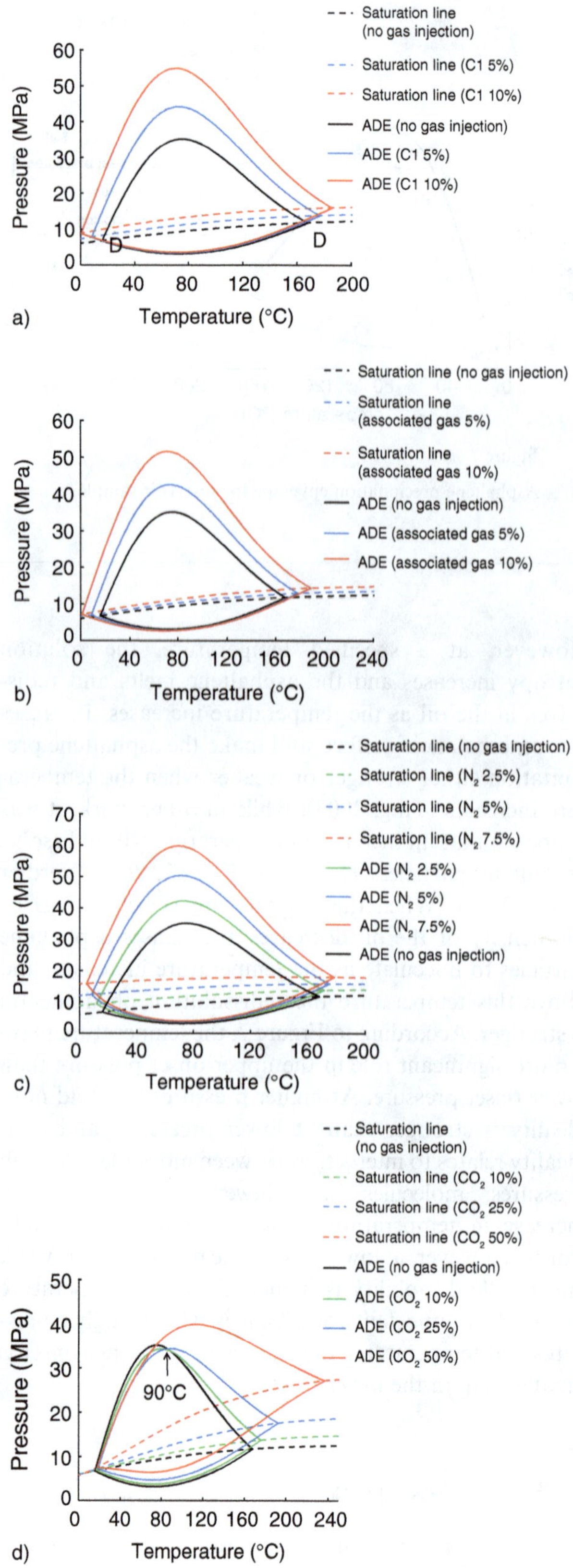

Figure 8

Effect of a) methane, b) associated gas, c) N_2, d) CO_2 on the APE.

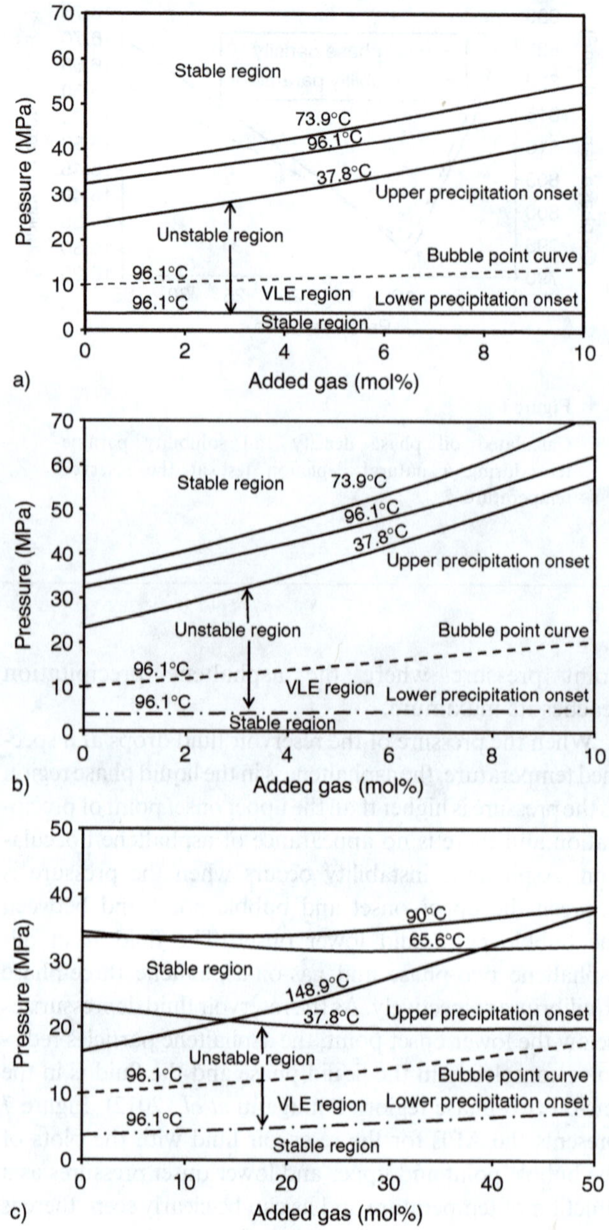

Figure 9

Temperature effect on asphaltene precipitation onset after the addition of a) methane, b) N_2, c) CO_2.

The bubble point curves and the asphaltene stability boundary increase when different compositions of methane, associated gas and N_2 are added to the original fluid, as shown in Figures 8a to 8c. A different behavior is observed upon the addition of CO_2. As shown in Figure 8d, for a known amount of injected CO_2, below a specific crossover temperature, the stability of asphaltene improves (*e.g.*, at 10 mol% of injected gas, the crossover temperature is about 90°C).

TABLE 4
Summary of composition for original fluid and mixtures

	Reservoir fluid	Mixture of reservoir fluid and associated gas			Injected gas
Injected gas added (mol%)	0	10	25	50	100
Component (mol%)					
N_2	0.39	0.442	0.520	0.650	0.91
CO_2	1.74	1.976	2.330	2.920	4.10
C1	20.55	23.333	27.508	34.465	48.38
C2	7.31	8.301	9.787	12.265	17.22
C3	5.34	6.049	7.113	8.885	12.43
i-C4	1.00	1.117	1.292	1.585	2.17
n-C4	3.65	3.919	4.323	4.995	6.34
i-C5	3.10	3.000	2.850	2.600	2.10
n-C5	4.75	4.520	4.175	3.600	2.45
C6	5.48	5.176	4.720	3.960	2.44
C7	3.23	3.014	2.690	2.150	1.07
C8	1.32	1.224	1.080	0.840	0.36
C9	2.27	2.046	1.710	1.150	0.03
C10	2.19	1.971	1.642	1.095	0
C11	1.81	1.629	1.358	0.905	0
C12+	35.87	32.283	26.902	17.935	0
Total	100	100	100	100	100

This behavior can be explained by the solubility parameter of CO_2. At temperatures below the crossover point, the solubility parameter of reservoir oil is lower than the solubility parameter of CO_2, so the probability of asphaltene precipitation in original oil is stronger than the CO_2-induced asphaltene. This phenomenon is in agreement with the one that Gonzalez *et al.* (2007) proved based on the PC-SAFT model. Above the crossover temperature, the asphaltene solubility decreases and the onset pressure increases at higher temperatures. The increase in the amount of gas at different temperatures was also calculated, as shown in Figure 9. The onset pressure of the mixture increases at any temperature when methane, N_2 and associated gases are added, which is why the upper onset curves in Figure 9 always have positive slopes. This behavior is completely different for CO_2. At temperatures below the crossover point, the CO_2 acts as an inhibitor and the upper onset curves have negative or zero slopes, whereas at temperatures above this point, it acts as a strong asphaltene precipitant and therefore the upper onset curves are ascending. The slope variation in the precipitation onset curves above and below the crossover point supports the specific behavior of the CO_2.

This gas can behave as an inhibitor or precursor of asphaltene precipitation, depending upon the temperature of the system for a fixed pressure and live oil composition. This dual effect is not observed with other gases, such as N_2 or methane, which always expand the ADE. Table 4 summarizes the fluid compositions at different mole percentages of associated gas based on the RKSA model.

Figure 10, that was calculated by Verdier in 2006, shows the solubility parameters of N_2, methane and CO_2 *versus* pressure, and among these gases N_2 has the lowest solubility. In a wide range of temperatures, different amounts of N_2 increase the upper onset pressures more than different percentages of methane and CO_2;

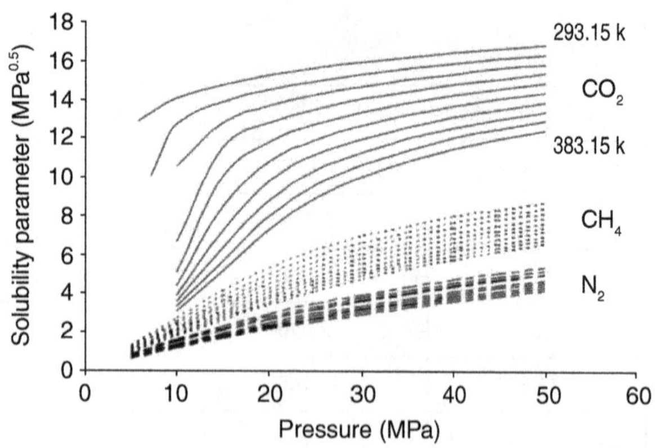

Figure 10

Solubility parameters of N_2, methane and CO_2 calculated every 10 K, starting at 293.15 K for the top curves by Verdier (2006).

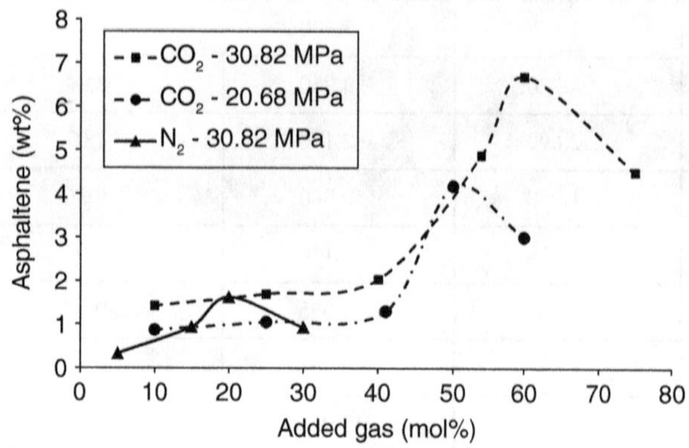

Figure 11

Experimental asphaltene precipitation due to CO_2 and N_2 gas injection at the reservoir temperature.

however, CO_2 is a stronger precipitant agent at a specified temperature and leads to a higher content of asphaltene to precipitate.

Figure 11 proves this behavior of CO_2 and N_2 based on experimental results. CO_2 and N_2 gas injection tests were carried out at different mole percentages. By adding more gas into the system, the saturation pressure of the mixture increases. Therefore, at each pressure point, the asphaltene precipitation will increase, as the saturation pressure is below the test pressure. This means that before the mixture reaches the saturation pressure, all injected gases mix with the oil and the density of the mixture decreases. Based on the solubility theory the asphaltene precipitation increases when density decreases. When the operating pressure is lower than the saturation pressure, the gas does not mix with the fluid, and some light components of the oil can travel to the mixture.

The result of this phenomenon is asphaltene dissolution and an increase in the density of the oil. For a specified amount of gas injection, N_2 increases the saturation pressure more than CO_2, and it reaches its maximum content at lower percentages of injected gas compared with CO_2. It is worth mentioning that if the results obtained here are confirmed by other authors, they need to be verified.

CONCLUSION

The general results are summarized as follows:

1. Depending on a counterbalancing effect of density and solution entropy, the temperature can cause an increase or a decrease in the amount of asphaltene precipitation;
2. Although temperature has a slight effect on the lower onset of asphaltene precipitation and the bubble point, the temperature can lead to a big change in the upper onset of asphaltene precipitation;
3. APE were gradually shifted to the high-pressure side and expanded as C1, N_2 and associated gas content increased. This phenomenon is stronger for N_2;
4. The shape of APE are similar to the original reservoir oil APE;
5. CO_2 injection causes asphaltene precipitation to increase at high temperatures but below a certain crossover, the asphaltene stability increases. Therefore, CO_2 can behave as an inhibitor or precursor of asphaltene precipitation;
6. The APE was gradually expanded to a wide range of pressures and temperatures in a CO_2 injection scenario, while the increase in the onset pressure was more significant in other gas injections;
7. Although at the same mol% of N_2 and CO_2 injection, the APE expands more with N_2 injection; CO_2 is a strong precipitant agent at a specified temperature, which leads to a higher content of asphaltene to precipitate.

REFERENCES

ASTM D6560–00, (2000) *Standard Test Method for Determination of Asphaltenes (Heptane Insolubles) in Crude Petroleum and Petroleum Products*, London, UK.

Akbarzadeh K., Hammami A., Kharrat A., Zhang A. (2007) Asphaltenes—Problematic but Rich in Potential, *Oilfield Review* **19**, 2, 22-43.

Andersen S.I. (1994) Effect of Precipitation Temperature on the Composition of *n*-Heptane Asphaltenes, *Petroleum Science and Technology* **12**, 1, 51-74.

Chung T.H. (1992) Thermodynamic Modeling for Organic Solid Precipitation, *SPE Annual Technical Conference and Exhibition*, 4-7 Oct, Washington, DC.

Creek J.L. (2005) Freedom of Action in the State of Asphaltenes: Escape from Conventional Wisdom, *Energy & Fuels* **19**, 4, 1212-1224.

de Boer R.B., Leerlooyer K., Eigner M.R.P., van Bergen A.R.D. (1995) Screening of Crude Oils for Asphalt Precipitation: Theory Practice, and the Selection of Inhibitors, *SPE Production & Facilities* **10**, 1, 55-61.

Edmonds B., Moorwood R.A.S., Szczepanski R., Zhang X., Heyward M., Hurle R. (1999a) A New Method to Give Quantitative Prediction of Asphaltene Deposition from Petroleum Fluids, *Third International Symposium on Colloid Chemistry in Oil Production. Asphaltene & Wax Deposition, ISCOP'99*, Huatelco, Mexico, 14-17 Nov.

Edmonds B., Moorwood R.A.S., Szczepanski R., Zhang X. (1999b) Latest Developments in Integrated Prediction Modelling Hydrates, Waxes and Asphaltenes, Presented at *IBC Forum Controlling Hydrates, Waxes and Asphaltenes*, Aberdeen, UK, 28-29 Oct.

Escobedo J., Mansoori G.A. (1997) Viscometric Principles of Onsets of Colloidal Asphaltene Flocculation in Paraffinic Oils and Asphaltene Micellization in Aromatics, *SPE Production & Facilities* **12**, 2, 116-122.

Gonzalez D.L., Ting P.D., Hirasaki G.J., Chapman W.G. (2005) Prediction of Asphaltene Instability under Gas Injection with the PC-SAFT Equation of State, *Energy & Fuels* **19**, 4, 1230-1234.

Gonzalez D.L., Vargas F.M., Hirasaki G.J., Chapman W.G. (2007) Modeling Study of CO_2-Induced Asphaltene Precipitation, *Energy & Fuels* **22**, 2, 757-762.

Gonzalez D.L., Garcia M.E., Diaz O. (2012) Unusual Asphaltene Phase Behavior of Fluids From Lake Maracaibo, Venezuela, *SPE Latin America and Caribbean Petroleum Engineering Conference*, Mexico City, Mexico, 16-18 April.

Hassanvand M., Shahsavani B., Anooshe A. (2012) Study of Temperature Effect on Asphaltene Precipitation by Visual and Quantitative Methods, *Journal of Petroleum Technology and Alternative Fuels* **3**, 2, 8-18.

Hirschberg A., de Jong L.N.J., Schipper B.A., Meijer J.G. (1984) Influence of Temperature and Pressure on Asphaltene Flocculation, *SPE Journal* **24**, 3, 283-293.

Jamaluddin A.K.M., Creek J., Kabir C.S., McFadden J.D., Cruz D., Joseph M.T., Joshi N., Ross B. (2001) A Comparison of Various Laboratory Techniques to Measure Thermodynamic Asphaltene Instability, *SPE Asia Pacific Improved Oil Recovery Conference*, Kuala Lumpur, Malaysia, 6-9 Oct.

Jamaluddin A.K.M., Jushi N., Iwere F., Gurpinar O. (2002) An Investigation of Asphaltene Instability Under Nitrogen Injection, *SPE International Petroleum Conference and Exhibition*, Villahermosa, Mexico, 10-12 Feb.

Kesler M.G., Lee B.I. (1976) Improve Prediction of Enthalpy of Fractions, *Hydrocarbon Processing* **53**, 153-158.

Leontaritis K.J., Mansoori G.A. (1987) Asphaltene Flocculation During Oil Production and Processing: A Thermodynamic Collodial Model, *SPE International Symposium on Oilfield Chemistry*, San Antonio, Texas, 4-6 Feb.

Mansoori G.A., Vazquez D., Shariaty-Niassar M. (2007) Polydispersity of Heavy Organics in Crude Oils and Their Role in Oil Well Fouling, *Journal of Petroleum Science and Engineering* **58**, 3-4, 375-390.

Nakhli H., Alizedeh A., Moqadam M.S., Afshari S., Kharrat R., Ghazanfari M.H. (2011) Monitoring of Asphaltene Precipitation: Experimental and Modeling Study, *Journal of Petroleum Science and Engineering* **78**, 2, 384-395.

Nghiem L.X., Coombe D.A. (1997) Modeling Asphaltene Precipitation during Primary Depletion, *SPE Journal* **2**, 2, 170-176.

Nghiem L.X., Hassam M.S., Nutakki R., George A.E.D. (1993) Efficient Modelling of Asphaltene Precipitation, *SPE Annual Technical Conference and Exhibition*, Houston, Texas, 3-6 Oct.

Novosad Z., Costain T.G. (1990) Experimental and Modeling Studies of Asphaltene Equilibria for a Reservoir Under CO_2 Injection, *SPE Annual Technical Conference and Exhibition*, New Orleans, Louisiana, 23-26 Sept.

Pan H., Firoozabadi A. (1998) A Thermodynamic Micellization Model for Asphaltene Precipitation: Part I: Micellar Size and Growth, *SPE Production & Facilities* **13**, 2, 118-127.

Panuganti S.R., Vargas F.M., Gonzalez D.L., Kurup A.S., Chapman W.G. (2012) PC-SAFT Characterization of Crude Oils and Modeling of Asphaltene Phase Behavior, *Fuel* **93**, 1, 658-669.

Pedrosa N., Szczepanski R., Zhang X. (2013) Integrated Equation of State Modelling for Flow Assurance, *Fluid Phase Equilibria* **359**, 24-37.

Peneloux A., Rauzy E., Freze R. (1982) A Consistent Correction for Redlich-Kwong-Soave Volumes, *Fluid Phase Equilibria* **8**, 1, 7-23.

Riazi M.R., Al-Sahhaf T.A. (1996) Physical Properties of Heavy Petroleum Fractions and Crude Oils, *Fluid Phase Equilibria* **117**, 1-2, 217-224.

Soave G. (1972) Equilibrium Constants from a Modified Redlich-Kwong Equation of State, *Chemical Engineering Science* **27**, 6, 1197-1203.

Srivastava R.K., Huang S.S. (1997) Asphaltene Deposition during CO_2 Flooding: A Laboratory Assessment, *SPE Production Operations Symposium*, Oklahoma City, Oklahoma, 9-11 March.

Stankiewicz A.B., Flannery M.D., Fuex N.A., Bronze G., Couch J.L., Dubey S.T., Iyer S.D., Ratulowski J., Westerich J.T. (2002) Prediction of Asphaltene Deposition Risk in E&P Operations, *3rd International Conference on Petroleum Phase Behavior and Fouling/AIChE Spring National Meeting*, New Orleans, Louisiana, 10-14 March.

Takahashi S., Hayashi Y., Takahashi S., Yazawa N. (2003) Characteristics and Impact of Asphaltene Precipitation during CO_2 Injection in Sandstone and Carbonate Cores: An Investigative Analysis Through Laboratory Tests and Compositional Simulation, *SPE International Improved Oil Recovery Conference in Asia Pacific*, Kuala Lumpur, Malaysia, 20-21 Oct.

Ting P.D. (2003) Thermodynamic Stability and Phase Behavior of Asphaltenes in Oil and of Other Highly Asymmetric Mixtures, *PhD Thesis*, Rice University.

Vafaie-Sefti M., Mousavi-Dehghani S.A., Mohammad-Zadeh M. (2003) A simple Model for Asphaltene Deposition in Petroleum Mixtures, *Fluid Phase Equilibria* **206**, 1-2, 1-11.

Vazquez D., Mansoori G.A. (2000) Identification and Measurement of Petroleum Precipitates, *Journal of Petroleum Science and Engineering* **26**, 1-4, 49-55.

Verdier S. (2006) Experimental Study and Modelling of Asphaltene Precipitation Caused by Gas Injection, *PhD Thesis*, Technical University of Denmark.

Verdier S., Carrier H., Andersen S.I., Daridon J.L. (2006) Study of Pressure and Temperature Effects on Asphaltene Stability in Presence of CO_2, *Energy & Fuels* **20**, 4, 1584-1590.

Victorov A.I., Firoozabadi A. (1996) Thermodynamic Micellizatin Model of Asphaltene Precipitation from Petroleum Fluids, *AIChE Journal* **42**, 6, 1753-1764.

Wang J.X., Creek J.L., Buckley J.S. (2006) Screening for Potential Asphaltene Problems, *SPE Annual Technical Conference and Exhibition*, 24-27 Sept.

Whitson C.H. (1983) Characterizing Hydrocarbon Plus Fractions, *Society of Petroleum Engineering Journal* **23**, 4, 683-694.

Wu J., Prausnitz J.M., Firoozabadi A. (2000) Molecular Thermodynamics of Asphaltene Precipitation in Reservoir Fluids, *AIChE Journal* **46**, 1, 197-209.

Zhang X., Pedrosa N., Moorwood T. (2012) Modeling Asphaltene Phase Behavior: Comparison of Methods for Flow Assurance Studies, *Energy & Fuels* **26**, 2611-2620.

Zekri A.Y., Shedid S.A., Alkashef H. (2007) A New Technique for Treatment of Permeability Damage due to Asphaltene Deposition using Laser Technology, *Journal of Petroleum Science and Engineering* **59**, 3-4, 300-308.

Permissions

All chapters in this book were first published in OGST, by EDP Sciences; hereby published with permission under the Creative Commons Attribution License or equivalent. Every chapter published in this book has been scrutinized by our experts. Their significance has been extensively debated. The topics covered herein carry significant findings which will fuel the growth of the discipline. They may even be implemented as practical applications or may be referred to as a beginning point for another development.

The contributors of this book come from diverse backgrounds, making this book a truly international effort. This book will bring forth new frontiers with its revolutionizing research information and detailed analysis of the nascent developments around the world.

We would like to thank all the contributing authors for lending their expertise to make the book truly unique. They have played a crucial role in the development of this book. Without their invaluable contributions this book wouldn't have been possible. They have made vital efforts to compile up to date information on the varied aspects of this subject to make this book a valuable addition to the collection of many professionals and students.

This book was conceptualized with the vision of imparting up-to-date information and advanced data in this field. To ensure the same, a matchless editorial board was set up. Every individual on the board went through rigorous rounds of assessment to prove their worth. After which they invested a large part of their time researching and compiling the most relevant data for our readers.

The editorial board has been involved in producing this book since its inception. They have spent rigorous hours researching and exploring the diverse topics which have resulted in the successful publishing of this book. They have passed on their knowledge of decades through this book. To expedite this challenging task, the publisher supported the team at every step. A small team of assistant editors was also appointed to further simplify the editing procedure and attain best results for the readers.

Apart from the editorial board, the designing team has also invested a significant amount of their time in understanding the subject and creating the most relevant covers. They scrutinized every image to scout for the most suitable representation of the subject and create an appropriate cover for the book.

The publishing team has been an ardent support to the editorial, designing and production team. Their endless efforts to recruit the best for this project, has resulted in the accomplishment of this book. They are a veteran in the field of academics and their pool of knowledge is as vast as their experience in printing. Their expertise and guidance has proved useful at every step. Their uncompromising quality standards have made this book an exceptional effort. Their encouragement from time to time has been an inspiration for everyone.

The publisher and the editorial board hope that this book will prove to be a valuable piece of knowledge for researchers, students, practitioners and scholars across the globe.

List of Contributors

Yu Haiyang, Wang Yefei, Lv Peng and Shi Shenglong
School of Petroleum Engineering, China University of Petroleum (East China), Qingdao 266555 - China

Zhang Jian
Technology Research Dept., CNOOC Research Center, Beijing 100027 – China

Zhao Yu-long, Zhang Lie-Hui, Feng Guo-Qing and Zhang Bo-Ning
State Key Laboratory of Oil and Gas Reservoir Geology and Exploitation, Southwest Petroleum University, Chengdu, Sichuan 610500, P.R. China

Kang Bo
Chengdu North Petroleum Exploration and Development Technology LTD, Chengdu, Sichuan 610500, P.R. China

Chuang Wen
College of Pipeline and Civil Engineering, China University of Petroleum, 66 Changjiang West Road, Huangdao District, Qingdao, 266580 - China
CSIRO Computational Informatics, Clayton, Victoria 3169 - Australia

Xuewen Cao
College of Pipeline and Civil Engineering, China University of Petroleum, 66 Changjiang West Road, Huangdao District, Qingdao, 266580 - China

Yan Yang
School of Petroleum Engineering, Changzhou University, Zhonglou District, Changzhou, Jiangsu, 213016 - China

Yuqing Feng
CSIRO Computational Informatics, Clayton, Victoria 3169 – Australia

Hadi Saboorian-Jooybari
Reservoir Studies Division, Department of Petroleum Engineering, Main Office Building, National Iranian South Oil Company (NISOC), P.O. Box 61735-1333, Ahvaz – Iran

Vangala Dhanunjana Chari
Gas Hydrate Group, National Geophysical Research Institute (Council of Scientific & Industrial Research), Hyderabad – 500 007 - India
Department of Physics, Osmania University, Hyderabad – 500 007 - India

Pinnelli S.R. Prasad
Gas Hydrate Group, National Geophysical Research Institute (Council of Scientific & Industrial Research), Hyderabad – 500 007 - India

Sarabu Ramana Murthy
Department of Physics, Osmania University, Hyderabad – 500 007 – India

Laurie Starck, Ludivine Pidol and Nicolas Jeuland
IFP Energies nouvelles, 1-4 avenue de Bois-Préau, 92852 Rueil-Malmaison - France

Thierry Chapus
IFP Energies Nouvelles, Rond-point de l'échangeur de Solaize, BP 3, 69360 Solaize - France

Paul Bogers and Joanna Bauldreay
Shell Global Solutions Downstream, Shell Technology Centre Thornton, P.O. Box 1, Chester CH1 3SH - United Kingdom

Manuel Nepveu, Filip Neele, Cor Hofstee and Jens Wollenweber
TNO, Princetonlaan 6, 3584 CB Utrecht - The Netherlands

Florence Delprat-Jannaud, Olivier Vincké, Anne Battani, Axelle Baroni and Bruno Garcia
IFP Energies Nouvelles, Rueil-Malmaison - France

Jonathan Pearce and Maxine Akhurst
BGS, Nottingham - United Kingdom

Valentina Volpi
OGS, Trieste - Italy

Ane Lothe
SINTEF Petroleum Research, Trondheim - Norway

Suzanne Brunsting
ECN, Amsterdam - The Netherlands

X.F. Lv, B.H. Shi, Y. Wang, L.Y. Wang and J. Gong
Beijing key laboratory of Urban Oil and Gas Distribution Technology, China University of Petroleum (Beijing), Beijing 102249 - China

Y.X. Tang
SINOPEC Luoyang Petrochemical Engineering Corporation, Luoyang 471003 – China

Joonho Jeon
Department of Mechanical Convergence Engineering, Graduate School of Hanyang University, Seoul 133791 - Republic of Korea

Yong Hee Park and Sang Il Kwon
National Institute of Environmental Research, Incheon 404170 - Republic of Korea

Sungwook Park
School of Mechanical Engineering, Hanyang University, 17 Haengdang-dong, Seongdong-gu, Seoul 133791 - Republic of Korea

Laure Patouillard and Pierre Collet
IFP Energies nouvelles, 1-4 avenue de Bois-Préau, 92852 Rueil-Malmaison - France

Peng Zhang
Chongqing University of Science & Technology, School of Chemistry and Chemical Engineering, Chongqing 266555 - China
China University of Petroleum (East China), School of Petroleum Engineering, Qingdao 266555 - China

Yefei Wang, Yan Yang, Wuhua Chen and Shixun Bai
China University of Petroleum (East China), School of Petroleum Engineering, Qingdao 266555 - China

Abraham de Swaan
Chilpa 16-4, Mexico D.F. cp 03910 - Mexico

Veronica Uribe-Vargas, Bernardo Carreón-Calderón, Edgar Ramírez-Jaramillo and Mario Ramírez-de-Santiago
Mexican Institute of Petroleum, Mexico City - Mexico

Pavel Spirov and Svetlana Rudyk
Sultan Qaboos University - Department of Petroleum and Chemical Engineering, PO Box 34, Al-Khod 123, Muscat - Sultanate of Oman

Christophe Preux
IFP Energies nouvelles, 1-4 avenue de Bois-Préau, 92852 Rueil-Malmaison Cedex - France

Oscar Mauricio Forero Camacho
Electrical Engineering Department-DTU, Frederiksborgvej 399, 4000 Roskilde - Denmark

Lucian Mihet-Popa
Electrical Engineering Department-DTU, Frederiksborgvej 399, 4000 Roskilde - Denmark
Politehnica University of Timisoara, V. Parvan 2, 300223 Timisoara - Romania

P.P.C. Udani and M. Rønning
Department of Chemical Engineering, Norwegian University of Science and Technology, NO-7491 Trondheim - Norway

K. Karimpour, R. Zarghami, M.A. Moosavian and H. Bahmanyar
Process Design and Simulation Research Center, Oil and Gas Processing Centre of Excellence, School of Chemical Engineering, College of Engineering, University of Tehran, P.O. Box 11155/4563, Tehran - Iran

Peyman Bahrami
Department of Oil and Gas Engineering, Science and Research Branch, Islamic Azad University, Tehran - Iran

Riyaz Kharrat
Petroleum University of Technology, Tehran - Iran

Sedigheh Mahdavi
Memorial University of Newfoundland - Canada

Hamed Firoozinia
Research Institute of Petroleum Industry, Tehran - Iran

Index

www.ingramcontent.com/pod-product-compliance
Lightning Source LLC
Chambersburg PA
CBHW082339130326
41169CB00012B/5023